SETON HALL UNIVERSITY
QH505 .L24 1984
Equations of membrane biophysics
MAIN

EQUATIONS OF MEMBRANE BIOPHYSICS

EQUATIONS OF MEMBRANE BIOPHYSICS

N. LAKSHMINARAYANAIAH

Department of Pharmacology
Jefferson Medical College
Thomas Jefferson University
Philadelphia, Pennsylvania

1984

ACADEMIC PRESS, INC.
(Harcourt Brace Jovanovich, Publishers)

Orlando San Diego San Francisco New York London
Toronto Montreal Sydney Tokyo São Paulo

Copyright © 1984, by Academic Press, Inc.
ALL RIGHTS RESERVED.
NO PART OF THIS PUBLICATION MAY BE REPRODUCED OR
TRANSMITTED IN ANY FORM OR BY ANY MEANS, ELECTRONIC
OR MECHANICAL, INCLUDING PHOTOCOPY, RECORDING, OR ANY
INFORMATION STORAGE AND RETRIEVAL SYSTEM, WITHOUT
PERMISSION IN WRITING FROM THE PUBLISHER.

ACADEMIC PRESS, INC.
Orlando, Florida 32887

United Kingdom Edition published by
ACADEMIC PRESS, INC. (LONDON) LTD.
24/28 Oval Road, London NW1 7DX

Library of Congress Cataloging in Publication Data

Lakshminarayanaiah, N., Date
  Equations of membrane biophysics.

  Bibliography: p.
  Includes index.
  1. Biological transport--Mathematics. 2. Membranes
(Biology)--Mathematics. I. Title. [DNLM: 1. Membranes
--Physiology. 2. Biological transport. 3. Models,
Biological. QH 509 L192e]
QH505.L24 1984 574.87'5   83–3743
ISBN 0–12–434260–4

PRINTED IN THE UNITED STATES OF AMERICA

84 85 86 87   9 8 7 6 5 4 3 2 1

CONTENTS

Preface   ix

Chapter 1   INTRODUCTION   1
                 References   5

Chapter 2   BASIC PRINCIPLES
                 I. Thermodynamic Concepts   7
                 II. Electrostatics   17
                 III. Physical and Electrochemical Principles   37
                 References   63

Chapter 3   ELECTROCHEMISTRY OF SOLUTIONS AND MEMBRANES
                 I. The Debye–Hückel Theory   66
                 II. Debye–Hückel Theory and Activity Coefficients   70
                 III. Debye–Hückel Theory and Electrolyte Conductance   75
                 IV. Distribution of Ions and Potential Differences at Interfaces   78
                 V. Electrokinetic Phenomena   98
                 VI. Donnan Equilibrium   107
                 VII. Donnan Equilibrium in Charged Membranes   114
                 VIII. Membrane Potential   117
                 IX. Some Applications of the Double-Layer Theory   119
                 X. Model-System Approach to Evaluation of Surface Charge Density   123
                 References   126

Chapter 4 ELECTRICAL POTENTIALS ACROSS MEMBRANES

I. Bi- and Multi-Ionic Potentials	130
II. Determination of Selectivity Coefficients K_{ij}^{pot}	134
III. Integration of Nernst–Planck Flux Equation	137
IV. Other Models	151
V. Liquid Membranes	155
VI. Thermodynamic Approach to Isothermal Membrane Potential	159
VII. Kinetic Approach to Membrane Potentials	160
References	163

Chapter 5 KINETIC MODELS OF MEMBRANE TRANSPORT

I. Equations of Enzyme Kinetics	165
II. Schematic Method of Deriving Rate Equations	170
III. Enzyme Kinetics of Mediated Transport	173
IV. Eyring Model for Membrane Permeation	174
V. Eyring Model and Biological Membranes	178
VI. Model for Lipid-Soluble Ions	188
VII. Model for Carriers of Small Ions	210
VIII. Models for Channel-Forming Ionophores	229
References	265

Chapter 6 STEADY-STATE THERMODYNAMIC APPROACH TO MEMBRANE TRANSPORT

I. Basic Principles	269
II. Electrical Parameters	275
III. Electrokinetic Phenomena	277
IV. Transport of a Solution of Nonelectrolyte across a Simple Membrane	283
V. Permeation of Electrolyte Solution through a Membrane	296
VI. Nature of Water Flow across Membranes	314
References	325

Chapter 7 IMPEDANCE, CABLE THEORY, AND HODGKIN–HUXLEY EQUATIONS

I. Impedance	328
II. Elements of the Cable Theory	334
III. Models to Relate Input Impedance to Electrical Cell Constants	344
IV. Hodgkin–Huxley Equations	352
References	367

Chapter 8 FLUCTUATION ANALYSIS OF THE ELECTRICAL PROPERTIES OF THE MEMBRANE

I. Nonmathematical Description of Noise Analysis	370
II. Statistical Concepts	373

III. Mathematical Preliminaries	381
IV. Spectral Density and Rayleigh's Theorem	395
V. Spectral Density and Source Impedance	396
VI. Filters	399
VII. Correlation Function and Spectra	404
VIII. Types of Noise Sources	409
References	416

Index 419

PREFACE

In recent years several books related to transport phenomena in membranes edited by reputable scientists have been published. Necessarily, the different chapters were written by authorities who reviewed material pertaining to their own areas of expertise. Although these serve as good reference texts for the advanced student and the researcher, they are seldom used by the ordinary students of biology, microbiology, pharmacology, and physiology who would like to learn the general aspects of the several phenomena that arise across membranes. In giving a course on the physiology and the pharmacology of membranes to students with varied backgrounds, it became apparent to me that the course had to be nonmathematically descriptive to be understood. Consequently, several important aspects of membrane phenomena could not be adequately described. This became a serious problem since no one book containing the quantitative material could be prescribed. To alleviate this problem, an attempt has been made in this book to introduce the relevant principles of thermodynamics, kinetics, electricity, surface chemistry, electrochemistry, and other mathematical theorems so that the quantitative aspects of membrane phenomena in model and biological systems could be described.

The book begins with an introduction to several phenomena that arise across membranes, both artificial and biological, when different driving forces act across them. In Chapter 2, an outline of the thermodynamic principles related to properties of dilute aqueous electrolyte solutions and a review of the principles of electrostatics, electrochemical principles, Fick's laws of diffusion, and the rate theory of diffusion are presented. The chapter ends with an outline of the several quantitative relations describing flows across barriers in general. Chapter 3 contains a summary of the quantitative aspects of the electrochemistry of solutions and membranes followed by an outline of the quantitative relations between charges and electrostatic potentials related to surfaces and interfaces. Chapter 4 is devoted to a review of membrane theories pertaining to electrical potentials arising across a variety of membranes. Chapter 5 contains a summary of the several quantitative relationships that could be used to discuss a variety of kinetic models of transport. The steady-state thermodynamic approaches to several transport phenomena in membranes are described in Chapter 6 which is followed by outlines of tissue impedance, cable theory, and Hodgkin–Huxley equations in Chapter 7. Finally, the several mathematical concepts related to fluctuation analysis used to derive values for the unitary parameters of membrane channels in both biological and artificial lipid membranes are presented in Chapter 8. In all of these presentations the emphasis has been to show how the several membrane parameters are related to one another and how well the final equation predicts the particular membrane phenomenon.

Some general comments about the presentation of the material are in order. References are kept to a minimum, and not all the names of workers who have investigated several membrane phenomena and made important contributions have been mentioned in the text, but their important papers are given in the references. The second comment concerns details about presentation of a particular theme. Verbal description is kept to a minimum, but enough is given to draw attention to the main equations that describe the phenomena. The third comment is that the symbols used may cause confusion because the same symbol may represent different parameters in different chapters or in some cases in the same chapter. Care has been exercised to spell out what each symbol means when it is not particularly obvious in the context it is used. Some symbols, particularly k and K, may represent a variety of parameters such as a constant, distribution coefficient, rate constant, or dissociation constant. Often the meaning of a symbol is obvious if some attention is paid to the context in which it is used. Although it would have been better if different symbols were used, this was deliberately avoided because it was thought unwise to change notations of generally well-known parameters such as Boltzmann constant, gas constant, farad, and Faraday. In instances where two groups of investigators have made contributions toward exposition of a particular phenomenon, names and symbols used by one group only are used without mentioning the other group in the text. This was done for the sake of clarity of presentation. However, the important papers of the other group are given in the references. This in no way should be construed as minimizing or ignoring the contributions of the other group.

It is my hope that this attempt in assembling the material will prove useful as an introductory text for those interested in entering the fascinating and at the same time perplexing world of membrane science.

My thanks are due to Dr. C. Paul Bianchi for his interest and encouragement. In the belief I am raised in, this work is dedicated to Venkateswara.

Chapter 1

INTRODUCTION

A membrane, in simple terms, may be defined as a phase that acts as a barrier to prevent mass movement but allows restricted and/or regulated passage of one or several species through it. It can be a solid or liquid containing ionized or ionizable groups, or it can be completely un-ionized. Functionally, all membranes are active when used as barriers to separate two other phases unless they are too porous or too fragile.

The existence of ionogenic groups and pores (space occupied by water) in the membrane confers to the membrane characteristics described by the words *permselectivity* and *semipermeability*. The transport property that determines the former is the transport number $\bar{t}_i$. The latter is indicated by the Staverman reflection coefficient σ_i, which is usually given by the ratio of the actual hydrostatic pressure across the membrane required to give zero volume flow to that that would be required if the membrane were truly semipermeable.

Membranes may be broadly classified into artificial or man-made and natural or biological. Artificial membranes that are used in several technologies are generally thick, although thin membranes from Parlodion (nitrocellulose) have been prepared. Natural or biological membranes are thin (less than 10 nm) and contain a variety of structurally well-organized lipids and proteins. Although it has not been possible so far to form a membrane artificially from known components (lipids and proteins) of biological membranes, bilayer membranes resembling the physical structure of biological membranes have been prepared from several pure and mixed phospholipids. A rough classification of several types of membranes according to their physical dimensions and chemical structures is given in Fig. 1.

1. Introduction

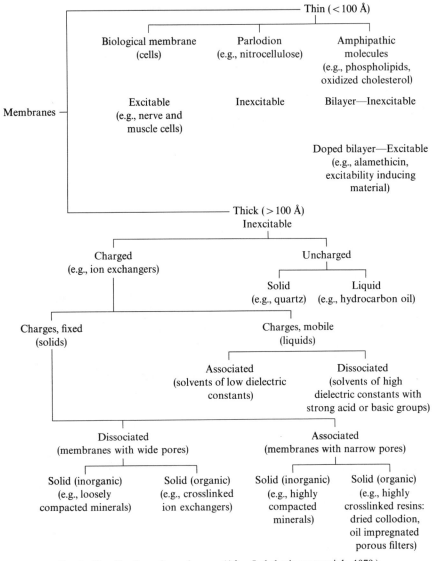

Fig. 1. Classification of membranes. (After Lakshminarayanaiah, 1979.)

When a membrane separates two aqueous phases, several driving forces may be made to act across it to cause a flow or flux of molecules or ionic species through it. The most common driving forces, in the absence of external magnetic and gravitational forces, are (a) difference of chemical potential $\Delta\mu$, (b) difference of electric potential ΔE, (c) difference of pressure ΔP, and

(d) difference of temperature ΔT. These forces, when they act severally or in combination across a membrane, produce a number of transport phenomena, and these are indicated in Table I.

Generation of a particular phenomenon across a membrane is dependent on the structural characteristics of the membrane. Several transport processes, such as dialysis, electrodialysis, and reverse osmosis, have been investigated extensively and used industrially to separate constituents of aqueous

TABLE I

Transport Phenomena in Membranes[a]

Driving force	Phenomena	Primary flow	Comments
Chemical or electrochemical potential $\Delta \mu$ or $\Delta \bar{\mu}$	Mixing or diffusion	Chemical constituents	Establishment of chemical equilibrium
	Osmosis	Solvent	Solvent enters concentrated solution (osmometry)
	Membrane potential	Ionic solute	Source of emf; negligible solvent flow
	Dialysis	Solute	Solute leaves concentrated solution (hemodialysis)
	Diasolysis	More mobile component	Selective flow of more mobile species
	Osmoionosis	Ionic solute	Three streams of different concentration produce driving force. Similar to electrodialysis without application of external electric field
	Dufour effect	Thermal	Gives rise to ΔT
Electric field ΔE	Electric conduction	Current	Evaluation of membrane resistance
	Transport number of species	Fraction of current carried by the species	Evaluation of membrane permselectivity
	Piezoelectricity	Residual polarization and storage of electricity	Production of electrets
	Electroosmosis	Solvent	Solvent transfer
	Electrodialysis	Ionic solute and/or solvent transfer	Solute removal

(Continued)

TABLE I (*Continued*)

Driving force	Phenomena	Primary flow	Comments
	Electrophoresis	Ionic solute	Separation of large molecules
	Transport depletion	Ionic solute	Simplified high current density electrodialysis
Pressure ΔP	Hydralic or mechanical permeability	Solvent	Relates to space available for laminar and/or diffusional flow
	Filtration	Solvent	Particulate matter retained by sieving
	Pressure permeation	Selective transport of most mobile component	Separation of liquids and/or gases
	Ultrafiltration or reverse osmosis	Solvent	Solvent leaves and solution concentrated
	Piezodialysis	Ionic solute	Product of reduced salinity
	Streaming potential	Solvent	Generation of emf
	Streaming current	Ionic solute	Current very small; studies rare
Vacuum	Pervaporation	Selective transport of most mobile component	Separation of liquids and/or gases
Temperature ΔT	Heat conduction	Heat	Thermal conductance
	Thermoosmosis	Solvent	Solvent may move from hot to cold side or vice versa
	Thermo-emf	Ionic solute	Source of emf; studies rare
	Soret effect	Ionic solute	Gives rise to $\Delta\mu$; difficult to measure
Electric field plus pressure ($\Delta E + \Delta P$)	Forced flow electrophoresis	Ionic solute and solvent	Purification of blood, sewage; in experimental stage
Electric field plus temperature ($\Delta E + \Delta T$)	Electrodecantation	Ionic solute and solvent	Electrical heating and/or cooling supplies ΔT; ionic matter concentrates downward and solvent concentrates upward

a From Lakshminarayanaiah (1979).

and nonaqueous solutions. The membranes used are mostly of the ion-exchange type. Membrane films have been used in the separation of mixtures of gases. Although special artificial membranes are required for performing special functions, such as the separation of constituents of gas mixtures, of biological fluids (artificial kidney), treatment of sewage, biological membranes perform several functions, such as maintenance of ionic gradients, cell volume, electrolyte levels in blood, secretion, excretion, at the same time to sustain life.

The many quantitative relations used to describe the several membrane phenomena listed in Table I together with the background material required to follow those equations are outlined in this book.

References

Buck, R. P. (1976). *CRC Crit. Rev. Anal. Chem.* **5**, 323.
Lakshminarayanaiah, N. (1969). "Transport Phenomena in Membranes." Academic Press, New York.
Lakshminarayanaiah, N. (1979). *Subcell. Biochem.* **6**, 401.

Chapter 2

BASIC PRINCIPLES

Some of the thermodynamic, physical and electrochemical principles that find application in the treatment of membrane phenomena are reviewed in this chapter.

I. Thermodynamic Concepts

A. Free Energy

The energy available for doing useful work denoted by the symbol G (free energy) is defined by the thermodynamic equation

$$G = H - TS = E + PV - TS, \qquad (1)$$

where H is the heat content or enthalpy and E is the energy content of the system whose temperature, pressure, volume, and entropy are denoted by T, P, V, and S, respectively.

The free-energy change in going from an initial to a final state is given by

$$\Delta G = \Delta E + \Delta(PV) - \Delta(TS). \qquad (2)$$

At constant temperature and pressure, Eq. (2) becomes

$$\Delta G = \Delta E + P\,\Delta V - T\,\Delta S; \qquad (3)$$

that is,

$$\Delta G = \Delta H - T\,\Delta S. \qquad (4)$$

According to first law of thermodynamics

$$\Delta E = q_{rev} - W_{max} = T \Delta S - W_{max} \tag{5}$$

holds for a reversible isothermal process, where q_{rev} is the heat supplied and W_{max} is the maximum work done since the process is performed reversibly.

Substitution of Eq. (5) into Eq. (3) gives

$$-\Delta G = W_{max} - P \Delta V. \tag{6}$$

The quantity $P \Delta V$ is the work done against external pressure, and so $W_{max} - P \Delta V$ is the net work. Consequently, the increase of free energy is a measure of the net work done on the system.

Differentiation of Eq. (1) gives

$$dG = dE + P\,dV + V\,dP - T\,dS - S\,dT. \tag{7}$$

For a reversible process in which work is restricted to the work of expansion ($W = P\,dV$), the quantity of heat absorbed is given by

$$q_{rev} = dE + P\,dV. \tag{8}$$

But $q_{rev}/T = dS$, and so Eq. (8) becomes

$$dE = T\,dS - P\,dV. \tag{9}$$

Substitution of Eq. (9) into Eq. (7) gives

$$dG = V\,dP - S\,dT. \tag{10}$$

For a reversible change, dE, dV, and dT are zero. Thus Eqs. (9) and (10) lead to the conditions

$$(\partial S)_{E,V} = 0 \tag{11}$$

and

$$(\partial G)_{T,P} = 0 \tag{12}$$

for a system at equilibrium.

At constant pressure, Eq. (10) becomes

$$dG = -S\,dT$$

and so

$$dG_1 = -S_1\,dT \quad \text{and} \quad dG_2 = -S_2\,dT.$$

Subtraction of one from the other gives

$$d(G_2 - G_1) = -dT(S_2 - S_1),$$
$$d\,\Delta G = -\Delta S\,dT,$$

or

$$\left(\frac{\partial(\Delta G)}{\partial T}\right)_P = -\Delta S. \quad (13)$$

Thus substituting from Eq. (4) for ΔS gives

$$T\left(\frac{\partial(\Delta G)}{\partial T}\right)_P = \Delta G - \Delta H. \quad (14)$$

This is called the Gibbs–Helmholtz equation.

B. Chemical Potential and Activity

The free energy of an open system can be written as a function of several different variables. Thus

$$G = f(T, P, n_1, n_2, \ldots, n_i),$$

where $n_1, n_2, \ldots, n_i$ are the numbers of moles of the respective constituents $1, 2, \ldots, i$ of the system. When all the variables change by infinitesimal amounts, then the change in free energy is given by

$$dG = \left(\frac{\partial G}{\partial T}\right)_{P, n_1, \ldots, n_i} dT + \left(\frac{\partial G}{\partial P}\right)_{T, n_1, \ldots, n_i} dP$$
$$+ \left(\frac{\partial G}{\partial n_1}\right)_{T, P, n_2, \ldots, n_i} dn_1 + \cdots + \left(\frac{\partial G}{\partial n_i}\right)_{T, P, n_1, \ldots, n_{i-1}} dn_i. \quad (15)$$

In general, the derivative $(\partial X/\partial n)$, where X is an extensive property of the system, is called the partial molar quantity. The partial molar free energy or chemical potential μ of the species i is given by

$$\mu_i = \left(\frac{\partial G}{\partial n_i}\right)_{T, P, n_j}. \quad (16)$$

Thus Eq. (15) may be written

$$dG = \left(\frac{\partial G}{\partial T}\right)_{P, n_1, \ldots, n_i} dT + \left(\frac{\partial G}{\partial P}\right)_{T, n_1, \ldots, n_i} dP + \mu_1 \, dn_1$$
$$+ \mu_2 \, dn_2 + \cdots + \mu_i \, dn_i. \quad (17)$$

For a closed system in which composition N stays constant, Eq. (17) becomes

$$dG = \left(\frac{\partial G}{\partial T}\right)_{P, N} dT + \left(\frac{\partial G}{\partial P}\right)_{T, N} dP. \quad (18)$$

Comparing Eq. (18) with Eq. (10) gives

$$\left(\frac{\partial G}{\partial T}\right)_P = -S, \tag{19}$$

$$\left(\frac{\partial G}{\partial P}\right)_T = V. \tag{20}$$

These two equations substituted into Eq. (17) give

$$dG = -S\,dT + V\,dP + \sum_i (\mu_i\,dn_i). \tag{21}$$

Equation (21), unlike Eq. (10), which is applicable to a closed system, is valid for any open system.

For isothermal and isobaric conditions, Eq. (21) becomes

$$(dG)_{T,P} = \sum_i (\mu_i\,dn_i). \tag{22}$$

Integration of Eq. (22) gives

$$(G)_{T,P,N} = \sum_i \mu_i n_i. \tag{23}$$

Differentiation of Eq. (23) gives

$$(dG)_{T,P} = \sum (\mu_i\,dn_i + n_i\,d\mu_i). \tag{24}$$

In view of Eq. (22), Eq. (24) becomes

$$\sum (n_i\,d\mu_i) = 0. \tag{25}$$

Equation (25) is called the Gibbs–Duhem equation and has an important place in the study of thermodynamics of dilute solutions.

Differentiation of Eq. (20) with respect to n_i, the number of moles of i in the system, gives

$$\frac{\partial^2 G}{\partial P\,\partial n_i} = \bar{V}_i, \tag{26}$$

where $\bar{V}_i$ is the partial molar volume of the constituent i of the system.

The effect of pressure on the chemical potential is given by the differentiation of Eq. (16). Thus

$$\left(\frac{\partial \mu_i}{\partial P}\right)_{T,N} = \frac{\partial^2 G}{\partial n_i\,\partial P}. \tag{27}$$

Equations (26) and (27) are equivalent according to the calculus principle that the second differential of G with respect to any pair of variables is independent of the order of differentiation. Thus

$$\left(\frac{\partial \mu_i}{\partial P}\right)_{T,N} = \bar{V}_i. \tag{28}$$

For a system of ideal gases

$$V_i = n_i RT/P_i, \tag{29}$$

and so

$$\bar{V}_i = RT/P_i. \tag{30}$$

Consequently, Eq. (28) can be written

$$\left(\frac{\partial \mu_i}{\partial P}\right)_{T,N} = \frac{RT}{P_i}. \tag{31}$$

Total pressure of a gaseous mixture is given by

$$P = P_1 + P_2 + \cdots + P_i = \sum P_i, \tag{32}$$

where the P_i are partial pressures. If P_1, P_2, etc., except P_i are constant, then

$$\partial P = \partial P_i, \tag{33}$$

and so Eq. (31) becomes

$$d\mu_i = \frac{RT\, dP_i}{P_i} = RT\, d(\ln P_i). \tag{34}$$

In the case of an ideal solution, the vapors of component at partial pressure P_i will be in equilibrium with the solution. If the solution is not in equilibrium with its vapors, then there will be flow of dn_i molecules from the solution phase (') to the vapor phase ('') to establish equilibrium. At constant temperature and pressure, the free-energy change accompanying this transfer is given by Eq. (22). Thus

$$(dG)_{T,P} = \mu_i' dn_i' + \mu_i'' dn_i'',$$

but $dn_i' = -dn_i''$ since loss from one phase equals gain in the other phase. According to Eq. (12), change in free energy is zero at equilibrium. Therefore

$$dn_i(\mu_i'' - \mu_i') = 0.$$

But $dn_i \neq 0$, and so

$$\mu_i' = \mu_i'', \tag{35}$$

2. Basic Principles

The condition for stable equilibrium so that no phase disappears is given by

$$d\mu'_i = d\mu''_i. \tag{36}$$

According to Raoult's law (i.e., $P_i = P_i^0 N_i$, where N_i is the mole fraction of i and P_i^0 is the vapor pressure in the pure phase), partial pressure of component i is proportional to the mole fraction of i in solution. Therefore Eq. (34) for an ideal solution can be written

$$\mu_i = \mu_N^0 + RT \ln N_i, \tag{37}$$

where μ_N^0 is a constant for the particular constituent and is independent of composition but depends on temperature and pressure. If the solution is not ideal, Eq. (37) is not applicable and so it is arbitrarily modified as

$$\mu_i = \mu_N^0 + RT \ln N_i f_i, \tag{38}$$

where f_i is a correction factor and is known as the activity coefficient. The product $N_i f_i$ is called activity a_i. Thus μ_N^0 is the chemical potential when the activity of $i(a_i)$ is unity. Activity is not an absolute parameter. It is defined with reference to an arbitrarily chosen standard state of unit activity. Any other convenient standard state may be chosen without paying attention to experimental difficulties.

Generally, in practice, two standard states are chosen. For solid, liquid, or gas and mixtures in which the solute is completely miscible with the solvent, the standard state used is usually the pure substance itself. In the case of dilute solutions, the standard state chosen is dependent on the system used to express concentration of the solute. The concentration of a solute species i may be expressed in three different ways—mole fraction (N_i, the ratio of the number of moles of i to the total number of moles in solution), molar (C_i, moles in a liter of solution), or molal (m_i, moles in 1000 g of solvent). For each concentration scale, the standard state is so chosen that the activity coefficient approaches unity when the concentration is made zero. This is applicable at every temperature and pressure. Equation (38) may therefore be rewritten

$$\mu_i = \mu_N^0 + RT \ln N_i (f_i)_N = \mu_N^0 + RT \ln(a_i)_N,$$
$$\mu_i = \mu_C^0 + RT \ln C_i (f_i)_C = \mu_C^0 + RT \ln(a_i)_C, \tag{39}$$
$$\mu_i = \mu_m^0 + RT \ln m_i (f_i)_m = \mu_m^0 + RT \ln(a_i)_m.$$

$(f_i)_N$, $(f_i)_C$, and $(f_i)_m$ are called rational, molar, and molal activity coefficients. The practical activity coefficients are $(f_i)_C$ and $(f_i)_m$. Usually $(f_i)_m$ is represented by γ_i, which is sometimes called stoichiometric activity coefficient when the total molality of the electrolyte is considered without correction for its incomplete dissociation.

The interrelations between the concentrations and activity coefficients expressed in three different ways are

$$N = \frac{mM_1}{1000 + mM_1} = \frac{CM_1}{1000\rho - C(M_2 - M_1)}, \tag{40}$$

$$C = \frac{\rho m}{1 + 0.001 mM_2} \quad \text{or} \quad m = \frac{c}{\rho - 0.001 CM_2}, \tag{41}$$

$$f_N = f_C \frac{1000\rho - C(M_2 - M_1)}{1000\rho_0} = f_m \frac{1000 + mM_1}{1000}, \tag{42}$$

where M_1 and M_2 are the molecular weights of solvent and solute, respectively, and ρ and ρ_0 are the densities of the solution and solvent, respectively.

When electrolyte solutes are considered there are difficulties in defining mole fraction and activity of the solute as a whole.

An electrolyte $M_{v_+} A_{v_-}$ which dissociates to yield v_+ positive ions (M^{z+}) of valence z_+ and v_- negative ions (A^{z-}) of valence z_- may be considered. Mole fraction of solute may be expressed either as

$$N_2 = \frac{vmM_1}{vmM_1 + 1000}$$

or as

$$N_2 = \frac{m}{m + 1000/vM_1}, \tag{43}$$

where $v = v_+ + v_-$.

In the case of the activity of an electrolyte, one can write for the ions according to Eq. (39)

$$\mu_+ = \mu_+^0 + RT \ln a_+, \tag{44}$$

$$\mu_- = \mu_-^0 + RT \ln a_-, \tag{45}$$

and for the electrolyte as a whole

$$\mu_2 = \mu_2^0 + RT \ln a_2. \tag{46}$$

But

$$\mu_2 = v_+ \mu_+ + v_- \mu_- \tag{47}$$

and

$$\mu_2^0 = v_+ \mu_+^0 + v_- \mu_-^0. \tag{48}$$

Using Eqs. (44), (45), (47), and (48) in Eq. (46) gives on simplification

$$a_2 = a_+^{v_+} a_-^{v_-}. \tag{49}$$

Mean activity $a_\pm$ of an electrolyte is defined by
$$a_\pm^\nu = a_+^{\nu_+} a_-^{\nu_-} \tag{50}$$
Thus
$$a_2 = a_\pm^\nu. \tag{51}$$
The activity of each ion may be written
$$a_+ = m_+ \gamma_+ \quad \text{and} \quad a_- = m_- \gamma_-. \tag{52}$$
The mean activity coefficient according to Eq. (50) is defined as
$$\gamma_\pm^\nu = \gamma_+^{\nu_+} \gamma_-^{\nu_-}. \tag{53}$$
Thus
$$\gamma_\pm^\nu = \frac{a_+^{\nu_+}}{m_+^{\nu_+}} \frac{a_-^{\nu_-}}{m_-^{\nu_-}} = \frac{a_\pm^\nu}{m_\pm^\nu}$$
or
$$\gamma_\pm = \frac{a_\pm}{m_\pm}. \tag{54}$$
If the molality of solution is m, then
$$m_+ = \nu_+ m \quad \text{and} \quad m_- = \nu_- m. \tag{55}$$
The mean ionic molality is given by
$$m_\pm^\nu = m_+^{\nu_+} m_-^{\nu_-}. \tag{56}$$
Substitution from Eq. (55) into Eq. (56) gives
$$m_\pm^\nu = (\nu_+ m)^{\nu_+} \cdot (\nu_- m)^{\nu_-} = m^\nu(\nu_+^{\nu_+} \nu_-^{\nu_-})$$
or
$$m_\pm = m(\nu_+^{\nu_+} \nu_-^{\nu_-})^{1/\nu}. \tag{57}$$
In view of these equations, Eq. (51) may be rewritten
$$a_2 = (\nu_+^{\nu_+} \nu_-^{\nu_-})(m\gamma_\pm)^\nu. \tag{58}$$
The interrelations between mean ionic activity coefficients are as follows [see also Eq. (42)]:
$$f_N = f_C[(\rho + 0.001\nu C M_1 - 0.001 C M_2)/\rho_0]$$
$$= \gamma(1 + 0.001\nu m M_1), \tag{59}$$
$$\gamma = \frac{\rho - 0.001 C M_2}{\rho_0} f_C = \frac{C}{m\rho_0} f_C \tag{60}$$

or

$$f_C = (1 + 0.001mM_2)\frac{\rho_0}{\rho}\gamma = \frac{m\rho_0}{C}\gamma, \quad (61)$$

where $\pm$ subscripts have been dropped from the mean activity coefficients.

C. Osmotic Coefficient and van't Hoff Factor

Nonideal behavior of electrolyte solutions may be expressed in terms of the properties of the solvent. One such property is the osmotic coefficient, which is the ratio of the observed (Π_{obs}) to the ideal (Π_{id}) osmotic pressure. The rational osmotic coefficient g may therefore be written

$$g = \Pi_{obs}/\Pi_{id}. \quad (62)$$

When a semipermeable membrane separates the solvent phase (') from the solution phase ("), only the solvent molecules move across the membrane. At equilibrium, in accordance with Eq. (35), one can write

$$\mu'_{1(P^0)} = \mu''_{1(P^0+\Pi)} = [\mu_1^{0''} + RT \ln a_1]_{(P^0+\Pi)}, \quad (63)$$

where P^0 is the pressure at which solvent represented by 1 exists. Equation (36) applied to Eq. (63) gives

$$d\mu'_1 = d\mu''_1 = d\mu_1^{0''} + RT\, d\ln a_1. \quad (64)$$

As $d\mu'_1 = 0$ since the pressure P^0 is constant, Eq. (64) becomes

$$d\mu_1^{0''} = -RT\, d\ln a_1. \quad (65)$$

But $d\mu_1^{0''} = -(\partial \mu_1^{0''}/\partial P)\, dP$, and according to Eq. (28) $[\partial \mu_1^{0''}/\partial P]$ is equal to $\bar{V}_1$. Thus

$$\bar{V}_1\, dP = -RT\, d\ln a_1. \quad (66)$$

Integration of Eq. (66) gives

$$[\bar{V}_1 P]_{P=P^0}^{P=P^0+\Pi} = -[RT \ln a_1]_1^{a_1}.$$

Thus

$$\Pi = -\frac{RT}{\bar{V}_1}\ln a_1 = -\frac{RT}{\bar{V}_1}\ln N_1 f_1. \quad (67)$$

For the ideal case when $f_1 = 1$, Eq. (67) becomes

$$\Pi_{id} = -\frac{RT}{\bar{V}_1}\ln N_1. \quad (68)$$

Thus Eq. (62) becomes
$$g = (\ln a_1)/(\ln N_1). \tag{69}$$
But
$$N_1 = \frac{1000/M_1}{1000/M_1 + vm} = \frac{1}{1 + 0.001vmM_1}.$$
Equation (69) therefore becomes
$$\begin{aligned}\ln a_1 &= -g\ln(1 + 0.001vmM_1) \\ &= -g[0.001vmM_1 - \tfrac{1}{2}(0.001vmM_1)^2 + \cdots].\end{aligned} \tag{70}$$
The practical osmotic coefficient ϕ is defined by
$$\ln a_1 = -0.001vmM_1\phi. \tag{71}$$
Thus Eqs. (70) and (71) lead to an approximate relation between the two osmotic coefficients. Thus
$$\phi \approx g[1 - \tfrac{1}{2}(0.001vmM_1)]. \tag{72}$$
The relation between osmotic coefficient and osmotic pressure is given by equating Eqs. (67) and (71). Thus
$$\Pi = \frac{vRTM_1}{1000\bar{V}_1}m\phi, \tag{73}$$
whereas for a dilute solution, Eq. (73) approximates to
$$\Pi \approx \frac{vRTM_1}{1000\bar{V}_1}mg \approx vgRTC. \tag{74}$$
The osmotic coefficient is thus related to the van't Hoff factor i by $vg \approx i$.

The van't Hoff factor i, introduced by van't Hoff to cover all types of deviation from ideal behavior without any question about their origins, is related to the degree of dissociation (α) of an electrolyte. If an electrolyte splits into v ions and the solution contains n molecules in a given volume, there will be present in solution $n(1 - \alpha)$ undissociated molecules and $v\alpha n$ ions. The total number of particles is $n(1 - \alpha + v\alpha)$. This number determines the osmotic pressure Π. If there is no dissociation, Π is determined by n. Thus
$$i = \frac{n(1 - \alpha + v\alpha)}{n} = 1 - \alpha + v\alpha$$
and
$$\alpha = (i - 1)/(v - 1). \tag{75}$$
The Gibbs–Duhem equation (25) may be used to establish the relation between molal osmotic coefficient and the activity coefficient. Equation (25)

may be written in the form

$$(1000/M_1)\, d\ln a_1 + vm\, d\ln(m\gamma_\pm) = 0. \tag{76}$$

Differentiation of Eq. (71) and substitution for $d\ln a_1$ in Eq. (76) gives on simplification the relation

$$d[m(1-\phi)] + m\, d\ln\gamma_\pm = 0. \tag{77}$$

In integrated form, Eq. (77) is written

$$\phi = 1 + \frac{1}{m}\int_0^m m\, d\ln\gamma_\pm. \tag{78}$$

Equation (77) can also be written

$$(\phi - 1)\frac{dm}{m} + d\phi = d\ln\gamma_\pm, \tag{79}$$

which on integration gives

$$\ln\gamma_\pm = [\phi]_{m=0,\phi=1}^{m,\phi} + \int_0^m (\phi - 1)\, d\ln m. \tag{80}$$

Thus the limits of ϕ yield

$$\ln\gamma_\pm = (\phi - 1) + \int_0^m (\phi - 1)\, d\ln m. \tag{81}$$

The activity coefficient can thus be determined provided any colligative property (osmotic pressure, depression of freezing point, or elevation of boiling point) is measured as a function of m.

II. Electrostatics

Matter is composed of negative and positive electricity. The elementary negatively and positively charged particles are known as electron and proton. The charge of the proton is equal in magnitude and opposite in sign to that of the electron, but its mass, which is the same as that of the hydrogen atom, is 1836 times greater than that of the electron.

A. Coulomb's Law

The force of attraction or repulsion between two point charges (q, q') is directly proportional to the product of the charges and inversely proportional to the square of the distance between them. Thus

$$\mathscr{F} = kqq'/r^2, \tag{82}$$

where k is a proportionality constant whose magnitude depends on the units in which the several terms are expressed. In cgs units taking k to be unity, Eq. (82) becomes

$$\mathscr{F} = qq'/r^2. \tag{83}$$

While the effect of air existing between charges on the magnitude of the force is very small, Eq. (82) applies exactly only to the case of charges *in vacuo*.

B. System of Units

The units of force, distance, and quantity of charge may be so chosen as to get Eq. (83), i.e., $k = 1$. There are some systems in which units of force, distance, and charge are defined independently of Coulomb's law, with the result that k is not unity. The equations of electrostatics have been written using units of the electrostatic system (esu) and the equations pertaining to electromagnetic phenomena have been written using the electromagnetic system (emu). The ratio between any two is equal to the numerical value of the speed of light (3×10^{10} cm/s) or in some cases to the square of it (see Table I). The same quantity has different units in the two systems. The units

TABLE I

Conversion of Units

Quantity	Symbol	esu	emu[a]	Practical
Capacitance	C	1	v^{-2}	$10^9 \, v^{-2}$ F
Charge	q	1	v^{-1}	$10 \, v^{-1}$ C
Conductance	G	1	v^{-2}	$10^9 \, v^{-2}$ S
Current	i	1	v^{-1}	$10 \, v^{-1}$ A
Dielectric constant	ε	1	v^{-2}	$10^9 \, v^{-2}(4\pi)^{-1}$ F/cm
Displacement	D	1	v^{-1}	$10(4\pi v)^{-1}$ C/cm^2
Electromotive force	E	1	v	$10^{-8} \, v$ V
Electric field intensity	E	1	v	$10^{-8} \, v$ V/cm
Impedance	Z	1	v^2	$10^{-9} \, v^2 \, \Omega$
Potential	V	1	v	$10^{-8} \, v$ V
Resistance	R	1	v^2	$10^{-9} \, v^2 \, \Omega$
		mks to cgs		
Energy	W Joule	$10^7 \, W$ erg		
Force	$\mathscr{F}$ newton	$10^5 \, \mathscr{F}$ dyne		
Length	l meter	$10^2 \, l$ centimeter		
Mass	m kilogram	$10^3 \, m$ gram		
Power	P watt	$10^7 \, P$ erg per second		

[a] v is the speed of light = 3×10^8 m/s.

II. Electrostatics

of the practical system (see Table I) are based on the emu and differ from it in that units of more convenient size are used. Here two things are kept in mind: (1) the change is always made by some multiple of 10; and (2) the changes so made do not affect the fundamental equations.

In the esu system, forces are expressed in dynes and distances in centimeters, and the unit of charge is so chosen as to make the magnitude of k in Eq. (82) equal to unity. This unit of charge is called the statcoulomb. In the mks system, forces are expressed in newtons and distances in meters. The quantity of charge is defined in terms of the unit of current, the ampere. The mks unit of charge is the coulomb (i.e., ampere multiplied by second), and the value is

$$1 \text{ coulomb} = 2.996 \times 10^9 \text{ statcoulombs.}$$

The natural unit of electric charge is carried by an electron or proton, and exact measurements made find this charge to be

$$e = 1.602 \times 10^{-19} \text{ C}$$
$$= 4.802 \times 10^{-10} \text{ statcoulombs.}$$

In the mks system, the value of k in Eq. (82) is found to be

$$k = 8.988 \times 10^9.$$

In any system of units, the units of k are those of

$$(\text{force})(\text{distance})^2/(\text{charge})^2.$$

In the esu system

$$k = 1\left(\frac{\text{dyn cm}^2}{\text{statcoulomb}^2}\right).$$

In the mks system

$$k = 9 \times 10^9 \left(\frac{\text{N m}^2}{\text{C}^2}\right),$$

and so

$$1 \text{ N} = 10^5 \text{ dyn.}$$

In a number of equations derived from Coulomb's law, the factor 4π appears very frequently. A number of authors have avoided this appearance by introducing a new constant k_0 by the relation $k_0 = 1/4\pi k$, and so Coulomb's law is written

$$\mathscr{F} = \frac{1}{4\pi k_0} \frac{qq'}{r^2}. \tag{84}$$

k_0 is called the permittivity of free space and is given by $k_0 = 10^7/4\pi v^2$ where v is the velocity of light. When $k = 1$ [see Eq. (83)], $k_0 = 1/4\pi$.

From Eq. (83) both the unit of charge and electric field strength E (electric intensity) can be defined. The unit of electric charge is the charge that when placed 1 cm from an equal charge in vacuo repels or attracts it with a force of 1 dyn. E, the electric intensity at a point, is defined as the force in dynes that would act on a unit positive charge at that point. It therefore follows that the electric field strength at a point due to a charge q' at a distance r from the point is

$$E = q'/r^2. \tag{85}$$

From Eqs. (83) and (85) it follows that

$$\mathscr{F} = Eq. \tag{86}$$

That is, charge q in field E experiences a force that is given by Eq. (86).

The experimental test for the existence of an electric field at any point is to place a test charge q at the point. If a force is exerted on the test charge, then an electric field, whose magnitude is given by Eq. (86), exists at that point. In the mks system, electric intensity is measured in newtons per coulomb.

C. *Electrostatic Potential*

The potential V at a point in an electrostatic field is the work done in bringing a unit positive charge from infinity, i.e., from outside the field, to the point in question, the charges producing the field being held rigidly in position during the process. It represents physically the potential energy of a unit charge placed at a given point. Its unit is the joule per coulomb, i.e., the volt.

Consider two points P and Q (Fig. 1), a distance dl apart. Let the potentials at P and Q be V and $V + dV$, respectively. The drop in potential in going from P to Q is the work done, i.e., force (electric intensity) times distance moved. Therefore

$$V_P - V_Q = -dV = E(\cos \alpha)\, dl. \tag{87}$$

Hence the component of E in the direction of displacement PQ is E_l and

$$E_l = E \cos \alpha = -\frac{\partial V}{\partial l}. \tag{88}$$

E_l is greatest for α equal to zero. If dl is parallel to the X axis, it becomes dX, and $E \cos \alpha = E_X$, the X component of electric intensity. Consequently,

$$E_X = -\frac{\partial V}{\partial X}. \tag{89}$$

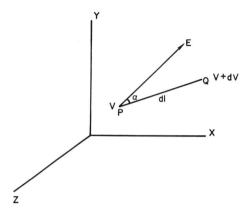

Fig. 1. Electric intensity E as a function of coordinates X, Y, and Z.

Similar expressions hold for the Y and Z components of E. If E is expressed as a function of coordinates X, Y, Z, then

$$E_X = -\frac{\partial V}{\partial X}, \quad E_Y = -\frac{\partial V}{\partial Y}, \quad \text{and} \quad E_Z = -\frac{\partial V}{\partial Z}. \tag{90}$$

If the potential function is known, electric intensity may be obtained from these relations. Since electric intensity is equal to the space rate of decrease of potential, a positive charge placed in an electric field experiences a force making it move from a region of higher to a region of lower potential, while a negative charge moves from a region of lower to a region of higher potential.

Work done in moving unit electric charge from a point A to a point B in the X direction is according to Eq. (89)

$$\int_a^b dV = -\int_a^b E_X \, dX, \tag{91}$$

where a and b are the distances of points A and B from charge q.

The value of the integral is independent of the path from A to B. If this were not so, the passage of a charge along the different paths between A and B would give rise to a net amount of work. That is, A would have more than one potential. This is not so by definition.

According to Eq. (91), the integral of the field strength from A to B is the potential difference between A and B.

The potential at a distance x from a charge q can be obtained from Eqs. (85) and (91) as

$$V_A - V_B = -\int_a^b \frac{q}{x^2} \, dx = \frac{q}{a} - \frac{q}{b}. \tag{92}$$

There is no absolute zero of potentials. When $b \to \infty$, V_B is arbitrarily taken to be zero. Therefore the potential at A is

$$V_A = q/a. \tag{93}$$

An equipotential surface is a surface all points of which are at the same potential. The electric field at any point is at right angles to the equipotential surface passing through that point, or else the potential would vary from point to point.

D. Electrostatic Energy

The electrostatic energy of a charge q is the amount of work done in bringing it from infinity to that point where the potential is V. The force on charge q at any point is Eq [Eq. (86)]. The work done in moving the charge through a distance dx is $E_x q\, dx$. Total work done is given by

$$\int_\infty^a E_x q\, dx,$$

that is,

$$q \int_\infty^a E_x\, dx = qV \tag{94}$$

according to Eq. (91). It is assumed that bringing the charge q from infinity to the point at which the potential is V had little effect on the potential V. This would be so if there were a large number of charges already present. Entry of a further charge into the medium would not affect the electric potential there. A different case of electrostatic energy is brought up in charging a body in such a way that the potential of the body is affected.

Let the body under consideration be at a potential V_1. During charging, V_1 changes from 0 to V. From Eq. (93), it follows that V_1 is proportional to q, and therefore $V_1 = \text{const}\, q$. When the charge on the body is increased from 0 to q, the potential changes from 0 to V, and the work done in charging is given by

$$\int_0^q \text{const}\, q\, dq.$$

Therefore the energy of charging of a body from 0 to q is

$$\text{const}\, \tfrac{1}{2} q^2 = \tfrac{1}{2} qV. \tag{95}$$

The difference between Eqs. (94) and (95) should be clearly understood.

E. Gauss's Theorem; Electric Flux Associated with Charge

To explain this concept of electric flux in terms of an analogy, imagine the space occupied by an electric field to be filled with an imaginary fluid whose velocity at every point is equal to the electric field in both magnitude and direction. The volume of fluid passing through any surface per unit time is equal to the electric flux through that surface.

The electric flux through a spherical surface of radius r due to a charge q at its center may be found as follows.

According to Coulomb's law [Eq. (85)], the field in this case is $E = q/r^2$. The velocity of the imaginary fluid is also q/r^2, and its flux therefore is q/r^2 multiplied by the area of the sphere, $4\pi r^2$. Thus the flux J is given by

$$J = 4\pi q. \tag{96}$$

If there are a number of charges in the closed surface, the total flux is given by 4π multiplied by the algebraic sum of the charges within the surface.

F. Poisson's and Laplace's Equations

Let P and Q (see Fig. 2) be two large plane parallel conducting plates between which there is an electric field directed from P to Q. Suppose there are electric charges distributed in the place between the plates. This charge is called space charge. Let this space charge density be ρ (i.e., charge per unit volume).

Apply Gauss's law to a thin volume element of cross-sectional area A and thickness Δx. E is the electric field having the same value at the left plate P and $E + \Delta E$ at the right plate Q. If dE/dx is the rate of increase of electric intensity with increasing x, the increase in intensity is given by

$$\Delta E = \frac{\partial E}{\partial x} \Delta x.$$

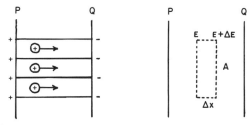

Fig. 2. Two large parallel conducting plates P and Q containing electric charges.

Therefore at face Q

$$E + \Delta E = E + \frac{\partial E}{\partial x} \Delta x.$$

The surface integral of E at the right hand face is

$$\left(E + \frac{\partial E}{\partial x} \Delta x \right) A.$$

Over the left face, it is $-EA$ since E is directed inward. The surface integral of E over the entire surface of the element is

$$\left(E + \frac{\partial E}{\partial x} \Delta x \right) A - EA = A \Delta x \frac{\partial E}{\partial x}.$$

The volume of the element is $A \Delta x$, and the charges enclosed are $\rho A \Delta x$. Hence according to Gauss's law

$$A \Delta x \frac{\partial E}{\partial x} = 4\pi \rho A \Delta x,$$

$$\frac{\partial E}{\partial x} = 4\pi \rho.$$

But

$$E = -\frac{\partial V}{\partial x}$$

[that is, Eq. (89)], and so

$$\frac{\partial^2 V}{\partial x^2} = -4\pi \rho. \tag{97}$$

In mks unit, Eq. (97) becomes $\partial^2 V/\partial x^2 = -\rho/kk_0$. This is Poisson's equation in one dimension where the potential varies only with x. Its general form where V depends on X, Y, and Z is found to be

$$\frac{\partial^2 V}{\partial X^2} + \frac{\partial^2 V}{\partial Y^2} + \frac{\partial^2 V}{\partial Z^2} = -4\pi \rho. \tag{98}$$

In the special case where charge density is zero,

$$\frac{\partial^2 V}{\partial X^2} = 0,$$

of which the general form is

$$\frac{\partial^2 V}{\partial X^2} + \frac{\partial^2 V}{\partial Y^2} + \frac{\partial^2 V}{\partial Z^2} = 0. \tag{99}$$

Equation (99) is called Laplace's equation.

G. *Electrostatic Capacity and Dielectric Constant*

A conductor is a material within which there are charges free to move provided a force is exerted on them by an electric field. If the field is maintained within the conductor, there will be continuous motion of its free charges. This motion is called a current. An insulator or dielectric is a substance within which there are no or relatively few charged particles free to move continuously under the influence of an electric field. When an uncharged body of any sort, conductor or dielectric, is brought into an electric field, a rearrangement of charges in the body always results. If the body is a conductor, the free electrons within it move in such a way as to make the interior of the body a field-free equipotential volume. If the body is a nonconductor, the electrons and positive nucleus in such a molecule are displaced by the field. Since they are not free to move indefinitely, the interior of the body does not become an equipotential region. The net charge remains zero, but certain regions of the body acquire excess positive or negative charges, called induced charges, when an uncharged conductor is introduced in the field between two plane parallel conductors having equal and opposite charges. If the fringe effect is neglected, the field is uniform in the region between the charged plates, as shown in Fig. 3. The motion of charges continues within the conductor until all the surface charge becomes equal and opposite to the original field.

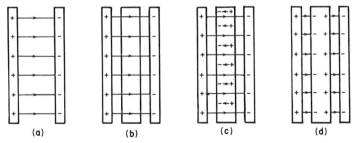

Fig. 3. (a) Electric field between two charged plates; (b) uncharged conductor introduced into the field without touching the charged plates; (c) induced charges and field; (d) resultant field induced. The field in the conductor is zero.

The case of a dielectric in the field may be considered. The molecules of the dielectric may be polar or nonpolar. A nonpolar molecule is one in which the centers of gravity of protons and electrons normally coincide, whereas a polar molecule is one in which they do not. Nonpolar molecules under the influence of a field are polarized and become induced dipoles of dipole moment equal to the product of both charges and the distance between them. The effect of the field will be to orient the dipole in the direction of the field. A polar molecule or a permanent dipole is also oriented in the direction of the external field.

When a nonpolar molecule is polarized, restoring forces come into play on the displaced charges. These are the interparticle binding forces that hold the molecule together. Under the influence of the field, the charges separate until the binding force is equal and opposite to the forces exerted by the field on the charges. The entire dielectric becomes polarized. The distribution of charges within the molecules of the dielectric will be as shown in Fig. 4. The two extreme layers indicated by the dotted lines have excess charges, one layer containing negative and the other layer containing positive excess charge. These are the induced surface charges.

The internal condition of a polarized dielectric is characterized not by an excess charge but by the relative displacement of the charges within it. The alignment of the induced charges will be the same as in a conductor, but the field created in the dielectric by the induced surface charges that are not free to move is weakened because of the restricted movement of the charges. The field in the dielectric is not reduced to zero, but in a conductor it is zero. The conditions existing in a polarized dielectric subject to an external field are indicated in Fig. 5.

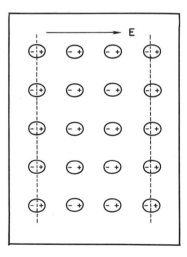

Fig. 4. Orientation of induced dipoles in the field.

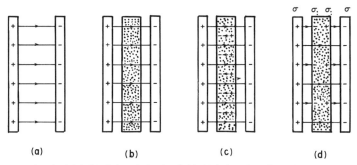

Fig. 5. (a) Electric field; (b) dielectric in the field; (c) induced surface charges and their field; (d) the resultant field. Induced charges on the faces of a dielectric in an electric field.

Let us consider the state of the polarized dielectric depicted in Fig. 5d. The edge or fringe effects may be neglected. The surface charge density of the induced charges will be uniform due to symmetry.

Let σ_i be the surface density of induced or bound charges, and let σ be the surface density of free charges on the plates. The induced charges reduce the effective surface density from σ to $\sigma - \sigma_i$. The electric field within the dielectric is

$$E = 4\pi(\sigma - \sigma_i) = 4\pi\sigma - 4\pi\sigma_i. \tag{100}$$

A new vector called displacement D is defined by $4\pi\sigma$, which represents that component of the resultant field set up by the free charges on the plates. $4\pi\sigma_i$ is the reversed field set up by the induced charges. Since induced charges are created by the field E, their magnitude will depend on the field E and also on the material of the dielectric. The ratio of induced charge density to the electric intensity is called the susceptibility of the material and is indicated by χ. Thus

$$\chi = \sigma_i/E \quad \text{or} \quad \sigma_i = \chi E. \tag{101}$$

A more general definition of electric susceptibility arises from the fact that surface effect is in reality a volume effect. The dipole moment of a polarized molecule is ql (charge times distance). If there are n such molecules per unit volume, the extent of polarization is $P = nql$. If q_i is the induced charge at each surface of a polarized slab of area A and thickness d, considering the entire slab as a large dipole, its dipole moment is $q_i d$ and its dipole moment per unit volume is

$$P = \frac{q_i d}{Ad} = \frac{q_i}{A} = \sigma_i. \tag{102}$$

In this special case dipole moment per unit volume is equal to surface density of induced charge. The effect of the field is considered as a surface phenomenon using σ_i. If it is considered as a volume phenomenon, the polarization P, dipole moment per unit volume, is used. If the two are numerically equal, they are expressed as charge per unit area. Therefore susceptibility in general terms is

$$\chi = P/E \quad \text{or} \quad P = \chi E. \tag{103}$$

The greater the susceptibility, the greater the induced charge in a given field, but the susceptibility of a given material is constant at constant temperature and moderate fields. So surface density of induced charges becomes proportional to resultant field.

In view of Eqs. (102) and (103), Eq. (100) may be rewritten

$$D = E + 4\pi P. \tag{104}$$

Also, Eq. (100) transforms in terms of susceptibility χ [Eq. (101)] to

$$E = \frac{4\pi\sigma}{1 + 4\pi\chi}. \tag{105}$$

If $1 + 4\pi\chi = \varepsilon$, then Eq. (105) reduces to

$$E = D/\varepsilon \quad \text{or} \quad \varepsilon = D/E. \tag{106}$$

ε is called the dielectric coefficient. It is also called by other names—specific inductive capacity or dielectric constant. It is a pure number since $4\pi\chi$ is a pure number. The quantities electric susceptibility χ, dielectric constant ε ($\varepsilon = 1 + 4\pi\chi$), and permittivity k ($k = \varepsilon k_0$) are different ways of describing the relative displacement or orientation of positive and negative charges within a substance when it is under the influence of an external field.

In terms of dielectric constant ε of the medium in which charges q, q' exist, the force acting between charges given by Eq. (84) will be reduced to

$$\mathscr{F} = \frac{qq'}{4\pi k_0 \varepsilon r^2},$$

the electric field to

$$E = q/\varepsilon r^2,$$

and the Poisson equation to

$$\frac{\partial^2 V}{dX^2} + \frac{\partial^2 V}{\partial Y^2} + \frac{\partial^2 V}{\partial Z^2} = -\frac{4\pi\rho}{\varepsilon}.$$

H. Capacitance of an Isolated Conductor

The potential V of an isolated charged sphere of radius a in empty space is given by Eq. (93), i.e., $q = Va$. The charge of a sphere is proportional to the potential. It can also be shown that charge on an isolated charged body is also proportional to its potential, i.e.,

$$q \propto V \quad \text{or} \quad q = CV, \quad (107)$$
$$C = q/V.$$

C is proportionality constant that depends on the size and shape of the conductor and is called its capacitance. It is expressed in coulombs per volt (farad).

The capacitance of an isolated spherical conductor is

$$C = q/V = aV/V = a.$$

The capacitance of a spherical conductor is equal to its radius. The capacitance of a spherical conductor immersed in a dielectric is given by εa, where ε is the dielectric constant.

I. The Parallel Plate Capacitor

The most common type of capacitor consists of two conducting plates of area A parallel to each other (see Fig. 6) and separated by a distance d that is small compared with the linear dimensions of the plates.

The entire field is localized in the region between the plates. The space between the plates, let us assume, is a vacuum. Then

$$E = 4\pi\sigma = 4\pi q/A.$$

The potential difference between the plates is V_{ab}, and

$$V_{ab} = Ed.$$

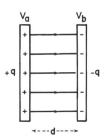

Fig. 6. Capacitor with two conducting plates separated by a small distance of d cm. qs are charges and Vs are voltages.

The capacitance therefore is given by

$$C = \frac{q}{V_{ab}} = \frac{EA}{4\pi}\frac{1}{Ed} = \frac{A}{4\pi d}. \qquad (108)$$

If the space is occupied by a substance of dielectric constant ε, then

$$C = \varepsilon A/4\pi d. \qquad (109)$$

If $C_0 = A/4\pi d$, the capacitance is increased by a factor of ε when a dielectric is inserted between the plates. The equation

$$\varepsilon = C/C_0. \qquad (110)$$

is used in defining dielectric constant and affords a direct, simple, and accurate method for determining dielectric constants as capacitance can be measured with high precision.

The unit of capacity

$$C_0 = \frac{A}{4\pi d}\frac{\text{cm}^2}{\text{cm}}$$

can simply be called centimeter. This is related to the farad by the relations

$$1\text{ V} = \tfrac{1}{300}\text{ esu} \quad\text{and}\quad 1\text{ C} = 3 \times 10^9\text{ esu}.$$

Thus

$$1\text{ F} = \frac{1\text{ C}}{1\text{ V}} = \frac{3 \times 10^9\text{ esu}}{\tfrac{1}{300}\text{ esu (statvolt)}}$$

$$= 9 \times 10^{11}\text{ cm}.$$

J. Combination of Capacitors

Capacitors can be arranged in parallel (Fig. 7a) or in series (Fig. 7b). In the former case, total charge q is given by

$$\begin{aligned}q &= q_1 + q_2 + \cdots + q_n \\ &= C_1(V_1 - V_2) + C_2(V_1 - V_2) + \cdots + C_n(V_1 - V_2) \\ &= (V_1 - V_2)(C_1 + C_2 + \cdots + C_n).\end{aligned}$$

Thus the resultant capacitance is given by

$$C = C_1 + C_2 + \cdots + C_n = \sum_i C_i. \qquad (111)$$

In the latter case with capacitors in series, the same quantity of charge exists

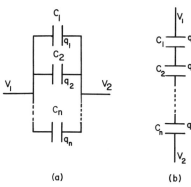

Fig. 7. (a) Capacitors in parallel; (b) capacitors in series.

in every capacitor. Thus

$$V_1 - V_2 = \frac{q}{C_1} + \frac{q}{C_2} + \cdots + \frac{q}{C_n} = q\left(\frac{1}{C_1} + \frac{1}{C_2} + \cdots + \frac{1}{C_n}\right).$$

In this case, C is given by

$$\frac{1}{C} = \frac{1}{C_1} + \frac{1}{C_2} + \cdots + \frac{1}{C_n} = \sum_i \frac{1}{C_i}. \tag{112}$$

K. Energy Stored in a Capacitor

Consider a capacitor of capacitance C with a potential difference $\Delta V = V_1 - V_2$ between the plates. The charge q is equal to $C \Delta V$. There is q charge on one plate and $-q$ charge on the other. If positive dq charge is transported from the negative to the positive plate, the work that has to be done is $dW = \Delta V \, dq = q \, dq / C$. Therefore to charge the capacitor from uncharged state to some final charge q_f, the work that must be done is

$$W = \frac{1}{C} \int_{q=0}^{q=q_f} q \, dq = \frac{1}{2} \frac{q_f^2}{C}. \tag{113}$$

This is the energy U stored in the capacitor. It can also be expressed by

$$U = \tfrac{1}{2} C (\Delta V)^2. \tag{114}$$

In the case of the parallel plate capacitor with plate area A and separation d, the capacitance is given by Eq. (107) and the electric field is $E = \Delta V/d$. Hence Eq. (114) is equivalent to

$$U = \frac{1}{2} \frac{A}{4\pi d} (Ed)^2 = \frac{E^2}{8\pi} \text{ (volume)}.$$

L. Current, Resistance, and Electromotive Force

Current across an area is defined quantitatively as the net charge flowing across the area in unit time. Thus if dq charge flows across an area in a time dt, current i across the area is

$$i = \frac{dq}{dt}. \tag{115}$$

The current per unit cross-sectional area is called the current density I. That is

$$I = i/A. \tag{116}$$

One of the earliest discoveries about electric currents in matter is contained in Ohm's law, which is

$$i = V/R, \tag{117}$$

where V is the voltage and R the resistance of the material of which the conductor is made. The resistance of a homogeneous conductor of length l and area of cross section A is given by

$$R = \rho(l/A), \tag{118}$$

where ρ is called the specific resistance or resistivity (Ω cm) and the reciprocal is called specific conductivity $[(\Omega\,\text{cm})^{-1}]$.

A device such as a galvanic cell (source) is schematically shown in Fig. 8. The source is said to be open since there is no external conducting path between the terminals a and b. The terminal a (+) is maintained by the source at a higher potential than the terminal b (−). Therefore there is an electrostatic field E_e at all points between and around the terminals, both inside and outside the source. The source is itself a conductor (electrolytic or metallic). If the force on the free charges within the source were only E_e, then positive charges would move from a to b and negative charges from b to a. The excess charges on terminals would decrease and the potential difference between the terminals would decrease and eventually become zero. As this does not happen, there must exist at every point within the source a force $\mathscr{F}_n$ of nonelectrostatic origin acting on every charged particle. This $\mathscr{F}_n$ must be equal and opposite to the electrostatic force $\mathscr{F}_e = qE_e$.

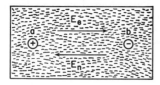

Fig. 8. Schematic representation of a galvanic cell.

The origin of $\mathscr{F}_n$ depends on the nature of the source. Irrespective of the origin of the source, an equivalent nonelectrostatic field can be defined by the equation

$$\mathscr{F}_n = qE_n \quad \text{or} \quad E_n = \mathscr{F}_n/q.$$

When the source is on open circuit as in Fig. 8, the charges are in equilibrium and the resultant field E, the sum of E_e and E_n, must be zero at every point. Thus

$$E = E_e + E_n = 0 \quad \text{or} \quad E_e = -E_n.$$

Hence on open circuit,

$$\int_a^b E_e \, dl = -\int_a^b E_n \, dl = \int_b^a E_n \, dl.$$

The first integral is the potential difference $(V_a - V_b)$. The last integral, the nonelectrostatic field from b to a (line integral), is called the electromotive force (emf) $\tilde{E}$. Thus for a source on open circuit, the potential difference $V_{ab} = V_a - V_b$ or open circuit terminal voltage is equal to the emf. That is

$$V_{ab} = \tilde{E}. \tag{119}$$

The electrostatic field within the source (i.e., $V_a - V_b$) depends on the current in the source, whereas the nonelectrostatic field (i.e., emf of source), in many cases a constant, is independent of the current. Thus emf represents a definite property of the source.

When the terminals of a source are connected by a wire, both the source and the wire form a closed circuit. Electrostatic field E_e sets up a current in the wire from a toward b. Both the charges on the terminals and the electrostatic fields within the wire and the source decrease. Consequently the electrostatic field within the source becomes smaller than the constant nonelectrostatic field. Thus positive charges within the source are driven toward the positive terminal, and there is a current within the source from b to a. The circuit settles down to a steady state in which the current is the same at every cross section.

Application of Eq. (117) to the wire and to the source gives, respectively,

$$V_{ab} = iR \tag{120}$$

and

$$\int_b^a E \, dl = ir, \tag{121}$$

where R is the resistance of the wire and r the internal resistance of the source within the source. The resultant field E is the vector sum of E_n and E_e, that is, $E = E_n + E_e$. Hence

$$\int_b^a E\,dl = \int_b^a E_n\,dl + \int_b^a E_e\,dl.$$

The first integral on the right-hand side is the emf $\tilde{E}$, and the second integral equals $-V_{ab}$. Therefore

$$\tilde{E} - V_{ab} = ir \quad \text{or} \quad V_{ab} = \tilde{E} - ir. \tag{122}$$

Thus when there is a current in the source from the negative to the positive terminal, the terminal voltage V_{ab} is smaller than the emf by the product ir. When $i = 0$, Eq. (122) reduces to Eq. (119).

Eliminating V_{ab} from Eqs. (120) and (122) gives

$$i = \tilde{E}/(R + r). \tag{123}$$

Substituting for i from Eq. (120) in Eq. (123) gives

$$V_{ab}/\tilde{E} = R/(R + r).$$

When $R \gg r$, the terminal voltage V_{ab} is apparently equal to the emf.

M. Resistors in Series and in Parallel

Most electrical circuits are composed of a number of sources, resistors, capacitors, etc., interconnected in a complicated manner. It is always possible to reduce the combination into a simple equivalent circuit.

In the case of resistors arranged in series (see Fig. 9a), the current is the same in each element, and is equal to the line current.

If the equivalent resistance of the series is R and V_{ab} is the potential difference between the terminals of the circuit, then

$$V_{ab} = V_{ax} + V_{xy} + V_{yb} = i(R_1 + R_2 + R_3).$$

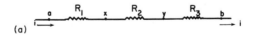

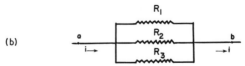

Fig. 9. (a) Series combination of resistors; (b) parallel combination of resistors.

Thus

$$V_{ab}/i = R_1 + R_2 + R_3.$$

The equivalent resistance is given by $R = V_{ab}/i$; hence

$$R = R_1 + R_2 + R_3. \tag{124}$$

In general, for any number of resistances in series,

$$R = \sum_i R_i.$$

If the resistors are in parallel (Fig. 9b), the potential between the terminals of each must be the same and equal to V_{ab}. If the currents in each are i_1, i_2, i_3, then

$$i_1 = \frac{V_{ab}}{R_1}, \quad i_2 = \frac{V_{ab}}{R_2}, \quad \text{and} \quad i_3 = \frac{V_{ab}}{R_3}.$$

As all the charge delivered at a is removed from a (no accumulation anywhere), it follows that

$$i = i_1 + i_2 + i_3 = V_{ab}\left(\frac{1}{R_1} + \frac{1}{R_2} + \frac{1}{R_3}\right).$$

As $1/R = i/V_{ab}$, we have

$$\frac{1}{R} = \frac{1}{R_1} + \frac{1}{R_2} + \frac{1}{R_3}. \tag{125}$$

In general for parallel combination of resistors,

$$\frac{1}{R} = \sum_i \frac{1}{R_i}.$$

When it is not possible to reduce networks into simple series–parallel combinations (see Fig. 10), special methods are used for handling such networks. One such method was developed by Kirchhoff.

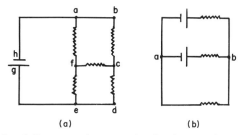

Fig. 10. (a) and (b) are complex networks of resistors and sources of emf.

N. Kirchhoff's Laws

A branch point in a network is a point where three or more conductors are joined. In Fig. 10a, a, f, c, and e are branch points, but b and d are not. In Fig. 10b, there are only two points a and b. A loop or mesh is any closed conducting path. Possible loops in Fig. 10a are *abcfa*, *fcdef*, *hafegh*, and *hafcdegh*.

LAW 1: The algebraic sum of all currents meeting at a junction is zero.

LAW 2: The algebraic sum of the Ri terms around any closed path is equal to the algebraic sum of the applied emfs in the given path.

Application of these rules to the network of Fig. 10b may be considered. In Fig. 11 are given the magnitudes and directions of the emfs and resistances: at point b, $i_1 + i_2 + i_3 = 0$; at point a, $-i_1 - i_2 - i_3 = 0$. These two equations are identical. Taking clockwise direction of current in the loop to be positive, Law 2 gives for the upper loop

$$\tilde{E}_1 - \tilde{E}_2 = i_1 r_1 + i_1 R_1 - i_2 r_2 - i_2 R_2,$$

and for the lower loop

$$\tilde{E}_2 = i_2 r_2 + i_2 R_2 - i_3 R_3.$$

Hence the three independent equations may be solved for the three unknown currents.

If n galvanic cells are arranged in series, the emf increases n times and the internal resistances must be added so that the current in a given external

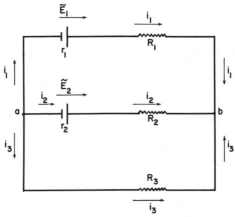

Fig. 11. The network of Fig. 10b giving magnitudes and directions of resistors and emfs to illustrate Kirchhoff's laws.

resistance is given by
$$i = n\tilde{E}/(R + nr).$$
When $R \gg nr$, $i = n\tilde{E}/R$. That is, the current is n times greater than that of a single cell. When $r \gg R$, $i = \tilde{E}/r$. This means that the current is approximately the same as in a single cell.

If n cells are placed in parallel in the circuit of external resistance R, the emf remains the same. The internal resistance is reduced n times. Thus
$$i = \frac{\tilde{E}}{R + (r/n)}.$$
When $R \gg r$, $i = \tilde{E}/R$; the current is the same as for a single cell. When $r \gg R$, $i = n\tilde{E}/r$; the current increases with the number of cells in the parallel circuit.

III. Physical and Electrochemical Principles

A. Relation between Free–Energy Change and the Potential of a Galvanic Cell

For a galvanic cell whose emf is $\tilde{E}$ volts and which is operated reversibly at constant temperature and pressure, the electrical work done on the passage of an infinitesimal quantity of electricity (δq, C) is $\tilde{E}\,\delta q$, V C (or joules). If $-\Delta G$ is the decrease in free energy of the reacting species in the cell due to passage of n Faradays of electricity, the change in free energy brought about by δq C of electricity is $-\Delta G\,\delta q/nF$, and this is equivalent to electrical work $\tilde{E}\,\delta q$. Therefore
$$-\Delta G = n\tilde{E}F. \tag{126}$$

B. Equilibrium of Ions in Different Phases

Equation (35) is valid for electrically charged ions of electrolyte solutions since single ions cannot be transferred without transferring an equivalent quantity of oppositely charged ions. But the electrical field arising from this movement of oppositely charged ions of different mobilities must be taken into account. If V' and V'' are the electrical potentials in the two phases (') and (''), and the valence of the ion is z_i, the electrical work done in moving a charge $z_i F\,dn_i$ from a region of potential V' to a region of potential V'' is $(V'' - V')z_i F\,dn_i$. The chemical work, i.e., change in free energy, is
$$(\mu_i'' - \mu_i')\,dn_i.$$
The total work done in the transfer of dn_i moles is
$$(\mu_i'' - \mu')\,dn_i + (V'' - V')z_i F\,dn_i, \tag{127}$$

and according to Eq. (12), Eq. (127) is zero for equilibrium conditions. As $dn_i \neq 0$, it follows that

$$\mu_i'' + z_i F V'' = \mu_i' + z_i F V'. \qquad (128)$$

The quantity $\mu + zFV$ is called the electrochemical potential $\bar{\mu}$ and has the same property for ions as chemical potential has for neutral species.

C. Diffusion in Aqueous Solutions

Diffusion is one of the most fundamental irreversible processes by which a difference of concentration is reduced by spontaneous flow of matter. In a solution, flow of solute occurs in the direction of lower concentration and flow of solvent occurs in the reverse direction. The rate of flow would be approximately proportional to the difference in concentration between the two solutions.

The flux of matter J is defined as the quantity (moles, for example) of material crossing unit area of a plane perpendicular to the direction of flow in unit time (mol cm^{-2} s^{-1}).

The concentration gradient $\partial C/\partial x$ is the rate of increase of concentration with distance measured in the direction of flow. The diffusion coefficient D is defined by Fick's first law,

$$J = -D \frac{\partial C}{\partial x}. \qquad (129)$$

The partial differential is necessary, as the concentration is dependent both on time and distance. The negative sign is required to make D a positive quantity. The diffusion flux of species i is caused by the chemical potential gradient $(d\mu_i/dx)$ (see Fig. 12).

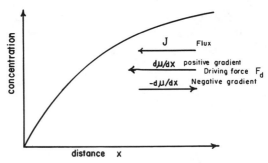

Fig. 12. Schematic illustration of the distance variation of concentration of a species and relative directions of the diffusion flux, etc.

In the steady state, both the driving force $\mathscr{F}_d$ and the flux J reach values that remain constant with time. The relation between the two can be expressed by a power series

$$J = A + B\mathscr{F}_d + C\mathscr{F}_d^2 + \cdots.$$

A, B, and C are constants. If $\mathscr{F}_d$ is sufficiently small, power terms in $\mathscr{F}_d$ can be neglected. Thus

$$J = A + B\mathscr{F}_d. \tag{130}$$

The constant A must be zero since when $\mathscr{F}_d$ is zero, there can be no diffusional flux. Hence for sufficiently small driving force, flux is linearly related to the driving force

$$J = B\mathscr{F}_d. \tag{131}$$

The driving force on 1 mol of a species i is $-(d\mu_i/dx)$. If the concentration of i is C_i (moles per unit volume) adjacent to the transit plane, the driving force at the plane is $-C_i(d\mu_i/dx)$. Substituting this value in Eq. (131) gives

$$J = -BC_i \frac{d\mu_i}{dx}. \tag{132}$$

As $\mu_i = \mu_i^0 + RT \ln C_i$ for an ideal solution (activity coefficient 1), Eq. (132) becomes

$$J = -BC_i RT \frac{d \ln C_i}{dx} = -BRT \frac{dC_i}{dx}. \tag{133}$$

Equating Eqs. (129) and (133) gives the relation

$$D = BRT. \tag{134}$$

Fick's first law is of importance in the study of diffusion by steady-state methods in which the concentration gradient does not change with time. However, in many methods currently used in the experimental study of diffusion, variation of concentration with both distance and time is of importance. For these methods, Eq. (129) may be transformed into a second-order differential equation connecting C, x, and t.

Consider a parallelepiped (Fig. 13) of unit area and length dx. Molecules diffuse in through face A and diffuse out through face B. If C is the concentration of the species at the face A, the concentration at face B is $C + (dc/dx)\, dx$. The flux J_A at face A is given by Eq. (129). Thus

$$J_A = -D \frac{\partial C}{\partial x}, \tag{129}$$

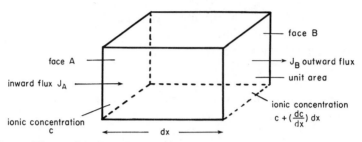

Fig. 13. The parallelepiped of electrolyte solution used to derive Fick's second law.

and the flux J_B out of face B is given by

$$J_B = -D \frac{\partial}{\partial x}\left(C + \frac{\partial C}{\partial x} dx\right)$$

$$= -D \frac{\partial C}{dx} - D \frac{\partial^2 C}{\partial x^2} dx. \qquad (135)$$

The net flow of material from the parallelepiped of volume dx is

$$J_A - J_B = D \frac{\partial^2 C}{\partial x^2} dx. \qquad (136)$$

Thus the net flow per unit volume per unit time is

$$\frac{J_A - J_B}{dx} = D \frac{\partial^2 C}{\partial x^2}.$$

As $(J_A - J_B)/dx$ is the rate of change of concentration with time, it follows that

$$\frac{\partial C}{\partial t} = D \frac{\partial^2 C}{\partial x^2} \qquad (137)$$

For diffusion in three dimensions, Eqs. (129) and (137) take the forms

$$J = -D \text{ grad } C$$

and

$$\frac{\partial C}{\partial t} = \text{div}(D \text{ grad } C).$$

D. Relation of Diffusion Coefficient D to Other Molecular Quantities

From a kinetic point of view, one can consider the molecules in solution (solute molecules or ions) as being in constant haphazard zigzag random motion in three-dimensional space. For the sake of simplicity, the special case of one-dimensional random walk (that of a drunken sailor) may be considered. Starting from $x = 0$ in the x direction, the forward motion (a step that the drunken sailor takes) is as likely as a backward step. In the first try, if N steps are taken, let x_1 be the distance covered from the origin. In the ith attempt, let x_i be the distance from the origin transversed. Then the average distance $\langle x \rangle$ from the origin is

$$\langle x \rangle = \left(\sum_i x_i \right) \bigg/ \left(\sum i \right) = \frac{\text{sum of distances from origin}}{\text{number of attempts}}.$$

Since the overall distance traversed is as likely to be in the plus direction as in the minus x direction, it is obvious that the mean distance traversed in a large number of times will be $\langle x \rangle = 0$. Consequently, computing the mean distance from the origin is not useful. Instead, if the square of the distance traversed (minus or plus, the square will always be positive) is considered, the mean square distance $\langle x^2 \rangle$ is given by

$$\langle x^2 \rangle = \left(\sum_i (x_i)^2 \right) \bigg/ \left(\sum i \right) = \frac{\text{sum of distances squared}}{\text{number of trials}}.$$

Thus $\langle x^2 \rangle^{1/2}$ is the linear distance covered by a molecule in time t.

Consider two compartments each of unit cross-sectional area and length $\langle x^2 \rangle^{1/2}$ on either side of the plane of transit (Fig. 14). Let C_L and C_R be the concentration of molecules in the two compartments. The average rate of movement of molecules in either compartment is $\langle x^2 \rangle^{1/2}/t$. As both forward and backward movements are likely, the number of molecules crossing the plane of transit per second from the left compartment is $\frac{1}{2} C_L \langle x^2 \rangle^{1/2}/t$. Similarly, the number crossing from the right compartment is $\frac{1}{2} C_R \langle x^2 \rangle^{1/2}/t$. Thus the net flux across the transit plane is given by

$$J = \tfrac{1}{2}(C_L - C_R)\langle x^2 \rangle^{1/2}/t. \tag{138}$$

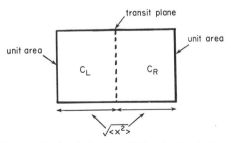

Fig. 14. Schematic diagram for the derivation of Einstein–Smoluchowski equation showing the transit plane in between two imaginary compartments.

But the concentration gradient is $(C_R - C_L)/\langle x^2 \rangle^{1/2}$. Thus Eq. (138) may be rewritten

$$J = -\frac{1}{2} \frac{C_R - C_L}{\langle x^2 \rangle^{1/2}} \frac{\langle x^2 \rangle}{t}. \tag{139}$$

Equating Eqs. (139) and (129) gives

$$D = \langle x^2 \rangle / 2t. \tag{140}$$

Equation (140) is the Einstein–Smoluchowski equation. It bridges the microscopic view of random walking molecules and the diffusion coefficient D of the macroscopic Fick's law.

The mean square distance $\langle x^2 \rangle$ traversed by a molecule is related to the number of jumps N the molecule makes and the mean jump distance l.

After $N - 1$ jumps, let x_{N-1} be the distance covered by the molecule. On taking one more jump, the distance x_N from the origin is either

$$x_N = x_{N-1} + l \tag{141}$$

or

$$x_N = x_{N-1} - l. \tag{142}$$

Squaring Eqs. (141) and (142) and taking the average gives

$$x_N^2 = x_{N-1}^2 + l^2, \tag{143}$$

which is true when the distance traveled after $N - 1$ jumps is exactly x_{N-1}. However, one can only expect an averaged value $\langle x_{N-1}^2 \rangle$ for the value of the square of the distance at the $(N - 1)$th jump. Thus Eq. (143) must be written

$$\langle x_N^2 \rangle = \langle x_{N-1}^2 \rangle + l^2.$$

At the start of the jump at x_0,

$$\langle x_0^2 \rangle = 0.$$

After jump one, the distance covered is

$$\langle x_1^2 \rangle = l^2.$$

After jump two, it is

$$\langle x_2^2 \rangle = \langle x_1^2 \rangle + l^2 = l^2 + l^2 = 2l^2.$$

After jump three, it is

$$\langle x_3^2 \rangle = \langle x_2^2 \rangle + l^2 = 2l^2 + l^2 = 3l^2.$$

In general, after N jumps, the distance covered is

$$\langle x_N^2 \rangle = Nl^2. \tag{144}$$

Substituting Eq. (144) in Eq. (140) gives

$$D = Nl^2/2t,$$

when $N = 1$. $D = l^2/2\tau$, where τ is the mean jump time in seconds required to traverse the mean jump distance l. The number of jumps per second or the rate constant k for transfer is given by $1/\tau$. Thus

$$D = \tfrac{1}{2}kl^2 \tag{145}$$

The number $\tfrac{1}{2}$ comes in Eq. (140) on the consideration that transfer of molecules occurred by one-dimensional random walk. In general, it should be related to the probability of molecular jumps taking place in several directions, not just forward and backward. So for convenience, the coefficient could be taken as unity and so

$$D \approx kl^2. \tag{146}$$

E. Rate Theory of Diffusion

Molecules jump from one site to another in a liquid. If the free energy of the system is plotted as a function of the position of the jumping molecule, then the standard free energy of the system must attain a critical value (see Fig. 15)—activation free energy ΔG—for the process to be completed. That is, the system has to cross an energy barrier for the rate process to occur. The peak of the curve corresponds to the critical configuration of the system, the "activated complex" whose energy E is given by

$$E = h\nu, \tag{147}$$

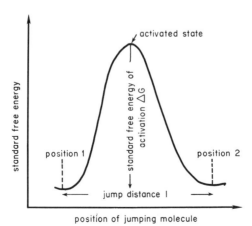

Fig. 15. Activation energy barrier.

where v is the vibration frequency of the complex and h the Planck constant. A vibration in the activated complex corresponds to a relative displacement (or translation) of the constitutents of the complex. The translational energy associated with an entity is $\tfrac{1}{2}\kappa T$ (κ is the Boltzmann constant) or κT for a pair of entities. Thus

$$E = \kappa T. \tag{148}$$

As Eqs. (147) and (148) are applicable to the same process, although looked at in two different ways, they must be equal. Thus

$$hv = \kappa T. \tag{149}$$

It is possible that not all systems reaching the activated state would attain the final state. Some of them may be turned back. So a factor called transmission coefficient ($\dot{k}$) is introduced to allow for this possibility. Generally it is assumed that $\dot{k}$ is close to unity.

The activated complexes are assumed to be in equilibrium with the reactants. For example,

jumping molecule + site ⇌ molecule-site
C_R (reactants) C_{AC} (activated complex)

Thus by the law of mass action

$$\frac{C_{AC}}{C_R} = K = \exp(-\Delta G/RT) \tag{150}$$

where K is the equilibrium constant and ΔG the standard free energy of activation. When C_R is at unit concentration

$$C_{AC} = \exp(-\Delta G/RT) \tag{151}$$

The number of times per second (k) that the rate process occurs is given by the product of the concentration of activated complex and its vibration frequency v. Thus

$$k = vC_{AC}. \tag{152}$$

Substitution of Eqs. (149) and (151) in Eq. (152) gives

$$k = \frac{\kappa T}{h} \exp(-\Delta G/RT). \tag{153}$$

This is the rate constant for the jumping of a molecule form site to site. Substituting Eq. (153) in Eq. (146) gives

$$D \approx l^2 \frac{\kappa T}{h} \exp(-\Delta G/RT). \tag{154}$$

F. Membrane Permeability

In the study of biological membranes whose thickness is not known with any degree of precision, diffusion of molecules across them has been described by a permeability coefficient P instead of a diffusion coefficient D.

Consider a membrane of thickness d cm interposed between two phases i (inside) and o (outside), where the concentrations of substance are measured at C_i and C_o (Fig. 16). In principle, the concentration profile through the membrane is determined by solving the diffusion equation (137). But as diffusion through the membrane reaches a steady state, $\partial C/\partial t = 0$. Consequently, $\partial^2 C/\partial x^2 = 0$. The solution of this equation is

$$C(x) = Ax + B, \tag{155}$$

where A and B are constants. It follows that the concentration profile is linear in a one-dimensional steady-state diffusion process. The constants A and B are found by satisfying the boundary conditions $C(x) = C_i$ for $x = 0$ and $C(x) = C_o$ for $x = d$. This gives

$$C(x) = \frac{C_o - C_i}{d} x + C_i, \tag{156}$$

which describes the concentration profile in the membrane during steady-state diffusion. The flux is given by Fick's law [Eq. (129)], so that

$$J = -D \frac{C_o - C_i}{d} = \frac{D}{d}(C_i - C_o). \tag{157}$$

This solution is based on the assumption that the substance has the same solubility in both the inside and outside phases and the membrane.

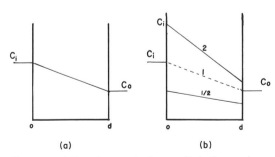

Fig. 16. Schematic representation of concentration profile in the membrane of thickness d cm: (a) linear profile; (b) linear profiles when the solubility of the permeating species in the membrane is greater, equal to, and less than unity.

If the solubility of the substance in the membrane differs from the solubility in the phases on either side of the membrane, the concentration profile is still determined by Eq. (155), but the boundary conditions for $C(x)$ are

$$C(x) = \begin{cases} \beta C_i & \text{for } x = 0 \\ \beta C_o & \text{for } x = d, \end{cases}$$

where β is the distribution or partition coefficient for the substance. Solving Eq. (155) for these boundary conditions gives

$$C(x) = \frac{\beta}{d}(C_o - C_i)x + \beta C_i. \tag{158}$$

Figure 16b shows the concentration profiles corresponding to $\beta > 1$ and $\beta < 1$. The flux now becomes

$$J = \frac{\beta D}{d}(C_i - C_o). \tag{159}$$

For a given C_i and C_o, the flux is thus determined by the term

$$P = \frac{\beta D}{d} = \frac{\beta BRT}{d}. \tag{160}$$

The last term in Eq. (160) is obtained by using the Nernst relation, Eq. (134). P is called the permeability coefficient or the permeability for the substance under consideration. Its units are centimeters per second or meters per second. When the thickness of the membrane is known approximately, Eq. (159) can be written in the form

$$J = P(C_i - C_o).$$

G. Stationary Diffusion through Composite Membranes

Two membranes of thickness d_1 and d_2 in which the diffusing substance has different diffusion coefficients may be considered. In the steady state $(\partial^2 C/\partial x^2) = 0$. Thus concentration profiles are linear in both the membranes. The flux across the total membrane unit is constant. Thus the flux is

$$J = P_1 \Delta C_1 = P_2 \Delta C_2,$$

where ΔC_1 and ΔC_2 are the concentration differences across membrane 1 and membrane 2 having permeabilities P_1 and P_2, respectively. But

$$\Delta C_1 + \Delta C_2 = \Delta C = J/\langle P \rangle,$$

where $\langle P \rangle$ is the equivalent permeability. From these considerations it follows that

$$\frac{J}{P_1} + \frac{J}{P_2} = \frac{J}{\langle P \rangle} \quad \text{or} \quad \frac{1}{\langle P \rangle} = \frac{1}{P_1} + \frac{1}{P_2}. \tag{161}$$

In general for a pack of membranes of n different permeabilities, one can write

$$\frac{1}{\langle P \rangle} = \frac{1}{P_1} + \frac{1}{P_2} + \cdots + \frac{1}{P_n} = \sum_i \frac{1}{P_i}. \tag{162}$$

The material transport through a composite membrane is limited by that membrane unit with the lowest permeability. Equation (162) is useful in evaluating the role of Nernst stationary or unstirred layers close to the membrane faces in determining the overall transport across the membrane.

H. Diffusion and Mobility

In the case of electrolyte solutions when random-walking ions are subject to a directed force $\mathscr{F}$, they acquire a nonrandom extra velocity in the direction of the force. This additional component of its velocity due to the force is called drift velocity v_d. Drift velocity when the force is 1 dyn is called absolute mobility. That is,

$$u_{abs} = v_d/\mathscr{F} \quad \text{cm s}^{-1}\text{dyn}^{-1}. \tag{163}$$

But mobility of ions are usually expressed per unit electric field (1 V cm^{-1}) acting on the ions and are conventionally referred to as electrochemical or electrical mobilities u_{conv}. Thus

$$u_{conv} = \frac{v_d}{E}\left(\frac{\text{cm s}^{-1}}{\text{V cm}^{-1}}\right). \tag{164}$$

The electric force $\mathscr{F}$ acting on the ion is equal to electric force per unit charge, i.e., electric field E multiplied by the charge ze on each ion (z is the valence and e the electronic charge). Thus

$$\mathscr{F} = zeE/300.$$

Equation (164) therefore becomes

$$u_{conv} = v_d ze/300\mathscr{F}.$$

Thus

$$u_{conv} = (ze/300)u_{abs} = 1.6 \times 10^{-12}\, zu_{abs}. \tag{165}$$

When an electric field is applied to an electrolyte solution, both positive and negative ions will drift across a chosen transit plane of unit area in opposite directions. If v_+ and v_- are the drift velocities of positive and negative ions, then all positive ions within a distance of v_+ cm from the transit plane will move across it in 1 s. Therefore, the flux J is given by

$$J_+ = C_+ v_+ \tag{166}$$

(C_+ = mol cm^{-3}). Current density I is given by

$$I_+ = z_+ F J_+ = z_+ F C_+ v_+.$$

In general, one can write

$$I_j = z_j F C_j v_j. \tag{167}$$

The total current density due to all ions will therefore be

$$I = \sum I_j = \sum z_j F C_j v_j. \tag{168}$$

Substituting for drift velocity v_j from Eq. (164) into Eq. (168) gives

$$I = \sum z_j F C_j u_{\text{conv}} E,$$

but Eqs. (116) and (117) give for the electrolyte solution

$$i = \frac{VA}{\rho l} \quad \text{or} \quad I = \frac{V}{\rho l} = \frac{E}{\rho}.$$

Specific conductance k_{sp} is given by

$$k_{\text{sp}} = \frac{1}{\rho} = \frac{I}{E} = \sum z_j F C_j u_{j(\text{conv})}. \tag{169}$$

Electrolyte solution resistance R may be considered to be given by Eq. (118) where l/A describes the geometry of the cell and is called the cell constant a, which in practice is determined by using a solution of known specific resistance. If a cube of the electrolyte solution is considered, the cell constant is unity, and so $R = 1/k_{\text{sp}}$. If the length of this cube in the direction of current is not unity but l, the resistance becomes

$$R = \rho l = l/k_{\text{sp}} \tag{170}$$

and the conductance becomes

$$G = 1/R = k_{\text{sp}}/l. \tag{171}$$

In the mks system, the units for ρ, R, and G are ohm meters (Ω m), ohm square meters (Ω m^2), and siemens per square meter (S m^{-2}).

The considerations discussed above are valid for a solution of uniform concentration, but in a case where the electrolyte concentration in the region under consideration is not uniform, the conductance can be calculated if the concentration profile $C_j(x)$ is known for the individual ions. For a layer of

III. Physical and Electrochemical Principles

infinitesimal thickness dx, Eq. (170) becomes

$$dR = \frac{dx}{k_{sp}}.$$

The total resistance through the entire layer of thickness l is, according to Eq. (169),

$$R = \int_0^l dR = \int_0^l dx/(\sum z_j F C_j u_j), \tag{172}$$

where for u_j, the subscript (conv) has been dropped.

The resistance calculated from Eq. (172) is called the integral resistance, and the conductance $G = 1/R$ is called the integral conductance. For individual ions, the resistance R_j or conductance G_j is given by

$$R_j = \int_0^l \frac{dx}{z_j F C_j u_j} = \frac{1}{G_j}. \tag{173}$$

The total conductance $G = \sum G_j$ may not be equal to that calculated from Eq. (172). In the case of a regime that is homogeneous, the total conductance is given by Eq. (172) whereas the single-ion conductances calculated from Eq. (173) have only a formal significance. But in the case of a mosaic regime where each ion has its own migration pathway or channel, the single-ion resistances or conductances calculated by using Eq. (173) have a reality and the total conductance is given by

$$G = \sum_j G_j.$$

In the case of $(z:z)$ valence electrolyte solution, Eq. (169) becomes

$$k_{sp} = zFC[u_+ + u_-].$$

The molar conductance λ_m of an electrolyte solution is given by

$$\lambda_m = k_{sp}/C = zF(u_+ + u_-). \tag{174}$$

Equivalent conductivity $\lambda = (\lambda_m/z)$ is given by

$$\lambda = F(u_+ + u_-), \tag{175}$$

which shows that equivalent conductivity is independent of concentration. This is only true at infinite dilution where ion–ion interactions are absent.

The equivalent conductance at infinite dilution λ_∞ may be considered as the sum of two ionic conductances λ_+ and λ_-. Thus Eq. (175) becomes

$$\lambda_\infty = \lambda_+ + \lambda_- = F(u_+ + u_-)$$

and so

$$u_+ = \lambda_+/F \quad \text{and} \quad u_- = \lambda_-/F. \tag{176}$$

In applying these equations care is needed with the units. For any given ion, u_{abs} cm s^{-1} dyn^{-1} = 6.469 × 10^6 $\lambda/|z|$ cm^2 Ω^{-1} equiv^{-1}.

Imposition of a driving force (e.g., electrochemical field) allows ions to migrate (conduction) with the drift velocity v_d. In addition, there is also diffusion arising from the random walk of ions. Einstein established the relation between drift u_{abs} and diffusion D. Diffusion flux J_d is given by Eq. (129). Conduction flux J_c is obtained by dividing conduction current density by the charge per mole of ions. Thus

$$J_c = \frac{I}{zF} = \frac{zFCv_d}{zF} = Cv_d$$

and so

$$J_c = Cu_{abs}\mathscr{F}.$$

If the applied electric field is adjusted to just compensate the diffusional flux, the net flux will be zero. Thus

$$J_d + J_c = 0$$

or

$$-D\frac{dC}{dx} + Cu_{abs}\mathscr{F} = 0,$$

$$\frac{dC}{dx} = \frac{Cu_{abs}\mathscr{F}}{D}$$

(177)

For these equilibrium or balanced conditions, the concentration C at any distance x, as the field varies in the x direction, can be determined by the application of the Boltzmann equation. Hence

$$C = C_0 \exp(-U/\kappa T) \tag{178}$$

where U is the potential energy of an ion in the electric field and C_0 is the concentration in the region where $x = 0$. Thus differentiation of Eq. (178) gives

$$\frac{dC}{dx} = -C_0 \exp(-U/\kappa T)\frac{1}{\kappa T}\frac{dU}{dx}$$

$$= -\frac{C}{\kappa T}\frac{dU}{dx}.$$

But $\mathscr{F} = -dU/dx$, and so

$$\frac{dC}{dx} = \frac{C}{\kappa T}\mathscr{F}. \tag{179}$$

Equating Eq. (179) to Eq. (177) yields

$$D = \kappa T u_{\text{abs}}. \tag{180}$$

This is the Einstein relation. On the other hand, Eq. (134) is due to phenomenological treatment of diffusion. Equating the two gives

$$BRT = \kappa T u_{\text{abs}}, \tag{181}$$

$$B = u_{\text{abs}}/N,$$

where N is Avogadro's number. The Einstein relation allows experiments on diffusion to be related to phenomena that produce drift velocities such as equivalent conductance (due to electric field) and viscosity (viscous drag).

I. Nernst–Einstein Relation

Equation (165) substituted into Eq. (176), omitting the conversion factor, gives

$$(u_{\text{abs}})_j z_j e = \lambda_j / F,$$

i.e.,

$$\lambda_j = (u_{\text{abs}})_j z_j e F. \tag{182}$$

Substituting the Einstein relation [Eq. (180)] into Eq. (182) gives

$$\lambda_j = \frac{z_j e F}{\kappa T} D_j \quad \text{or} \quad \lambda_j = \frac{z_j F^2}{RT} D_j. \tag{183}$$

From a knowledge of the diffusion coefficients of individual ions, a more useful form of Nernst–Einstein relation can be obtained, and it is

$$\lambda = \lambda_+ + \lambda_- = \frac{zF^2}{RT}[D_+ + D_-]. \tag{184}$$

An important limitation of this equation is that it is applicable only to an ideal case in which there are no ion–ion interactions of any kind.

J. Stokes–Einstein Relation

When a particle is moving in a zigzag fashion, it is subject to a viscous drag or friction force by the environment. Stokes showed that this frictional force is given by

$$\mathscr{F}_f = 6\pi \eta r v_d, \tag{185}$$

where η is the viscosity of the medium in which the particle of radius r is drifting with velocity v_d. This viscous force is equal to the diffusional force $-d\mu/dx$. Hence

$$-\frac{d\mu}{dx} = 6\pi\eta r v_d.$$

But

$$u_{abs} = \frac{v_d}{-d\mu/dx} = \frac{v_d}{6\pi\eta r v_d} = \frac{1}{6\pi\eta r}.$$

Substituting the Einstein relation [Eq. (180)] for u_{abs} gives

$$D = \frac{\kappa T}{6\pi\eta r}. \tag{186}$$

Thus the process of diffusion and viscous flow are linked. If the particle is equated to a sphere of molecular weight M, then r is proportional to $M^{1/3}$. Thus

$$DM^{1/3} = \text{const.} \tag{187}$$

When the particle radius is large compared to its jump distance l in water, this relation has been found to be approximately valid. At comparable values of r and l,

$$DM^{1/2} = \text{const} \tag{188}$$

is found to be valid.

Study of diffusion as a function of temperature yields the experimental energy for activation E_{expt}, which is given by the Arrhenius equation,

$$\frac{d\ln D}{dT} = \frac{E_{expt}}{RT^2}. \tag{189}$$

Equation (154) in logarithmic form on differentiation with respect to temperature gives

$$\frac{d\ln D}{dT} = \frac{1}{T} + \frac{\Delta G}{RT^2} - \frac{1}{RT}\frac{\partial(\Delta G)}{\partial T}. \tag{190}$$

Substitution of Eq. (14) into Eq. (190) on simplification gives

$$\frac{d\ln D}{dT} = \frac{1}{T} + \frac{\Delta H}{RT^2}. \tag{191}$$

Equating Eqs. (189) and (191) gives

$$E_{expt} = RT + \Delta H. \tag{192}$$

III. Physical and Electrochemical Principles

Thus by measuring D as a function of temperature, both experimental activation energy E_{expt} and enthalpy ΔH can be evaluated. Further, substitution of Eq. (4) in Eq. (154) gives

$$D = l^2 \frac{\kappa T}{h} \exp\left(-\frac{\Delta H}{RT}\right) \exp\left(\frac{\Delta S}{R}\right). \tag{193}$$

From Eq. (193) a value for $l^2 \exp(\Delta S/R)$ can be derived. In this manner the preceding equations have been used in the study of permeation of gases and nonelectrolytes through artificial membranes and of nonelectrolytes through biological membranes.

K. Transport Numbers

Transport number is defined as the fraction of the total current carried by the particular ionic species. Thus

$$t_i = I_i/(\sum I) \quad \text{and} \quad \sum t_i = 1.$$

As $I_i = z_i F C_i u_i$ (subscript conv for the mobility term dropped), t_i in terms of ion parameters can be written

$$t_i = z_i C_i u_i \bigg/ \left(\sum_j z_j C_j u_j\right). \tag{194}$$

L. Nernst–Planck Flux Equations

The driving forces on ions are both diffusional and electrical. Hence flux (mol cm^{-2} s^{-1}) of a charged species is caused by the gradient of electrochemical potential. As flux is given by

$$\text{Flux} = (\text{mobility})(\text{concentration})(\text{driving force}),$$

$$J_i = -u_{i(\text{abs})} C_i \frac{d}{dx} [\mu_i + z_i F V]$$

$$= -u_{i(\text{abs})} C_i \left[RT \frac{d \ln a_i}{dx} + z_i F \frac{dV}{dx}\right]$$

$$= -u_{i(\text{abs})} RT \left[\frac{dC_i}{dx} + C_i \frac{d \ln \gamma_i}{dx} + z_i \frac{F}{RT} C_i \frac{dV}{dx}\right].$$

If electrical mobility u_i is considered (subscript conv dropped), the preceding equation becomes

$$J_i = -\frac{RT}{z_i F} u_i \left[\frac{dC_i}{dx} + C_i \frac{d \ln \gamma_i}{dx} + z_i \frac{F}{RT} C_i \frac{dV}{dx}\right]. \tag{195}$$

This is the Nernst–Planck flux equation with the activity factor included. It is applicable to all mobile ions, and the set of flux equations, one for each species, is subject to the condition of electroneutrality, viz.,

$$\sum_i z_i C_i = 0.$$

The total current I carried by the charged species is given by

$$I = F \sum_i z_i J_i. \tag{196}$$

For the ideal case when $\gamma_i = 1$, Eq. (195) becomes

$$J_i = -\frac{RT}{z_i F} u_i \left[\frac{dC_i}{dx} + z_i \frac{F}{RT} C_i \frac{dV}{dx} \right],$$

which for current I_i carried by 1 g ion of i is

$$I_i = -RT u_i \left[\frac{dC_i}{dx} + z_i \frac{F}{RT} C_i \frac{dV}{dx} \right]. \tag{197}$$

M. Liquid Junction or Diffusion Potential

A liquid junction represents a boundary between two dissimilar solutions across which ions move by diffusion. The liquid junction potential (V_L) arises from the differences in the mobilities of positive and negative ions. If the cation has a higher mobility than the anion, the former will move ahead of the latter into dilute solution that will become positively charged with respect to the concentrated solution. If the anion moves faster, the dilute solution will be negatively charged. In either case, a "double layer" on a microscopic scale is produced at the junction between the two solutions. In other words, a gradient of electrical potential will be set up. This will oppose the diffusional flow of ions by increasing the speed of the slower ion and slowing that of the faster ion. In the steady state, the diffusional flux will be exactly balanced by the electric flux. The concentrations and the electrostatic potential through the boundary region become invariant with time. On this basis the sum of the electrical and diffusional work of transporting ions across a lamina dx in the boundary region becomes zero.

If one equivalent of charge (both positive and negative ions) is taken across the lamina, the electrical work is $F\,dV_L$. This one equivalent charge is due to transport of (t_i/z_i) g ion of each species. The diffusional work therefore is

III. Physical and Electrochemical Principles

given by $\sum_i (t_i/z_i)\,d\mu_i$. Hence

$$F\,dV_{\rm L} + \sum_i \frac{t_i}{z_i}\,d\mu_i = 0,$$

$$dV_{\rm L} = -\frac{RT}{F}\sum_i \frac{t_i}{z_i}\,d\ln a_i.$$

The diffusion potential $V_{\rm L}$ across the diffusion zone between two solution phases 1 and 2 of activity a_1 and a_2 is given by

$$V_{\rm L} = -\frac{RT}{F}\int_{a_1}^{a_2}\sum_i \frac{t_i}{z_i}\,d\ln a_i. \tag{198}$$

Integration of Eq. (198) is difficult. Approximations have to be made about a_i, a single ion property which is not directly measurable, about concentration dependence of t_i and about concentration profile of the species in the diffusion zone.

N. Diffusion Potential between Two Solutions of the Same Electrolyte

Application of Eq. (198) to the case of diffusion of a simple $(z_+:z_-)$ electrolyte in a concentration cell of the type

```
Electrolyte solution   │ │   Electrolyte solution
       (m₁)            │ │          (m₂)
                       │ │
                       │ │
   Mean activity, a₁   │ │    Mean activity, a₂
                       V_L
```

The total potential is given according to Eq. (198) by

$$V_{\rm L} = -\frac{RT}{F}\int_{(a_+)_1}^{(a_+)_2} \frac{t_+}{z_+}\,d\ln a_+ + \frac{RT}{F}\int_{(a_-)_1}^{(a_-)_2} \frac{t_-}{z_-}\,d\ln a_-,$$

where t_+ and t_- are assumed to be concentration independent. Then

$$V_{\rm L} = -\frac{RT}{F}\frac{t_+}{z_+}\ln\frac{(a_+)_2}{(a_+)_1} + \frac{RT}{F}\frac{t_-}{z_-}\ln\frac{(a_-)_2}{(a_-)_1}.$$

Since $t_+ + t_- = 1$ and $z_+ = z_- = z$,

$$V_{\rm L} = \frac{RT}{zF}\left[-2t_+\ln\frac{a_2}{a_1} + \ln\frac{(a_-)_2}{(a_-)_1}\right], \tag{199}$$

where a_2 and a_1 are the mean activities of the two solutions [see Eqs. (49)–(51)]. Further simplification of Eq. (199) is not possible without an extra nonthermodynamic assumption. Generally, it is assumed that the ratio of the single ion activities in the two solutions is equal to the ratio of their mean activities, i.e., $[(a_-)_2/(a_-)_1 = a_2/a_1]$. With this assumption, Eq. (199) becomes

$$V_L = (1 - 2t_+)\frac{RT}{zF}\ln\frac{a_2}{a_1}. \tag{200}$$

In terms of anion and cation mobilities u_- and u_+, Eq. (200) becomes

$$V_L = \frac{u_- - u_+}{u_- + u_+}\frac{RT}{zF}\ln\frac{a_2}{a_1}. \tag{201}$$

The sign of V_L depends on the relative values of u_- and u_+.

In the case of two solutions of the same electrolyte forming a liquid junction, the structure of the boundary is not important. The potential does not depend on the manner in which the boundary is constituted. At any point in the boundary layer the electrolyte solution is the same and has a definite activity and transference number. But if different electrolytes are present in the liquid junction, the ionic concentration at any point in it is determined by its structure and so t_i and a_i will depend on the nature of the boundary.

In general, liquid junctions can be grouped into four types: (1) continuous-mixture-type boundary; (2) constrained diffusion junction; (3) free diffusion junction; and (4) flowing junction. The first two cases have been considered theoretically by Henderson and Planck, and some satisfactory integrations of Eq. (198) have been given. Most widely used are types 3 and 4, and they are found to be too complex to be treated theoretically.

O. Henderson Equation

Henderson postulated that there existed in the diffusion zone a continuous series of mixtures of solutions 1 and 2 when they were brought in contact. At any point in the diffusion zone, if the concentration of the species j is C_j and if C'_j and C''_j are its concentrations in solution 1 and 2, then

$$C_j = C'_j(1 - x) + C''_j x,$$

where x is the fraction of solution 2 existing at the given point. $1 - x$ is therefore the fraction of solution 1 existing at the same point. Using this expression, Eq. (198), in which activities were replaced by concentrations, was integrated. The final result given in the more familiar form is

$$V_L = \frac{RT}{F}\frac{(U_1 - V_1) - (U_2 - V_2)}{(U'_1 + V'_1) - (U'_2 + V'_2)}\ln\frac{U'_1 + V'_1}{U'_2 + V'_2}, \tag{202}$$

where

$$U_1 = \sum_n (C_+ u_+)_1, \qquad V_1 = \sum_n (C_- u_-)_1,$$
$$U'_1 = \sum_n (C_+ u_+ z_+)_1, \qquad V'_1 = \sum_n (C_- u_- z_-)_1. \qquad (203)$$

Analogous relations apply to U_2, U'_2, V_2, and V'_2 for ions in solution 2.

P. Planck Equation

Planck considered univalent electrolytes and assumed a "constrained diffusion" boundary, which is equivalent to separating solutions 1 and 2 by a permeable membrane. The junction has a finite thickness across which diffusion reached a steady state giving a constant potential. Planck integration of Eq. (197) is complex and obtained the transcendental equation

$$\frac{\xi U_2 - U_1}{V_2 - \xi V_1} \frac{\ln(C_2/C_1) - \ln \xi}{\ln(C_2/C_1) + \ln \xi} \frac{\xi C_2 - C_1}{C_2 - \xi C_1}, \qquad (204)$$

where U_1, U_2, V_1, and V_2 are defined by Eq. (203) and $\xi = \exp(V_L F/RT)$.

Equations (202) and (204) are very useful in calculating liquid junction potentials existing at several interfaces when intracellular and extracellular capillary electrodes are used to measure bioelectric potentials existing across cell membranes.

Application of Eqs. (202) and (204) to two special cases—one where two solutions of the same (uni–uni valent) electrolyte at two different concentrations, and two where two (uni–uni valent) electrolytes with a common ion at the same concentration form the boundary—gives similar results. Both equations, that of Henderson and that of Planck, reduce to the form

$$V_L = \frac{RT}{F} \frac{u_+ - u_-}{u_+ + u_-} \ln \frac{C_2}{C_1}$$

for the first case and to the form

$$V_L = \frac{RT}{F} \ln \frac{u_{+(1)} + u_-}{u_{+(2)} + u_-}$$

for the second case.

Several investigators have integrated the flux equation (197) for several conditions existing across the diffusion zone by making appropriate and different assumptions. Among them, Behn integrated it to determine the flux across the boundary of any one particular species contained in the two solutions forming the boundary. His final equation is

$$I_i = -u_i \frac{RT}{d} \left[\frac{C_2 - C_1}{C_2\xi - C_1} \frac{\ln(C_2\xi/C_1)}{\ln(C_2/C_1)} \right] \{C_{i(2)}\xi - C_{i(1)}\}, \qquad (205)$$

where C_2 and C_1 are the total concentrations in the two solutions, and $C_{i(2)}$ and $C_{i(1)}$ are the concentrations of the species i in the two solutions. This equation is of considerable interest in biology since it enables evaluation of fluxes of particular species through the biological membrane whose intracellular and extracellular environments are composed of different ion species.

Q. Time-Dependent Diffusion

The diffusion Eq. (137) must be solved for the initial and boundary conditions to obtain general solutions. Generally, a solution has either a series of error functions or is in the form of a trigonometric series. When diffusion is confined to a cylinder the trigonometric series is replaced by a series of Bessel functions.

The spread by diffusion of an amount of substance C_0 deposited at time $t = 0$ in the plane $x = 0$ is given by

$$C(x,t) = \frac{C_0}{2(\pi Dt)^{1/2}} \exp(-x^2/4Dt). \qquad (206)$$

If at $x = 0$ an impermeable wall exists, all the diffusion occurs in the positive x direction. A solution for the negative x direction is that reflected from the plane $x = 0$ and superposed on the original distribution in the region $x > 0$. As the original distribution is symmetrical about $x = 0$, the concentration distribution is given by

$$C(x,t) = \frac{C_0}{(\pi Dt)^{1/2}} \exp(-x^2/4Dt). \qquad (207)$$

The equations relating the equilibration of three shapes (sheet, cylinder, and sphere) to time and diffusion coefficient are as follows.

For a plane sheet of thickness d whose both sides exposed,

Fractional equilibration, the ratio of the amount of isotope C_t present at time t to the amount C_0 present upon full equilibration:

$$\frac{C_t}{C_0} = 1 - \frac{8}{\pi^2} \left[\exp\left(\frac{-\pi^2 Dt}{d^2}\right) + \frac{1}{9}\exp\left(\frac{-9\pi^2 Dt}{d^2}\right) \right.$$
$$\left. + \frac{1}{25}\exp\left(\frac{-25\pi^2 Dt}{d^2}\right) + \cdots \right]. \qquad (208)$$

III. Physical and Electrochemical Principles

For a cylinder of radius r,

Fractional equilibration

$$\frac{C_t}{C_0} = 1 - 4\left[\frac{1}{\mu_1^2}\exp\left(\frac{-\mu_1^2 Dt}{r^2}\right) + \frac{1}{\mu_2^2}\exp\left(\frac{-\mu_2^2 Dt}{r^2}\right) + \cdots\right], \quad (209)$$

where the μs are the zeros of the Bessel function J_0, that is, the values of x for which the Bessel function $J_0(x)$ becomes zero. The first three of these roots have values $\mu_1 = 2.4048$, $\mu_2 = 5.5201$, $\mu_3 = 8.6537$.

For a sphere of radius, r,

Fractional equilibration

$$\frac{C_t}{C_0} = 1 - \frac{6}{\pi^2}\left[\exp\left(\frac{-\pi^2 Dt}{r^2}\right) + \frac{1}{4}\exp\left(\frac{-4\pi^2 Dt}{r^2}\right) + \frac{1}{9}\exp\left(\frac{-9\pi^2 Dt}{r^2}\right) + \cdots\right] \quad (210)$$

These equations describe the uptake of substances into spaces of appropriate shape provided there are no rate-limiting barriers or membranes. If such barriers exist, description of diffusion would simplify the mathematics as described below.

1. KINETICS OF EXCHANGE BETWEEN TWO PHASES SEPARATED BY A MEMBRANE

The flux in the direction phase (i) → phase (o) is given by

$$J(t) = P(C_i - C_o), \quad (211)$$

where $J(t)$ is the time-dependent flux since C_i and C_o are not at constant values but move toward the equilibrium state, i.e., $C_i = C_o$ for $t \to \infty$.

(a) If one of the phases is infinitely large, for example, V_o, the volume large compared to V_i, then C_o is practically constant and

$$J(t) = PC_i.$$

If a quantity of substance dm_i is moved in time dt, then

$$dm_i = JA\, dt,$$

where A is the area of the membrane. Expressing in concentration

$$-\frac{dm_i}{V_i} = -\frac{JA\, dt}{V_i} = -\frac{APC_i}{V_i}dt.$$

If the change in concentration in phase (i) is dC_i in time dt, then

$$dC_i = -\frac{APC_i}{V_i} dt$$

or

$$\frac{dC_i}{dt} = -kC_i, \tag{212}$$

where $k = AP/V_i$ is the rate constant (per second).

Integration of Eq. (212) gives $[\ln C_i]_{C_i^0}^{C_i} = -[kt]_0^t$, and so

$$C_i = C_i^0 \exp(-kt). \tag{213}$$

For practical purposes it is convenient to plot $C_i(t)$ against t on semi-logarithmic paper; then Eq. (213) gives a straight line, the slope of which gives the rate constant k.

The exponential process described by Eq. (213) is characterized by its time constant $\tau = 1/k$ (the time for the concentration to reach $1/e$ times its value) or $t_{1/2}$ (the time for concentration to reach half its value); that is,

$$\begin{aligned} (C_i^0/2) &= C_i^0 \exp(-kt_{1/2}), \\ \ln(\tfrac{1}{2}) &= -\ln 2 = -kt_{1/2}, \\ t_{1/2} &= \frac{\ln 2}{k} = \frac{0.693}{k}. \end{aligned} \tag{214}$$

(b) If the outer compartment is finite and the inside concentration is initially zero, the flux into phase (i) according to the above reasoning is given by

$$J = P(C_o - C_i) \quad \text{and} \quad dm_i = AP(C_o - C_i)\,dt = V_i\,dC_i,$$

and therefore

$$\frac{dC_i}{dt} = k(C_o - C_i) \quad \text{and} \quad k = \frac{AP}{V_i}.$$

Integration gives

$$\begin{aligned}{} [\ln(C_o - C_i)]_{C_i=0}^{C_i} &= -[kt]_0^t \\ C_i &= C_o[1 - \exp(-kt)]. \end{aligned} \tag{215}$$

(c) If both the phases are comparable in size, V_i and V_o, the flux in the direction phase (i) $\to$ phase (o) is given by

$$J(t) = AP(C_i - C_o).$$

At $t = 0$, m_o is dissolved in V_i and V_o has no substance, then

$$(dm)_o = AP(C_i - C_o)\,dt. \tag{216}$$

Conservation of mass gives

$$m = m_i + m_o.$$

If the values of C_i and C_o are inserted in Eq. (216), then

$$\frac{(dm)_o}{dt} = AP\left(\frac{m - m_o}{V_i} - \frac{m_o}{V_o}\right).$$

Rearrangement gives

$$\frac{(dm)_o}{[mV_o/(V_i + V_o)] - m_o} = AP\frac{V_i + V_o}{V_i V_o} dt.$$

Integration between the time $t = t_1$ and $t = t_2$ when m_o changed from $m_o^{t_1}$ to $m_o^{t_2}$ gives

$$\ln\left[\frac{mV_o}{V_i + V_o} - m_o\right]_{m_o^{t_1}}^{m_o^{t_2}} = -\left[AP\frac{V_i + V_o}{V_i V_o} t\right]_{t_1}^{t_2}.$$

Rearrangement gives

$$P = \frac{2.303 V_i V_o}{A(V_i + V_o)(t_2 - t_1)} \log \frac{mV_o - m_o^{t_1}(V_i + V_o)}{mV_o - m_o^{t_2}(V_i + V_o)}.$$

But $C_o^{t_1} = m_o^{t_1}/V_o$ and $C_o^{t_2} = m_o^{t_2}/V_o$. Thus

$$P = \frac{2.303 V_i V_o}{A(V_i + V_o)(t_2 - t_1)} \log \frac{m - C_o^{t_1}(V_i + V_o)}{m - C_o^{t_2}(V_i + V_o)}. \tag{217}$$

This equation was first derived by Northrop and Anson. Here m is the total mass initially in phase (i). As the total quantity of the substance is constant $m = C_i^{t_1} V_i + C_o^{t_1} V_o$, substitution of this gives

$$P = \frac{2.303 V_i V_o}{A(V_i + V_o)(t_2 - t_1)} \log \frac{V_i(C_i^{t_1} - C_o^{t_1})}{C_i^{t_1} V_i + C_o^{t_1} V_o - C_o^{t_2}(V_i + V_o)}. \tag{218}$$

But if the initial conditions at $t = 0$ were $C_i = C_i^{to}$ and $C_o = C_o^{to}$ then, as $m = C_i^{to} V_i + C_o^{to} V_o$, Eq. (217) becomes

$$P = \frac{2.303 V_i V_o}{A(V_i + V_o)(t_2 - t_1)} \log \frac{C_i^{to} V_i + C_o^{to} V_o - C_o^{t_1}(V_i + V_o)}{C_i^{to} V_i + C_o^{to} V_o - C_o^{t_2}(V_i + V_o)}. \tag{219}$$

2. Kinetics of Exchange in Biological Tissues

When a nerve or muscle cell is equilibrated with a substance (e.g., radio-isotope), the efflux from the cell is usually described by the sum of two or three exponentials. Thus

$$C(t) = a_\infty + a_1 \exp(-k_1 t) + a_2 \exp(-k_2 t) + \cdots + a_n \exp(-k_n t). \quad (220)$$

All the ks are positive corresponding to the fact that the curve approaches a_∞ as $t \to \infty$. If all the as are positive, then $C(t)$ decays from beginning to end.

Equation (220) is frequently used to derive physiological information from the curve. This is done by the curve-peeling method, which is as follows.

The experimental data (fraction of substance remaining in the fiber with time) are plotted on semilogarithmic paper. The tail-end portion of the $C(t)$ curve is extrapolated to zero time to give a_1. The $t_{1/2}$ of this straight line gives $k_1 = 0.693/t_{1/2}$. If the tail-end portion of the curve is curvilinear, that is, $a_\infty \neq 0$, the phase plane method (see later) is used to find a_∞ and k_1 simultaneously. In practice, a_∞ is known as the amount of substance in the fiber left after washout and is usually subtracted from the raw data to construct the $C(t)$-versus-t curve.

The calculated values of $a_1 \exp(-k_1 t)$ are subtracted from $C(t)$ and plotted on the same semilogarithmic paper. The tail-end portion of this plot is extrapolated to zero time and the preceeding steps repeated to yield a_2 and k_2. The whole process is repeated until there are no more points left.

The modern digital computer can be used to fit sums of exponentials to the experimental data. Manual curve peeling is often used as an initial approximation in such procedures.

The phase plane method referred to above is carried out as follows. If the curve actually is given by

$$C(t) = C_\infty + (C^0 - C_\infty)\exp(-kt), \quad (221)$$

then a semilogarithmic plot will not give a straight line, which, however, is obtained by plotting $C(t) = Y$ against $[dC(t)/dt] = X$. Differentiation of Eq. (221) gives

$$\frac{dC(t)}{dt} = -k(C^0 - C_\infty)\exp(-kt).$$

Substituting this in Eq. (221) yields

$$C(t) = C_\infty - \frac{1}{k}\frac{dC(t)}{dt}, \quad \text{i.e.,} \quad Y = C_\infty - \frac{1}{k}X.$$

This phase plane method involves more work to calculate slopes numerically or to measure them by drawing tangents to $C(t)$ versus t on a linear plot at various times. It has the advantage of being more sensitive to data at long times.

In the case of biological tissue (e.g., nerve or muscle) where radioisotopes are used, the tissue is assumed to consist of three compartments in series: (1)

bathing medium, (2) extracellular compartment in the tissue, and (3) intracellular compartment or compartments in which case there will be more than three compartments. In the case of the three-compartment system, the exchange usually consists of two phases, fast and slow, with two time constants, the fast corresponding to the extracellular compartment. It is generally assumed that the sizes of the two compartments will correspond exactly to the amounts of material present in the two phases of the cell. This means

$$C_2 + C_3 = A\exp(-k_1 t) + B\exp(-k_2 t) \tag{222}$$

and $C_2 = A\exp(-k_1 t)$ and $C_3 = B\exp(-k_2 t)$. This independence need not be true when both rate constants determine C_2 and/or C_3. That is,

$$C_2 = A_1 \exp(-k_1 t) + B_1 \exp(-k_2 t)$$

and

$$C_3 = A_2 \exp(-k_1 t) + B_2 \exp(-k_2 t).$$

Huxley (1960) has shown that C_3 at $t = 0$ is given by

$$C_{3(t=0)} = \frac{AB(k_1 - k_2)^2}{Ak_1^2 + Bk_2^2}. \tag{223}$$

On this basis, the assumption that $C_{3(t=0)}$ is given by B is an overestimate. Provided A, B, k_1 and k_2 are known, a value for $C_{3(t=0)}$ can be calculated from Eq. (223).

References

Bockris, J. O'M., and Reddy, A. K. N. (1970). "Modern Electrochemistry," Vol. 1. Plenum, New York.
Denbigh, K. (1961). "Principles of Chemical Equilibrium." Cambridge Univ. Press, London and New York.
Glasstone, S. (1947a). "Text Book of Physical Chemistry." Van Nostrand-Reinhold, Princeton, New Jersey.
Glasstone, S. (1947b). "Thermodynamics for Chemists." Van Nostrand-Reinhold, Princeton, New Jersey.
Glasstone, S., Laidler, K. J., and Eyring, H. (1941). "The Theory of Rate Processes." McGraw-Hill, New York.
Huxley, A. F. (1960). In "Mineral Metabolism" (C. L. Comar and F. Bronner, eds.), Vol. 1, Part A, p. 163. Academic Press, New York.
Kortum, G., and Bockris, J. O'M. (1951). "Text Book of Electrochemistry," Vol. 1. Am. Elsevier, New York.
Lakshminarayanaiah, N. (1969) "Transport Phenomena in Membranes." Academic Press, New York.

Lassen, N. A., and Perl, W. (1979). "Tracer Kinetics Methods in Medical Physiology." Raven Press, New York.
Page, L., and Adams, N. I. A. (1958). "Principles of Electricity," 3rd ed. Van Nostrand-Reinhold, Princeton, New Jersey.
Robinson, R. A., and Stokes, R. H. (1959). "Electrolyte Solutions," 2nd ed. Academic Press, New York.
Sears, F. W., and Zemansky, M. W. (1970). "University Physics," 4th ed. Addison-Wesley, Reading, Massachusetts.
Stein, W. D. (1967). "The Movement of Molecules across Cell Membranes." Academic Press, New York.
Sten-Knudsen, O. (1978). *In* "Membrane Transport in Biology" (G. Giebisch, D. C. Tosteson, and H. H. Ussing, eds.), Vol. 1, Chapter 2. Springer-Verlag, Berlin and New York.

Chapter **3**

ELECTROCHEMISTRY OF SOLUTIONS AND MEMBRANES

Solutions in general contain molecules of both solute and solvent. In the case of aqueous electrolyte solutions, charged particles (positive and negative ions) will be present in water. The symmetrical electrical field of the ion as it enters water will remove some of the water molecules from the water lattice and orient them in such a way that appropriate charged ends are facing one another. A definite number of water molecules get trapped and orient toward the ion. Kinetically they become part of the ion itself (primary ion hydration). Far away from the ion the normal structure of water (bulk water) prevails since the ionic field is attenuated. In the region between these extremes, some water molecules are neither close enough to feel the full field strength of the ion nor distant enough to be part of the bulk water. Depending on their distance from the ion, they break away from the water network and orient to varying degrees, so the water structure partially breaks down in this intermediate region (secondary ion hydration). As the magnitude of ion hydration is determined by the ion field strength, large ions (e.g., organic) are considered to be without significant hydration.

In addition to these ion–water interactions, there are ion–ion (repulsive or attractive) interactions. The modern theory of electrolyte solutions in its elementary form considers the latter as it relates to thermal motions of these ions (considered as point charges). Considerations are given to the former only in higher refinements of the theory.

I. The Debye–Hückel Theory

The important feature of the theory is the calculation of the electrical potential V at a point in the solution in terms of the charges and concentrations of the ions and of the properties of the solvent.

Consider a reference or central ion i. The water molecules surrounding it form a continuous dielectric medium. The remaining ions form an ion cloud or atmosphere (charges smeared out) around the central ion. The net charge in this ion cloud is equal and opposite to that of the central ion. The problem is to determine the time-average distribution of ions inside the solution. This reduces, according the Debye–Hückel model, to finding out how excess charge density ρ varies with distance r from the central ion.

Consider an infinitesimal volume element located at a distance r from the central ion, and let the net charge density and the average electrostatic potential in the volume element be ρ_r and V_r respectively. The relation between ρ and V is given by the Poisson equation [Chapter 2, Eq. (98)], which in polar coordinates can be written

$$\frac{1}{r^2}\frac{d}{dr}\left(r^2\frac{dV_r}{dr}\right) = -\frac{4\pi\rho_r}{\varepsilon}, \tag{1}$$

where ε is the dielectric constant of the medium (water).

The excess charge density in the volume element is given by the product of total number of ions per unit volume and the charge on these ions. Thus

$$\rho_r = n'_1 z_1 e + n'_2 z_2 e + \cdots + n'_i z_i e = \sum_i n'_i z_i e, \tag{2}$$

where n'_i is the number of ions of type i bearing charge $z_i e$. To link local concentrations of ions $(n'_1, n'_2, \ldots, n'_i)$ to the bulk concentrations n_i, the Boltzmann law is used. Thus

$$n'_i = n_i \exp\left(-\frac{z_i e V_r}{\kappa T}\right), \tag{3}$$

where $z_i e V_r$ is the electrical potential energy of the ion. The net charge density in the volume element is given by substitution of Eq. (3) in Eq. (2) for all ions. Thus

$$\rho_r = \sum_i n_i z_i e \exp\left(-\frac{z_i e V_r}{\kappa T}\right). \tag{4}$$

Expanding Eq. (4) gives

$$\exp\left(-\frac{z_i e V_r}{\kappa T}\right) = 1 - \frac{z_i e V_r}{\kappa T} + \frac{1}{2}\left(\frac{z_i e V_r}{\kappa T}\right)^2 - \cdots.$$

I. The Debye–Hückel Theory

When the average electrostatic potential is small so that $z_i e V_r \ll \kappa T$, i.e., coulombic potential energy negligible compared to kinetic energy, only the linear term in V_r is appreciable. Furthermore, $\sum z_i e n_i$ gives the charge on the electrolyte solution as a whole, and this is zero because the whole solution is electrically neutral, i.e., $\sum z_i e n_i = 0$. Thus Eq. (4) becomes

$$\rho_r = \sum_i z_i e n_i \left(1 - \frac{z_i e V_r}{\kappa T}\right) = -\sum_i z_i^2 e^2 n_i \frac{V_r}{\kappa T}. \tag{5}$$

Substituting Eq. (5) in Eq. (1) gives the linearized Poisson–Boltzmann equation

$$\frac{1}{r^2} \frac{d}{dr}\left(r^2 \frac{dV_r}{dr}\right) = \frac{4\pi e^2}{\varepsilon \kappa T} \sum n_i z_i^2 V_r,$$

$$= \varkappa^2 V_r, \tag{6}$$

where

$$\varkappa^2 = \frac{4\pi e^2}{\varepsilon \kappa T} \sum_i n_i z_i^2. \tag{7}$$

Equation (6) has the general solution

$$V_r = A \exp(-\varkappa r)/r + B \exp(\varkappa r)/r, \tag{8}$$

where A and B are constants of integration to be evaluated from the physical conditions of the problem. At high values of r, V_r must have a finite value, and for this to be realized B must be zero. Therefore

$$V_r = A \exp(-\varkappa r)/r. \tag{9}$$

To evaluate the constant A, a hypothetical situation is considered in which the solution is so dilute and ions on the average so far apart that there is little interionic field. The potential near the central ion is due to the point charge $z_i e$. Hence

$$V_r = z_i e / \varepsilon r. \tag{10}$$

In addition, as the concentration of the hypothetical solution tends to zero (i.e., $n_i \to 0$), it is seen from Eq. (7) that $\varkappa \to 0$. Thus in Eq. (9) $\exp(-\varkappa r) \to 1$, and so Eq. (9) becomes

$$V_r = A/r. \tag{11}$$

Combining Eqs. (10) and (11) gives

$$A = z_i e / \varepsilon. \tag{12}$$

Thus variation of electrostatic potential with distance r from an arbitrarily chosen central ion is described by Eq. (13) obtained by substitution of

Eq. (12) in Eq. (9). Hence

$$V_r = \frac{z_i e}{\varepsilon} \frac{\exp(-\varkappa r)}{r}. \tag{13}$$

The spatial distribution of the charge density with distance from the central ion is given by combining Eqs. (1) and (6). Thus

$$\rho_r = -(\varepsilon/4\pi)\varkappa^2 V_r. \tag{14}$$

Substituting for V_r from Eq. (13) gives

$$\rho_r = -\frac{z_i e}{4\pi} \varkappa^2 \frac{\exp(-\varkappa r)}{r}. \tag{15}$$

The net charge density contained in the ion atmosphere or cloud enveloping the central ion will be negative if the central ion is positive and vice versa. If a thin spherical shell of thickness dr at a distance r from the central ion is considered, the charge dq in the small shell is given by ρ_r multiplied by the volume of the shell. Thus

$$dq = \rho_r 4\pi r^2 \, dr. \tag{16}$$

The total charge in the ion atmosphere q_{cloud} is given by

$$q_{\text{cloud}} = \int_{r=0}^{r=\infty} dq = \int_{r=0}^{r=\infty} \rho_r 4\pi r^2 \, dr.$$

Substituting from Eq. (15) for ρ_r gives

$$q_{\text{cloud}} = -z_i e \int_{r=0}^{r=\infty} (\varkappa r) \exp(-\varkappa r) \, d(\varkappa r). \tag{17}$$

Integration by parts of Eq. (17) gives

$$\int_{r=0}^{r=\infty} (\varkappa r) \exp(-\varkappa r) \, d(\varkappa r) = [-\exp(-\varkappa r)(\varkappa r + 1)]_{r=0}^{r=\infty} = 1.$$

Thus

$$q_{\text{cloud}} = -z_i e. \tag{18}$$

This means that a central ion of charge $+z_i e$ is surrounded by a cloud of total charge $-z_i e$ and thereby maintaining electroneutrality.

Equations (15) and (16) yield

$$dq = -z_i e \exp(-\varkappa r) \varkappa^2 r \, dr. \tag{19}$$

This dq is maximum for a certain value of r, and the condition is $(dq/dr) = 0$. But

$$\frac{dq}{dr} = -z_i e \varkappa^2 \frac{d}{dr}[\exp(-\varkappa r) r]$$

$$= -z_i e \varkappa^2 [\exp(-\varkappa r) - r\varkappa \exp(-\varkappa r)]. \tag{20}$$

For Eq. (20) to be zero, the terms within the brackets must be zero as $z_i e \varkappa^2$ is finite. Thus

$$r = \varkappa^{-1}. \tag{21}$$

This means that maximum charge is contained in a spherical shell when that shell is at a distance $r = \varkappa^{-1}$ from the central ion. $\varkappa$ has the dimensions of reciprocal length; for this reason $\varkappa^{-1}$ is called the thickness of the ion atmosphere, and it is some times referred to as the Debye or Debye–Hückel length. Substituting for n_i ($= C_i N/1000$) and $I = \frac{1}{2}\sum_i C_i z_i^2$ (C_i is in moles per liter, N is Avogadro's number, and I is called ionic strength), Eq. (7) becomes

$$\varkappa = \left(\frac{8\pi N e^2}{1000 \varepsilon \kappa T}\right)^{1/2} I^{1/2} = B I^{1/2}, \tag{22}$$

where

$$B = \left(\frac{8\pi N e^2}{1000 \varepsilon \kappa T}\right)^{1/2}. \tag{23}$$

Introducing numerical values for the universal constants into Eq. (22) gives

$$1/\varkappa = 1.988 \times 10^{-10} \sqrt{\varepsilon T/I}. \tag{24}$$

The Debye length thus becomes bigger with a decrease in ionic strength and smaller with an increase in valence of ions.

An isolated ion (absence of ion cloud) of valence z_i in a medium of dielectric constant ε gives rise to a potential at a distance r. Thus

$$V_{r(\text{ion})} = z_i e / \varepsilon r. \tag{25}$$

Introduction of an ion cloud around the central ion now gives rise to the total potential V_r given by Eq. (13). Hence one can write

$$V_r = V_{r(\text{ion})} + V_{r(\text{cloud})}. \tag{26}$$

Thus substituting Eqs. (13) and (25) in Eq. (26) gives

$$V_{r(\text{cloud})} = \frac{z_i e}{\varepsilon r} [\exp(-\varkappa r) - 1]. \tag{27}$$

When $\varkappa r \ll 1$, i.e., in dilute solutions, $\exp(-\varkappa r) - 1$ becomes equal to $-\varkappa r$. For this condition, Eq. (27) becomes

$$V_{r(\text{cloud})} = -z_i e / \varepsilon \varkappa^{-1}. \tag{28}$$

from which it can be seen that $V_{r(\text{cloud})}$ is independent of r and that the effect of the ion cloud is equated to that of a single charge, equal in magnitude but opposite in sign to that of the central ion placed at a distance equal to $\varkappa^{-1}$ from the central ion. Consequently, it is meaningful to call $\varkappa^{-1}$ the effective thickness of the ion atmosphere surrounding the central ion.

II. Debye–Hückel Theory and Activity Coefficients

The ion cloud arises from the interaction of the central ion with all other ions of its environment. If there were no interactions (absence of coulombic forces) between ions, thermal forces would take over and distribute the ions in a random fashion and wipe out the ion cloud. Formation of the ion cloud may be considered as a charging process whereby the initial state of non-interacting ions are taken to a final state of electrostatic charging. Work of electrostatic charging is equal to the change in free energy due to ion–ion interactions. If only the reference ion i is taken, its partial free-energy change or change in chemical potential of i due to its interaction with all other ions must be considered. In other words, the central ion i is considered to be of zero charge. Then the work of charging this ion of radius r_i from the initial state of zero charge to the final state of charge $z_i e$ is given by W multiplied by N, Avogadro's number. This work must be equal to the change in chemical potential of i due to its interaction with all other ions. Thus

$$\Delta \mu_i = WN, \tag{29}$$

where W is given by

$$W = \int dW = \int_0^{z_i e} V_{r_i} \, dq. \tag{30}$$

dW is the infinitesimal work given by the product of the electrical potential and the small charge dq put into an uncharged sphere immersed in solvent of dielectric constant ε. But V_{r_i} is given by Eq. (93) of Chapter 2 when $\varepsilon = 1$. For the present case,

$$V_{r_i} = q/\varepsilon r_i.$$

Using this relation in Eq. (30) gives on integration

$$W = \frac{(z_i e)^2}{2\varepsilon r_i} = \frac{z_i e}{2} \frac{z_i e}{\varepsilon r_i}.$$

$z_i e / \varepsilon r_i$ is the electric potential V at the surface of the ion and therefore can be equated to V_{cloud}. Thus Eq. (29) may be rewritten

$$\Delta \mu_i = \frac{N z_i e}{2} V_{\text{cloud}}. \tag{31}$$

Substitution for V_{cloud} from Eq. (28) yields

$$\Delta \mu_i = -\frac{N}{2} \frac{(z_i e)^2}{\varepsilon} \varkappa. \tag{32}$$

II. Debye–Hückel Theory and Activity Coefficients

Equations (37) and (38) of Chapter 2 relate the change in free energy $[\Delta\mu_i = \mu_{i(\text{real})} - \mu_{i(\text{ideal})}]$ to the activity coefficient f_i. Thus

$$\Delta\mu_i = RT \ln f_i.$$

But $\Delta\mu_i$ arising from ion–ion interactions is given by Eq. (32). Thus combining these equations yields the result

$$\ln f_i = -\frac{N(z_i e)^2}{2\varepsilon RT}\varkappa. \tag{33}$$

The mean rational activity coefficient $f_\pm$ of an electrolyte dissociating into v_+ cations of valence z_+ and v_- anions of valence z_-, according to Eq. (53) of Chapter 2, is given by

$$\ln f_\pm = \frac{1}{v}[v_+ \ln f_+ + v_- \ln f_-]. \tag{34}$$

Substituting the value of f_i from Eq. (33) for each ion and employing Eq. (22) and the electroneutrality condition

$$v_+ z_+ = -v_- z_-$$

leads to

$$\ln f_\pm = -\frac{N(z_+ z_-)e^2}{2\varepsilon RT} BI^{1/2}$$

or

$$\log f_\pm = -\frac{1}{2.303}\frac{Ne^2}{2\varepsilon RT} B(z_+ z_-)I^{1/2}. \tag{35}$$

Substituting the relation

$$A = \frac{1}{2.303}\frac{Ne^2}{2\varepsilon RT} B,$$

Eq. (35) becomes

$$\log f_\pm = -A(z_+ z_-)I^{1/2}. \tag{36}$$

The value of A for water at 25°C is 0.512. Equation (36) is the Debye–Hückel limiting law and is a very useful guide to the behavior of the activity coefficient at high dilutions. In the use of ion-selective electrodes that sense the activity of single ions for which the electrode is selective, computation of single-ion activity coefficients can be made by using Eq. (36) in the form

$$\log f_i = -A z_i^2 I^{1/2}. \tag{37}$$

In the foregoing considerations ions are considered as point charges. This approximation becomes less valid as the electrolyte concentration is increased because the radius of the ion cloud becomes smaller and smaller with increase in concentration and comes closer to the size of the ion. In order to remove this approximation, an ion-size parameter a must be introduced into the lower limit of integration of Eq. (17). Instead of choosing $r = 0$, it should be $r = a$. Before carrying out this integration, the constant of integration in Eq. (9) must be evaluated in a different way.

Introducing Eq. (9) into Eq. (14) gives

$$\rho_r = -\frac{\varepsilon}{4\pi} \varkappa^2 A \frac{\exp(-\varkappa r)}{r}.$$

Substituting this in Eq. (16) gives

$$dq = -A\varkappa^2\varepsilon[\exp(-\varkappa r)r\, dr],$$
$$q_{\text{cloud}} = -z_i e = \int_a^\infty dq = -A\varepsilon \int_a^\infty \exp(-\varkappa r)\varkappa r\, d(\varkappa r). \tag{38}$$

Integration by parts gives the relation

$$-A\varepsilon \exp(-\varkappa a)(1 + \varkappa a) = -z_i e.$$

Thus

$$A = \frac{z_i e}{\varepsilon} \frac{\exp(\varkappa a)}{1 + \varkappa a}. \tag{39}$$

Substituting Eq. (39) in Eq. (9) gives

$$V_r = \frac{z_i e}{\varepsilon} \frac{\exp(\varkappa a)}{1 + \varkappa a} \frac{\exp(-\varkappa r)}{r}. \tag{40}$$

Once again, Eq. (26) is applicable. Thus

$$V_{\text{cloud}} = \frac{z_i e}{\varepsilon r} \left[\frac{\exp(\varkappa(a-r))}{1 + \varkappa a} - 1 \right]. \tag{41}$$

Equation (41) holds for all ions up to $r = a$. When $r < a$, no other ion can penetrate. Therefore for $r = a$, Eq. (41) becomes

$$V_{\text{cloud}} = -\frac{z_i e}{\varepsilon} \frac{\varkappa}{1 + \varkappa a}. \tag{42}$$

Substituting Eq. (42) in Eq. (31) gives

$$\Delta\mu_i = -\frac{N(z_i e)^2}{2\varepsilon} \frac{\varkappa}{1 + \varkappa a}$$

or

$$\ln f_i = \frac{N(z_i e)^2}{2\varepsilon RT} \frac{\varkappa}{1+\varkappa a}. \tag{43}$$

Equation (43) in terms of mean activity coefficient becomes

$$\log f_\pm = -\frac{A(z_+ z_-)I^{1/2}}{1+BaI^{1/2}}. \tag{44}$$

In Eq. (44) the numerator gives the effect of the long-range coulombic forces, and the denominator shows how these are affected by short-range interactions between ions. In any solution there will be short-range ion–solvent interactions to consider. These are assumed to give linear variation of $\log f_\pm$ with concentration. In an empirical way, these may be included by adding a term linear in concentration. Thus

$$\log f_\pm = -\frac{A(z_+ z_-)I^{1/2}}{1+BaI^{1/2}} + bI. \tag{45}$$

The parameters b and a can be adjusted to fit any experimental curve. Several authors have written these equations in different empirical forms to describe the behavior of electrolytes. These equations are useful in calculating activity coefficients for salts for which no experimental values are available. Other important empirical equations are the following:

$$\log f_\pm = -\frac{A(z_+ z_-)I^{1/2}}{1+I^{1/2}}, \tag{46}$$

due to Guntelberg (here a has a value of 3.04 Å for values of I up to 0.1 m);

$$\log f_\pm = -\frac{A(z_+ z_-)I^{1/2}}{1+I^{1/2}} + bI, \tag{47}$$

due to Guggenheim (here b is an adjustable parameter);

$$\log f_\pm = -\frac{A(z_+ z_-)I^{1/2}}{1+I^{1/2}} + b(z_+ z_-)I, \tag{48}$$

due to Davies (b has a value of 0.1 or 0.15).

The limiting law [Eq. (36)] at extremely low electrolyte concentrations predicts the experimental results well, but at higher concentrations it fails and the plot of $\log f_\pm$ versus $I^{1/2}$ is not a straight line but a curve. This deviation from a straight line up to a certain concentration can be accounted for by Eq. (44), which differs from the limiting law in having the term $1+BaI^{1/2}$. The parameter a assigns a finite size to the ion, which is unknown. However, a numerical value can be derived on the basis of an experiment. Equation

(44) can be approximately represented by

$$\log f_\pm \approx -A(z_+ z_-)I^{1/2}(1 - BaI^{1/2}). \tag{49}$$

The values of the ion-size parameter or closest distance of approach a calculated for several electrolytes are reasonable and lie between 3 and 5 Å, which is greater than the sum of the crystallographic radii of positive and negative ions and pertains to solvated ions. In many cases, Eq. (49) has been found to give a good fit with experiment by choosing a reasonable value for a, independent of concentration. The fit has been found for ionic strengths up to 0.1 m.

An unfortunate feature of a is that it is concentration dependent and is an adjustable parameter. For example, it is found in the case of NaCl that a value of 6 Å for a at $m = 0.1$ (m is molality) has to be changed to about 14 Å at $m = 1.0$ to fit the experimental data. In addition, a has to assume impossible values for some electrolytes to fit the theory to experimental data (e.g., $a = -411.2$ at 2m HCl). Obviously a has been forced to account for a number of short-range interactions. Ions in solution exist in several states of interaction with solvent molecules, some of which form part and parcel of the ions themselves. This has the effect of increasing the electrolyte concentration. This aspect of ion hydration has not been taken care of by the parameter a, although it accounted for the increase in size of the point charge because of association of some water molecules with the ion. Attempts have been made to incorporate this aspect into the theory. It is based on the consideration that the total free energy of a known quantity of solution is fixed regardless of how much water (solvent) is attached to the ion. If in a certain quantity of solution, 1 mol of the unhydrated solute 2 dissociates into v_+ moles of cations and v_- moles of anions and dissolves in s mol of solvent 1, the total free energy G can be calculated for the system in two ways, first when the solute exists unhydrated and second when the solute is hydrated with n moles of water. Hence

$$G = s\mu_1 + v_+\mu_+ + v_-\mu_- \quad \text{(unhydrated)},$$
$$G = (s - n)\mu_1 + v_+\mu_+^s + v_-\mu_-^s \quad \text{(hydrated)}. \tag{50}$$

Substitution from Eq. (39) of Chapter 2 into Eq. (50) gives on rearrangement

$$(n\mu_1^0/RT) + v_+(\mu_+^0 - \mu_+^{s0})/RT + v_-(\mu_-^0 - \mu_0^{s0})/RT$$
$$= -n \ln a_1 + v_+ \ln a_+^s - v_+ \ln a_+ + v_- \ln a_-^s - v_- \ln a_-. \tag{51}$$

At infinite dilution (i.e., $s \to \infty$) all a_i become unity, and thus all logarithmic terms become zero. This makes the left-hand side zero. Substituting mole fractions for cations and anions and using Eq. (53) of Chapter 2 gives

$$\ln f_\pm^s = \ln f_\pm + \frac{n}{v} \ln a_1 + \ln \frac{\text{mole fraction of solute (unhydrated)}}{\text{mole fraction of solute (hydrated)}},$$

where the mole fraction of solute unhydrated is $v/(s + v)$ and the mole fraction of solute hydrated is $v/(s - n + v)$.

Substitution of these values gives

$$\ln f^s_\pm = \ln f_\pm + \frac{n}{v} \ln a_1 + \ln \frac{s - n + v}{s + v}. \tag{52}$$

Expressing $f_\pm$ in terms of molal activity coefficient $\gamma_\pm$ [see Eq. (59) of Chapter 2] and substituting for s ($= 1000/mM_1$), Eq. (52) becomes

$$\ln f^s_\pm = \ln \gamma_\pm + \frac{n}{v} \ln a_1 + \ln[1 + 0.001 mM_1(v - n)]. \tag{53}$$

Substituting for a_1 in terms of the osmotic coefficient ϕ [see Eq. (71) of Chapter 2], Eq. (53) becomes

$$\ln f^s_\pm = \ln \gamma_\pm - 0.001 nmM_1 \phi + \ln[1 + 0.001 mM_1(v - n)]. \tag{54}$$

In these equations, it is assumed that n is independent of concentration. Assuming that Eq. (44) is applicable to hydrated ions and ignoring other effects such as molecular size and heat of mixing, Eq. (53) can be rewritten

$$\log \gamma_\pm = -\frac{A(z_+ z_-) I^{1/2}}{1 + BaI^{1/2}} - \frac{n}{v} \log a_1 - \log[1 + 0.001 mM_1(v - n)]. \tag{55}$$

This equation contains only two parameters a and n whose values are difficult to determine by experimentation. However, it has been successfully tested for a number of aqueous electrolyte solutions. For example, in the case of NaCl, the activity coefficients calculated by using Eq. (55) [$a = 3.97$ Å, $n = 3.5$] agrees with experimental values for solutions as concentrated as $5m$. In summary, incorporation of a term to describe the influence of ion–solvent interaction on ion–ion interaction has extended the range of electrolyte concentration over which the theory becomes applicable although the theory has two parameters that defy experimental solution.

III. Debye–Hückel Theory and Electrolyte Conductance

Equations (175) and (176) of Chapter 2 show that the equivalent conductivity at infinite dilution of an electrolyte solution is the sum of the equivalent conductivities at infinite dilution of the ions of the electrolyte. This is generally known as Kohlrausch's law of independent migration of ions. Also, there is an empirical relation between equivalent conductivity and the square root of concentration of an electrolyte solution named after Kohlrausch. This experimental relation is

$$\lambda = \lambda_0 - AC^{1/2}, \tag{56}$$

where λ_0 is the equivalent conductance at infinite dilution and A is a constant.

The theoretical basis for this relation is provided by Debye–Hückel and Onsager. The decrease in conductance (or ionic mobilities) is due to the interionic attractions and repulsions. These result in producing two effects, the electrophoretic effect and the relaxation effect.

In the presence of an applied electric field, both the central ion i and its atmosphere (or ion cloud) move in opposite directions because they are oppositely charged. The ion cloud has a radius given by $\varkappa^{-1}$. Migration of this cloud may be equated to the movement of a large colloidal particle. When the cloud moves, it tries to carry along with it the central ion also with a force $\mathscr{F}_e$ called the electrophoretic force. On the other hand, the central ion whose coulombic field has spherical symmetry in the absence of an electric field loses its symmetry (becomes asymmetrical) in the presence of the electric field and its ion cloud becomes egg-shaped. The central ion as it moves builds up an ion cloud in front of it while leaving behind an ion cloud to stand still and die out. As the center of charge on the moving ion and the center of charge of the oppositely charged ion cloud do not coincide, an electrical force exists between the ion and its cloud. This means that the ion is subject to an electric field that arises from the decay (or relaxation) of the cloud behind the ion and its formation in front of the ion. This field is called the relaxation field and acts in the direction opposite to the direction of the externally applied field. Thus the relaxation field decreases the mobility of the central ion.

In a simple treatment of ion migration, it was considered that the drift velocity arose from an applied electric force $\mathscr{F}$. Because of interactions between ion and its cloud, two other forces—electrophoretic $\mathscr{F}_e$ and relaxation $\mathscr{F}_r$—operate on the ion. Thus the net force on the ion is given by

$$\mathscr{F}_{\text{net}} = \mathscr{F} - \mathscr{F}_e - \mathscr{F}_r. \tag{57}$$

In accordance with Eq. (57) the overall drift velocity v_d is given by

$$v_d = v_0 - (v_e + v_r). \tag{58}$$

v_0 is the drift velocity due to externally applied field only. v_e and v_r are the drift velocity components due to electrophoretic and relaxation fields, respectively.

In a straightforward, simple way, v_e can be evaluated by equating the applied electric force ($zeE/300$) to the Stokes viscous force retarding the rate of movement of the ion cloud. In computing the viscous force acting on the ion cloud, $r = \varkappa^{-1}$ and $v = v_e$ in Stokes formula. Thus

$$\frac{zeE}{300} = 6\pi\eta\varkappa^{-1}v_e \quad \text{or} \quad v_e = \frac{ze}{6\pi\eta\varkappa^{-1}}\frac{E}{300}. \tag{59}$$

III. Debye–Hückel Theory and Electrolyte Conductance

The relaxation component v_r of the drift velocity is given by [see Eq. (163) of Chapter 2]

$$v_r = u_{abs}^0 \mathscr{F}_r. \tag{60}$$

The relaxation force $\mathscr{F}_r$ was evaluated by Onsager, who showed that

$$\mathscr{F}_r = \frac{z_i e^3 \varkappa (z_+ z_-)}{3\varepsilon \kappa T} \frac{q}{1+\sqrt{q}} \frac{E}{300}, \tag{61}$$

where

$$q = \frac{z_+ z_-}{z_+ + z_-} \frac{\lambda_+ + \lambda_-}{z_-\lambda_+ + z_+\lambda_-}.$$

Substitution of Eq. (61) in Eq. (60) gives

$$v_r = u_{abs}^0 \frac{z_i e^3 \varkappa (z_+ z_-)}{3\varepsilon \kappa T} \frac{q}{1+\sqrt{q}} \frac{E}{300}.$$

Further substitution of Eq. (165) of Chapter 2 gives

$$v_r = \frac{u^0(z_+ z_-)e^2}{3\varepsilon \kappa T} \varkappa \frac{q}{1+\sqrt{q}} E. \tag{62}$$

Substituting Eqs. (59) and (62) in Eq. (58) gives the value for the overall drift velocity v_d. Thus

$$\frac{v_d}{E} = \frac{v_0}{E} - \left[\frac{z_i e}{6\pi\eta 300} + \frac{u^0(z_+ z_-)e^2}{3\varepsilon \kappa T}\frac{q}{1+\sqrt{q}}\right]\varkappa. \tag{63}$$

As v_d/E and v_0/E are equal to mobilities [see Eq. (164) of Chapter 2], Eq. (63) becomes

$$u = u^0 - \left[\frac{z_i e}{1800\pi\eta} + \frac{u^0(z_+ z_-)e^2}{3\varepsilon \kappa T}\frac{q}{1+\sqrt{q}}\right]\varkappa. \tag{64}$$

At infinite dilution, $1/\varkappa \to \infty$ or $\varkappa = 0$. Thus

$$u_{\lim, c \to 0} = u^0.$$

Equivalent conductivity λ is given by Eq. (175) of Chapter 2:

$$\lambda = F(u_+ + u_-).$$

Substituting for u_+ and u_- from Eq. (64) gives on rearrangement

$$\lambda = [u_+^0 + u_-^0]F - \left[\frac{\varkappa e F}{1800\pi\eta}(z_+ + z_-) + \frac{q}{1+\sqrt{q}}\varkappa \frac{e^2}{3\varepsilon \kappa T}z_+ z_- F(u_+^0 + u_-^0)\right].$$

Substituting for $\lambda_0 = (u_+^0 + u_-^0)F$ gives

$$\lambda = \lambda_0 - \left[\frac{\varkappa eF}{1800\pi\eta}(z_+ + z_-) + (z_+z_-)\frac{q}{1+\sqrt{q}}\frac{e^2\varkappa}{3\varepsilon\kappa T}\lambda_0\right]. \quad (65)$$

Substituting the values for the constants, Eq. (65) becomes

$$\lambda = \lambda_0 - \left[\frac{41.25(|z_+| + |z_-|)}{\eta(\varepsilon T)^{1/2}} + \frac{2.801 \times 10^6 |z_+z_-|q}{(\varepsilon T)^{3/2}(1+\sqrt{q})}\lambda_0\right]I^{1/2}. \quad (66)$$

Equation (66) for 1:1 electrolyte reduces to

$$\lambda = \lambda_0 - (B_1 + B_2\lambda_0)C^{1/2}, \quad (67)$$

where B_1 and B_2 are constants given by

$$B_1 = [82.5/\eta(\varepsilon T)^{1/2}],$$

$$B_2 = 8.204 \times 10^5(\varepsilon T)^{-3/2}.$$

Equation (67) is similar to Eq. (56). This law is valid for dilute solutions. For moderate concentrations, Eq. (66) may be rewritten

$$\lambda = \lambda_0 - \left[\frac{41.25(|z_+| + |z_-|)}{\eta(\varepsilon T)^{1/2}} + \frac{2.801 \times 10^6 |z_+z_-|q}{(\varepsilon T)^{3/2}(1+\sqrt{q})}\lambda_0\right]\frac{I^{1/2}}{(1+\varkappa a)}. \quad (68)$$

IV. Distribution of Ions and Potential Differences at Interfaces

An interface or boundary between two homogeneous phases shows properties differing from those of the materials in the contiguous homogeneous phases. It is a film of characteristic thickness. When polar oils, such as salicylaldehyde and O-toluidine, are placed in a tube between two aqueous phases, an oil or liquid membrane is formed (see Fig. 1). When organic salts (e.g., tetramethyl ammonium salts or picrates) are added to one of the aqueous phases, a difference of potential can be measured between the two

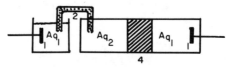

Fig. 1. Schematic representation of cell to measure potential differences across an oil membrane (4). (1) is reversible Ag–AgCl electrodes, (2) is agar-KCl salt bridge, Aq_1 is aqueous salt solution, Aq_2 is solution containing added organic salt.

IV. Distribution of Ions and Potential Differences at Interfaces

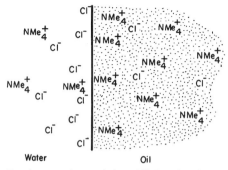

Fig. 2. Immiscible oil and water phases. Anions (Cl$^-$) and cations (tetramethylammonium, NMe$_4^+$) have different solubilities in the oil and water phases.

electrodes in the aqueous phases. Because of the unequal distribution of cations and anions across the interface (see Fig. 2), a potential difference exists across the interface. This potential difference is called by Lange the distribution potential or outer volta potential ψ. This was the view of Beutner. However, Baur thought that the potentials arose from selective adsorption of a monolayer of organic ions at the oil–water interface (see Fig. 3). Such potentials are called by Lange χ potentials or, as often called, contact potentials V. The sum of these potentials, i.e.,

$$\phi = \psi + V, \qquad (69)$$

is called the inner or Galvani potential of the oil phase relative to the water phase (see Fig. 4). The changes in ψ only may be brought about by adding organic salts (including surface active agents) to either phase. On the other hand, changes in χ can be measured between water and air or water and nonpolar oil when any surface active agent is allowed to adsorb at the interface. Changes in ψ may be called distribution potential while changes in χ or V may be called interfacial potentials or surface or dipole potentials.

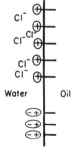

Fig. 3. Selective adsorption of a monolayer of organic ions at the oil–water interface. Adsorbable organic ions are shown by charged "head" groups and hydrocarbon "tails." Lower portion shows a monolayer of permanent dipoles. This would occur in case of adsorption of a long-chain alcohol.

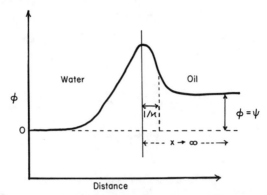

Fig. 4. Variation of Galvani potential ϕ relative to the water phase at a water–oil interface. Oil is considered polar. Distance from the interface is represented by x. Potential at $x \to \infty$ is compared with the ionic double layer thickness $1/\varkappa$. (After Davis and Rideal, 1961.)

A. Distribution Potentials

Distribution potentials arise from the differences in the solubilities of positive and negative ions in the two phases. As an example, the distribution of tetramethylammonium chloride in nitrobenzene and water may be considered. At equilibrium, the distribution of ions are defined by

$$RT \ln S_+ = \mu^0_{+(w)} - \mu^0_{+(o)}, \tag{70}$$

$$RT \ln S_- = \mu^0_{-(w)} - \mu^0_{-(o)}, \tag{71}$$

where S_+ and S_- are the cation and anion distribution or partition coefficients, μ^0 is the standard chemical potential, and subscripts w and o refer to water and oil phases respectively.

As tetramethyl ion will be in the oil phase and chloride ion in the water phase, the oil phase will be positive with respect to the water phase.

When changes in ϕ are measured, it will indicate changes in ψ and in χ. But since the adsorbed monolayer is thin compared to the bulk phases, changes in χ at the moment of formation of monolayer, although nonzero, will decay to zero, and so the measured changes in ϕ will be due to changes in ψ. The relation between ϕ, S_+, and S_- may be derived as follows.

The electrochemical potential of cation in the two phases is given by

$$\bar{\mu}_{+(w)} = \mu^0_{+(w)} + RT \ln C_{+(w)} \tag{72}$$

(ϕ assumed zero),

$$\bar{\mu}_{+(o)} = \mu^0_{+(o)} + RT \ln C_{+(o)} + \phi F. \tag{73}$$

IV. Distribution of Ions and Potential Differences at Interfaces

At equilibrium
$$\bar{\mu}_{+(w)} = \bar{\mu}_{+(o)}. \tag{74}$$

Similarly, for the anion,
$$\bar{\mu}_{-(w)} = \mu^0_{-(w)} + RT \ln C_{-(w)}, \tag{75}$$
$$\bar{\mu}_{-(o)} = \mu^0_{-(o)} + RT \ln C_{-(o)} - \phi F, \tag{76}$$

and
$$\bar{\mu}_{-(w)} = \bar{\mu}_{-(o)}. \tag{77}$$

Each bulk phase is electrically neutral, and so
$$C_{+(w)} = C_{-(w)}, \tag{78}$$
$$C_{+(o)} = C_{-(o)}. \tag{79}$$

Combinations of Eqs. (72) and (73) in view of Eq. (74) and of Eqs. (75) and (76) in view of Eq. (77) give
$$\mu^0_{+(w)} + RT \ln C_{+(w)} = \mu^0_{+(o)} + RT \ln C_{+(o)} + \phi F, \tag{80}$$
$$\mu^0_{-(w)} + RT \ln C_{-(w)} = \mu^0_{-(o)} + RT \ln C_{-(o)} - \phi F. \tag{81}$$

In view of Eqs. (78) and (79), subtraction of Eq. (81) from Eq. (80) gives
$$[\mu^0_{+(w)} - \mu^0_{+(o)}] - [\mu^0_{-(w)} - \mu^0_{-(o)}] = 2\phi F. \tag{82}$$

Substituting from Eqs. (70) and (71) into Eq. (82) gives
$$RT \ln S_+ - RT \ln S_- = 2\phi F$$

or
$$\phi = (RT/2F) \ln(S_+/S_-). \tag{83}$$

Overall distribution of salt S, i.e., (concentration of salt in oil membrane)/(concentration in water), according to Eq. (50) of Chapter 2, is given by
$$S = (S_+ S_-)^{1/2}. \tag{84}$$

Thus Eq. (83) can be expressed as
$$\phi = (RT/2F) \ln(S^2/S_-^2), \tag{85}$$

which, as it stands, cannot be verified experimentally, because the absolute potential ϕ cannot be measured. Changes in ϕ, however, can be measured. Thus, if NaI and KI are added to nitrobenzene and water, respectively, then according to Eq. (85)

$$\phi_{\text{NaI}} = \frac{RT}{F} \ln \frac{S_{\text{NaI}}}{S_{\text{I}^-}^2} \quad \text{and} \quad \phi_{\text{KI}} = \frac{RT}{F} \ln \frac{S_{\text{KI}}}{S_{\text{I}^-}^2}.$$

Thus

$$\Delta\psi = \Delta\phi = \phi_{NaI} - \phi_{KI} = \frac{RT}{F} \ln \frac{S_{NaI}}{S_{KI}}. \tag{86}$$

Experiments using salts of Na and K and tetraethylammonium, and chloride, and iodide ions in a precision cell of the type

Hg, Hg$_2$Cl$_2$	KCl 3 M water	M$^+$X$^-$ water	M$^+$X$^-$ Nitrobenzene	(Et)$_4$ N-picrate di-isopropyl ketone	(Et)$_4$N Pi	HgPi, Hg
	ϕ_1	ϕ		ϕ_2	ϕ_3	

[Cell-forming $\Delta\phi$s are at interface of water and a polar oil such as nitrobenzene. Salt bridges are chosen so that when salt MX is changed, the measured potential change is $\Delta\phi$, i.e., ϕ_1, ϕ_2, and ϕ_3 remain constant. $\Delta\phi$ is measured across the thickness of the solvent, which is great in relation to the thickness of the interfacial electrical double layers.] have shown that the measured $\Delta\phi$ values were in agreement with those calculated from Eq. (86).

If a layer of polar oil separates two aqueous phases of different salt concentrations, a transient $\Delta\phi$ will be measured across the electrodes in the aqueous phases. This $\Delta\phi$ arises from the differences in the transport numbers of anions and cations diffusing through the oil layer.

B. Interfacial and Surface Potentials

When an interfacial film is formed from an oil in which little salt dissolves, a double layer cannot form in the thickness of oil available. Thus Eq. (79) no longer applies. However, other equations [(69)–(78)] will apply. It can be shown that when the oil solubilities of salt and the interfacial film are very small, $\Delta\chi$ alone contributes to $\Delta\phi$.

Equations (80) and (81) give

$$[\mu^0_{+(w)} - \mu^0_{+(o)}] - [\mu^0_{-(w)} - \mu^0_{-(o)}] - RT \ln \frac{C_{+(o)}}{C_{-(o)}} = 2\phi F. \tag{87}$$

Thus

$$\phi = \frac{RT}{2F} \ln \frac{S_+}{S_-} - \frac{RT}{2F} \ln \frac{C_{+(o)}}{C_{-(o)}}, \tag{88}$$

where $C_{+(o)}$ and $C_{-(o)}$ refer to ionic concentrations in the oil at the point where ϕ is measured.

IV. Distribution of Ions and Potential Differences at Interfaces

If the contact potential is V, then applying Boltzmann equation to the distribution of ions near the interface gives

$$C_{+(o)} = C_{+(o)_{x=\infty}} \exp[-F(V - \psi_{x=\infty})/RT], \quad (89)$$

$$C_{-(o)} = C_{-(o)_{x=\infty}} \exp[F(V - \psi_{x=\infty})/RT], \quad (90)$$

and at $x = \infty$

$$C_{+(o)_{x=\infty}} = C_{-(o)_{x=\infty}}. \quad (91)$$

Under these conditions, Eq. (88) becomes

$$\phi = \frac{RT}{2F} \ln \frac{S_+}{S_-} - \frac{RT}{2F} \ln \exp[-2F(V - \psi_{x=\infty})/RT]. \quad (92)$$

The double layer at $x = \infty$ is collapsed and so $\phi_{x=\infty} = \psi_{x=\infty}$ (see Fig. 5). According to Eq. (83)

$$\psi_{x=\infty} = (RT/2F) \ln(S_+/S_-),$$

and so Eq. (92) becomes

$$\phi = \psi_{x=\infty} + \frac{RT}{2F} \frac{2F}{RT} (V - \psi_{x=\infty}) = \psi_{x=\infty} + (V - \psi_{x=\infty}),$$

$$\phi = V. \quad (93)$$

Thus for nonpolar oil or for air in contact with water, ϕ is independent of both S_+ and S_-, as opposed to Eq. (86). As nonpolar oils are nonconducting, a direct current method such as the one used to test Eq. (86) cannot be used

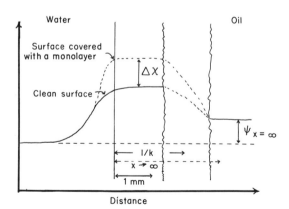

Fig. 5. Variation of ϕ relative to the water phase at the paraffinic–oil interface. Values correspond to $x \approx 1$ mm and less than the double-layer thickness. (After Davis and Rideal, 1961.)

to test Eq. (93). A capacity method is warranted. Special methods used to measure the surface potentials of monolayers have confirmed the validity of Eq. (93).

C. Monolayers and Components of ΔV

The surface of pure water has no charges, and so ψ is zero. The absolute potential ϕ of water is χ or V, the potential due to water dipoles at the surface. The absolute potential V has been estimated to be between -100 and -200 mV.

If a monolayer is spread on a clean water surface, the water dipoles will reorient about the film-forming molecules (see Fig. 6). Let this change be μ_1. The dipoles of the film-forming molecules (e.g., long-chain amine → $C{-}NH_2$) will contribute to ΔV by an amount determined by the group dipole moment μ_2. A third component of ΔV arises from the moment μ_3 of the bond at the upper limit of the monolayer (see Fig. 6). If there are an array of n dipoles per square centimeter, then

$$\Delta V = 4\pi n\mu_1 + 4\pi n\mu_2 + 4\pi n\mu_3, \tag{94}$$

provided they are all additive in the vertical direction. Thus

$$\Delta V = 4\pi n\mu_D, \tag{95}$$

where $\mu_D = \mu_1 + \mu_2 + \mu_3$ is the characteristic dipole moment of the head group of the molecules of the monolayer. When the film is compressed, μ_D will change if the dipolar head is unsymmetrical and reorient itself as shown for long-chain esters (see Fig. 7)

The typical values of bond moments are given in the figure. Resolution of these moments in the vertical direction gives the overall moments in milli-

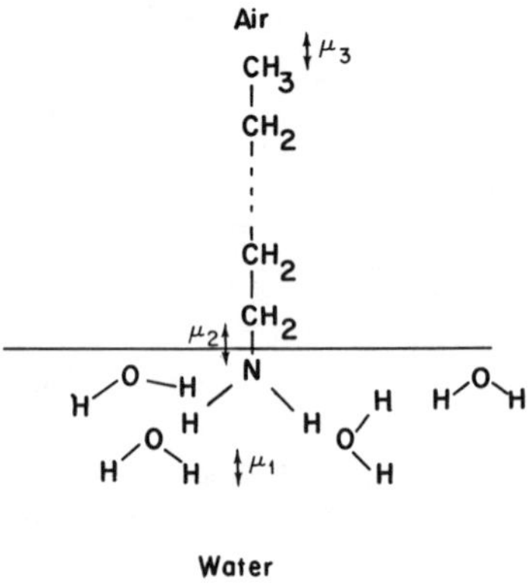

Fig. 6. Components μ_1, μ_2, and μ_3 of the surface potential for a neutral monolayer. (After Davis and Rideal, 1961.)

IV. Distribution of Ions and Potential Differences at Interfaces

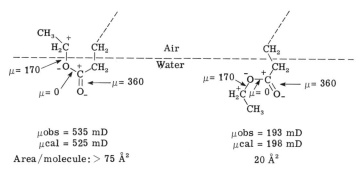

Fig. 7. Orientation and bond moments in the head groups of ethyl palmitate spread at the oil–water surface. At large areas, the vertical component of the moments is high, but is reduced on compression (low areas) when the ethyl chain is forced below the water surface. (After Davis and Rideal, 1961.)

debyes (mD). When the area per molecule is high, the vertical components of the moments is high. At low areas, the ethyl chain is forced to exist below the surface of water and the vertical component is lower.

If a monolayer of long-chain ions [e.g., $C_{18}H_{37}N(CH_3)_3^+$] is considered, the monolayer contributes not only a dipole component but also an electrostatic term to $\Delta\phi$ (see Fig. 8). The electrostatic potential arises from the distribution of ions as shown in Fig. 9. As the potential arises from the presence of the monolayer, the ψ potential is included with ΔV. This ψ potential is called ψ_0 (at $x = 0$). Thus Eq. (95) can be rewritten

$$\Delta V = 4\pi n\mu_D + \psi_0 = 4\pi n\mu, \qquad (96)$$

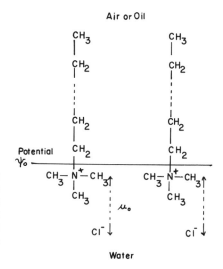

Fig. 8. A monolayer of $C_{18}H_{37}N(CH_3)_3^+$ at air–water (or paraffin oil–water) interface. The electric potential ψ_0 at the interface is relative to the bulk aqueous phase. The Cl^- counterions are at a mean depth of $1/\varkappa$. At low ionic strength, large dipole contributions are possible (see Table I). (After Davis and Rideal, 1961.)

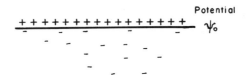

Fig. 9. Idealized double layer. Counterions behave as point charges and approach the plane of the charges.

where $\mu = \mu_D + \mu_0$, the overall dipole contribution, μ_0 is the dipole contribution of the electrical double layer, which at low surface pressure is given by

$$\mu_{0(n\to 0)} = \frac{\varkappa^{-1}(\text{cm}) \cdot 4.8 \times 10^{-10}(\text{esu})}{\varepsilon \times 10^{-18}} \quad \text{(D)}.$$

Thus

$$\mu = \mu_D + \frac{\varkappa^{-1} 4.8 \times 10^{-10}}{\varepsilon \times 10^{-18}} \quad \text{(D)}.$$

As the value of μ_D has been found to be constant at 450 mD, values for μ can be calculated as a function of ionic strength assuming a value of 80 for ε in the double layer. These values are shown in Table I.

TABLE I

Calculated and Measured Dipole Moments μ for Ionized Films of $C_{18}H_{37}N(CH_3)_3^+$ at the Air–Water or Decane–Water Interfaces.[a,b]

NaCl C (N)	$(1/\varkappa)$ (Å)	μ (mD) calculated from $\mu = 450 + \left[\dfrac{(1/\varkappa)4.8 \times 10^3}{80}\right]$	Measured[c]	ΔV (mV) at $n = 10^{13}$ 1000 Å² per chain
2	2.1	576	500	22.5
10^{-1}	9.5	1020	1026	43
10^{-2}	30	2250	2700	86
10^{-3}	95	6150	6170	124
10^{-4}	300	18450	19300	220

[a] From Davis and Rideal (1961, p. 78).
[b] The Debye length $1/\varkappa = 3/\sqrt{C}$ Å, where C is the salt concentration in the subphase.
[c] Surface potential was measured as a function of n and extrapolated to $n = 0$. μ was calculated from $\Delta V(\text{v})/4\pi n \times 300$ (esu) and $10^3 \text{ esu}/10^{-18}$ (mD).

The calculated values of μ are in good agreement with those observed by extrapolation ($n \to 0$) from data on films of very low charge density. These dipole moments are very large, whereas the potentials remain moderate, as shown in last column of Table I.

Equation (96) can be tested provided ψ_0 can be evaluated. This is indicated below.

D. Interfacial Potential: Gouy–Chapman and Stern Theories

At the membrane–solution interface, the polar groups of the membrane constituents (phospholipids and proteins) exist at physiological pH in different states of ionization, depending on the pK_a of the groups concerned. In general, it is considered that the groups will be negatively charged. Some water molecules will be associated with them. To maintain electroneutrality, cations (counterions) that are probably hydrated will exist close to the negative groups and form an electrical double layer, which is characteristic of all phase boundaries. A crude extrapolation to the membrane–solution interface of the concepts that have evolved over the years by electrochemists for the double layer existing at a metal–solution interface is illustrated in Fig. 10.

Helmholtz proposed the simplest model of distribution of ions at interfaces and considered the behavior of the double layer of charges on the surface and those in solution to be approximately equivalent to that of a parallel plate capacitor (Fig. 10a). Gouy regarded the Helmholtz model to neglect the thermal distribution of ions. He considered the distribution of ions analogous to that considered in the Debye–Hückel theory; that is, the distributed space charge near the membrane interface was considered to correspond to that of the ion atmosphere of the charged membrane surface (Fig. 10b). The mathematical theory was developed by Chapman following Gouy's treatment and was based on two assumptions: (1) ions are point charges and (2) the charge (σ cm^{-2}) is uniformly spread over the membrane surface. A physically realistic model is due to Stern, who considered the finite sizes of ions and the possibility of their specific adsorption.

The space charge density ρ_x at a distance x from the membrane surface is given by Eq. (4). ρ_x is related to ψ by the Poisson equation, which for a flat double layer takes the form

$$\frac{d^2\psi}{dx^2} = -\frac{4\pi\rho_x}{\varepsilon} = -\frac{4\pi}{\varepsilon} e \sum_i n_i z_i \exp\left(-\frac{z_i e \psi}{\kappa T}\right). \tag{97}$$

Using the identity

$$2\frac{d^2\psi}{dx^2} = \frac{d}{d\psi}\left[\frac{d\psi}{dx}\right]^2 \tag{98}$$

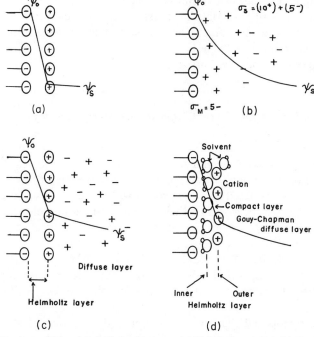

Fig. 10. Structure of the electrical double layer: (a) Helmholtz model; (b) Gouy model. Ion considered as a point charge. Surface charge per unit area indicated by cations (+) and anions (−). (c) Stern model. Finite size of the ion with thermal distribution considered. (d) General representation of the double layer with adsorption of water dipoles.

in Eq. (97) gives on integration

$$\frac{d\psi}{dx} = \left[\frac{8\pi\kappa T}{\varepsilon}\sum n_i\left\{\exp\left(-\frac{z_i e\psi}{\kappa T}\right) - 1\right\}\right]^{1/2} \quad (99)$$

for the boundary condition $\psi = 0$, $d\psi/dx = 0$ at $x = \infty$. The surface charge density σ on the membrane is related to ρ_x by

$$\sigma = -\int_0^\infty \rho_x \, dx. \quad (100)$$

Integration of Eq. (97) and substitution into Eq. (100) gives

$$\sigma = \frac{\varepsilon}{4\pi}\left(\frac{d\psi}{dx}\right). \quad (101)$$

Substituting Eq. (101) into Eq. (99) gives on rearrangement

$$\sigma = \left[\frac{\varepsilon\kappa T}{2\pi}\sum_i n_i\left\{\exp\left(\frac{-z_i e\psi}{\kappa T}\right) - 1\right\}\right]^{1/2}. \quad (102)$$

IV. Distribution of Ions and Potential Differences at Interfaces

Expressing n_i as $(C_iN/1000)$, where N is Avogadro's number and C_i is the bulk concentration (moles per liter), σ as the surface charge density in electronic charges per square angstrom, and $\psi = \psi_0$ at $x = 0$, Eq. (102) at 22°C becomes

$$\sigma = \frac{1}{272}\left[\sum_i C_i\left\{\exp\left(\frac{-z_ie\psi_0}{\kappa T}\right) - 1\right\}\right]^{1/2}. \tag{103}$$

In the case of a mixture of (1:1) and (2:1) electrolytes, Eq. (103) simplifies to

$$\sigma = \tfrac{1}{272}[C^+\{\exp(F\psi_0/RT) + \exp(-F\psi_0/RT) - 2\} + C^{2+}\{2\exp(F\psi_0/RT) + \exp(-2F\psi_0/RT) - 3\}]^{1/2}, \tag{104}$$

and in the case of symmetrical $(z:z)$ electrolyte simplifies to

$$\sigma = \tfrac{1}{272}C^{1/2}[\exp(zF\psi_0/RT) - 1 + \exp(-zF\psi_0/RT) - 1]^{1/2}$$

$$= \tfrac{1}{272}C^{1/2}[\exp(zF\psi_0/RT) - 2\exp(zF\psi_0/2RT)$$
$$\times \exp(-zF\psi_0/2RT) + \exp(-zF\psi_0/RT)]^{1/2}$$

$$= \tfrac{1}{272} C^{1/2}\left[2\sinh\frac{zF\psi_0}{2RT}\right] = \frac{C^{1/2}}{136}\sinh\frac{zF\psi^0}{2RT}. \tag{105}$$

Alternatively, Eq. (105) is usually written

$$\psi_0 = \frac{2RT}{zF}\sinh^{-1}\left(\frac{136\sigma}{C^{1/2}}\right). \tag{106}$$

Equation (96) can now be tested by writing Eq. (106) for (1:1) electrolyte as

$$\psi_0 = \frac{2RT}{F}\sinh^{-1}\frac{136n \times 10^{-16}}{C^{1/2}} \quad \left(\sigma = \frac{n}{10^{16}}\right). \tag{107}$$

Substituting Eq. (107) into Eq. (96) gives

$$\Delta V = 4\pi n\mu_D + \frac{2RT}{F}\sinh^{-1}(136n \times 10^{-16}/C^{1/2}). \tag{108}$$

When ψ_0 is high, Eq. (108) may be written

$$\Delta V = 4\pi n\mu_D + \frac{RT}{F}\ln\left(136^2 \times 4 \times 10^{-32}\frac{n^2}{C}\right) \tag{109}$$

$$[2\sinh^{-1}x = \ln(x + \sqrt{x^2 + 1})^2 \approx \ln 4x^2].$$

Differentiating Eq. (109) for each of the two conditions (constant n and constant C) gives

$$\left(\frac{\partial \Delta V}{\partial \log C}\right)_n = -\frac{2.303RT}{F} = -58 \text{ mV} \quad \text{(at 20°C)} \tag{110}$$

if C is not too high or

$$\left\{\frac{\partial[\Delta V - 4\pi n \cdot 450]}{\partial \log n}\right\}_C = \frac{2 \times 2.303 RT}{F} = 118 \text{ mV} \quad (111)$$

if n is not too large.

Equations (110) and (111) have been found to agree with experimental results. Also, Eq. (108) was found to be valid since it was found that the difference between the values of ΔV measured for films as a function of C at constant n and the values of ψ_G calculated by Eq. (107) was a constant and equal to $4\pi n \mu_D$. This means that change in μ_1 is small and ψ_G can be identified with ψ_0.

Equation (99) can be simplified for a single binary electrolyte of valence z in solution to

$$\left(\frac{d\psi}{dx}\right)^2 = \frac{8\pi n \kappa T}{\varepsilon}\left[\exp\left(\frac{ze\psi}{2\kappa T}\right) - \exp\left(-\frac{ze\psi}{2\kappa T}\right)\right]^2$$

or

$$\frac{d\psi}{dx} = -\sqrt{\frac{8\pi n \kappa T}{\varepsilon}}\left[\exp\left(\frac{ze\psi}{2\kappa T}\right) - \exp\left(-\frac{ze\psi}{2\kappa T}\right)\right]. \quad (112)$$

(The negative root is taken since at the positively charged electrode $\psi > 0$, but $d\psi/dx < 0$, while at the negatively charged electrode, it is reversed, corresponding to the physical situation.)

Integration of Eq. (112) gives an explicit relation between the potential and the coordinate of the double layer. Equation (112) can be rewritten

$$\int_{ze\psi/2\kappa T}^{ze\psi/2\kappa T} \frac{2d(ze\psi/2\kappa T)}{\exp(ze\psi/2\kappa T) - \exp(-ze\psi/2\kappa T)} = -\int_0^x \frac{8\pi n z^2 e^2}{\varepsilon \kappa T} dx = -\int_0^x d(\kappa x).$$

Integrating this equation for the condition $\psi = \psi_0$ for $x = 0$ gives

$$\left[\int \frac{dx}{\exp(x) - \exp(-x)} = \tfrac{1}{2}\ln \frac{\exp(x) - 1}{\exp(x) + 1}\right],$$

$$\kappa x = \ln \frac{[\exp(ze\psi/2\kappa T) + 1][\exp(ze\psi_0/2\kappa T) - 1]}{[\exp(ze\psi/2\kappa T) - 1][\exp(ze\psi_0/2\kappa T) + 1]}. \quad (113)$$

For small potentials, the exponentials may be expanded, and so Eq. (113) simplifies to

$$\kappa x = \ln(\psi_0/\psi) \quad \text{or} \quad \psi = \psi_0 \exp(-\kappa x). \quad (114)$$

This equation shows that the potential decreases exponentially to zero over a distance of the order of $1/\kappa$, the Debye length. The variation of potential in the double layer according to Eqs. (113) and (114) is shown in Fig. 11.

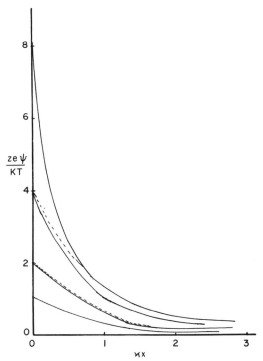

Fig. 11. Variation of electric potential in the double layer according to Gouy–Chapman theory. The solid curves correspond to Eq. (113) for $ze\psi_0/\kappa T = 1, 2, 4, 8$. The dotted curves are according to the approximate equation (114). (After Kruyt, 1952.)

Another approximation is obtained when $ze\psi_0/\kappa T \gg 1$ and $ze\psi/\kappa T \ll 1$. This is an approximation applicable to a surface at high potential. For this case, the relation is

$$\frac{ze\psi}{\kappa T} = 4\alpha \exp(-\varkappa x) \quad \text{or} \quad \psi = \frac{4\kappa T}{ze}\exp(-\varkappa x)\alpha, \tag{115}$$

where

$$\alpha = \frac{\exp(ze\psi_0/2\kappa T) - 1}{\exp(ze\psi_0/2\kappa T) + 1} \approx 1.$$

This means that the potential some distance away from a surface appears to follow Eq. (114), but with an apparent ψ_0 value of $4\kappa T/ze$, which is a constant independent of the actual value. For monovalent ions at room temperature, this apparent ψ_0 would be 100 mV and for divalent ions, 50 mV.

In the Debye–Hückel ion–cloud model, the electrical effect of the cloud on the central ion could be equated to placing the entire charge of the cloud $-z_i e$ at a distance $1/\varkappa$ from the central ion. In analogy with this, the Gouy–Chapman charge σ_d can be placed on a plane parallel to the membrane and at a distance equal to $1/\varkappa$. If this is done, one has a parallel-plate capacitor situation, i.e., at $x = 0$ (plane of the membrane), a charge σ and a diffuse charge $-\sigma_d$ at $x = 1/\varkappa$ plane. The potential across this parallel-plate capacitor is given by Eqs. (107) and (108) of Chapter 2.

σ is given by Eq. (102), which for a symmetrical electrolyte of valence z can be written for the condition $\psi = \psi_0$ at $x = 0$

$$\sigma = -\sigma_d = -\left(\frac{\varepsilon n \kappa T}{2\pi}\right)^{1/2}\left[\exp\left(\frac{ze\psi_0}{2\kappa T}\right) - \exp\left(-\frac{ze\psi_0}{2\kappa T}\right)\right]$$

$$= -2\left(\frac{\varepsilon n \kappa T}{2\pi}\right)^{1/2} \sinh\frac{ze\psi_0}{2\kappa T}.$$

At low potentials,

$$\sigma_d = \left(\frac{2\varepsilon n \kappa T}{\pi}\right)^{1/2}\frac{ze\psi_0}{2\kappa T} \quad \text{and} \quad \varkappa = \frac{1}{d} = \left(\frac{8\pi n z^2 e^2}{\varepsilon \kappa T}\right)^{1/2},$$

$$\sigma_d = (\varepsilon \varkappa \psi_0)/4\pi.$$

For large values of the potential, σ increases more than linearly with ψ_0 and so does the capacity $(C = \varepsilon \varkappa / 4\pi)$ of the double layer.

Several features of the behavior of the Gouy–Chapman equations are illustrated in Fig. 12.

At room temperature, $\varkappa$ is about $3 \times 10^7 zC^{1/2}$, where C is the concentration (moles per liter). Therefore 1-N solution of a monovalent electrolyte gives a calculated capacity of $3 \times 10^7 \times 80/4\pi = 200 \ \mu\text{F/cm}^2$. Experimentally the capacities found are about one-tenth of this amount. Evidently the description of the diffuse double layer is too simple. This simplification leads to other trouble when a dilute solution (0.1 N) is considered. If the potential at the surface is, say, 200 mV, then the concentration of the electrolyte near the membrane surface will be $0.1 = C_0 \exp(-\psi_0 F/RT) = C_0 \exp(-200/25)$, i.e., $C_0 = 300 \ N$, an impossible value. This is due to the assumption of point charge and consequent neglect of ionic diameters.

Stern suggested that the above difficulty could be handled by treating the region near the surface into two parts, the first consisting of a layer of ions adsorbed and forming an inner compact double layer, and the second consisting of the diffuse Gouy–Chapman layer. Roughly, the potential is

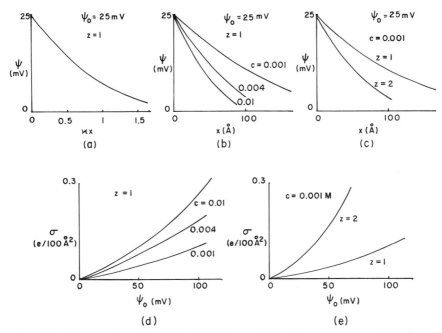

Fig. 12. Characteristics of the diffuse double layer. (a) Variation of ψ with $\varkappa x$. (b) Effect of electrolyte concentration on the variation in potential with distance, the fall in potential being sharp at higher electrolyte concentration. (c) Effect of valence on double-layer thickness, decreasing with increasing valence. (d) Relation between surface charge density σ and surface potential ψ_0, being proportional to each other at low values of ψ_0. Also shows the effect of electrolyte concentration on the relation between σ and ψ_0. (e) Effect of valence on charge–potential curve. (After Adamson, 1960.)

assumed to vary with distance, as shown in Fig. 13. If σ_s is the charge on the solution side of the double layer, then it is considered to be made up of two contributions σ_1 and σ_2, the first arising from charges associated with adsorption and the second from space charge associated with the ionic atmosphere. Thus

$$\sigma = \sigma_1 + \sigma_2. \qquad (116)$$

The total potential drop ψ_0 is divided into ψ_δ over the diffuse part and $(\psi_0 - \psi_\delta)$ over the molecular condenser of thickness δ. Thus

$$\sigma_1 = \frac{\varepsilon'}{4\pi\delta}(\psi_0 - \psi_\delta). \qquad (117)$$

The capacity of the molecular condenser is given by $C_1 = \varepsilon'/4\pi\delta$, where ε' is the local dielectric constant, which may be different from that of the solvent. The compact layer is followed by the diffuse layer with ψ_0 replaced

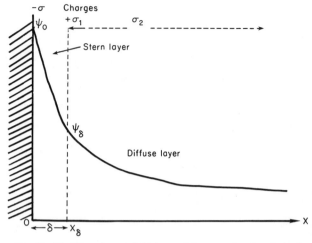

Fig. 13. The Stern layer of thickness δ (see text for description).

by ψ_δ. Thus the total electrical capacity may be considered as given by the formula for capacities in series [see Eq. (111) of Chapter 2]:

$$\frac{1}{C} = \frac{1}{C_1} + \frac{1}{C_2}.$$

C_1 is nearly constant, but C_2 depends on the electrolyte concentration ($C_2 \approx \varkappa$). So when ψ_0, the surface potential, is constant, ψ_δ, which controls the diffuse double layer, decreases with increasing $\varkappa$ and thus plays a part in electrokinetics.

In dilute solutions, $C \approx C_2$, and in concentrated solutions $C \approx C_1$.

The charge σ_2 is given by Eq. (102). The charge σ_1 is obtained in terms of site fraction statistics for solution and surface phases by applying a form of Langmuir isotherm.

If θ is the fraction of sites on the membrane occupied by adsorbed molecules, then the chemical potential of the adsorbed species i is given by

$$\mu_i = \mu_i^{0,A} + \kappa T \ln[\theta/(1-\theta)]. \tag{118}$$

Considering gas phase adsorption, μ_i is the chemical potential of the species in the gas phase. If the gas is ideal, then

$$\mu_i = \mu_i^{0,g} + \kappa T \ln P. \tag{119}$$

Equating Eqs. (118) and (119) gives

$$\exp\left(-\frac{\mu_i^{0,A} - \mu_i^{0,g}}{\kappa T}\right) = \frac{\theta}{1-\theta}\frac{1}{P}, \tag{120}$$

$$\theta/(1-\theta) = bP,$$

where $b = \exp(-\Delta G^0_{ads}/\kappa T)$ and $\Delta G^0_{ads} = (\mu_i^{0,A} - \mu_i^{0,g})$ is the standard free energy of adsorption, the standard states being unit pressure in the gas phase and half coverage ($\theta = \frac{1}{2}$) in the adsorbed layer.

The same considerations may be applied to a solution regarding it as a three-dimensional lattice of independent sites. The electrochemical potential of the adsorbate in solution can be written

$$\bar{\mu}_i = \bar{\mu}_i^{0,s} + \kappa T \ln \frac{\theta'}{1 - \theta'}, \tag{121}$$

where θ' is the fraction of sites occupied.

Equation (121) may be equated to Eq. (118) provided electrochemical potentials are considered in Eq. (118). Under these conditions, the adsorption isotherm obtained is

$$\frac{\theta}{1-\theta} = \frac{\theta'}{1-\theta'} \exp\left(-\frac{\overline{\Delta G}^0_{ads}}{\kappa T}\right), \tag{122}$$

where $\overline{\Delta G}^0_{ads}$ is the standard electrochemical free energy of adsorption.

Let n_m be the number of ions adsorbed per square centimeter of membrane surface, n_s the number of ions of the same type present in 1 ml of the bulk solution, Z_m the maximum number of ions that can be adsorbed on 1 cm² of membrane surface, and Z_s the maximum number of ions per milliliter for which space is available in the bulk solution. Then

$$\theta = n_m/Z_m \quad \text{and} \quad \theta' = n_s/Z_s. \tag{123}$$

Substitution of Eq. (123) into Eq. (122) gives on rearrangement

$$\frac{Z_m - n_m}{Z_s - n_s} \exp\left(-\frac{\overline{\Delta G}^0_{ads}}{\kappa T}\right) = \frac{n_m}{n_s}. \tag{124}$$

Solving Eq. (124) for n_m gives

$$n_m = \frac{Z_m}{1 + [(Z_s - n_s)/n_s]\exp(\overline{\Delta G}^0_{ads}/\kappa T)}. \tag{125}$$

Stern assumed that for dilute solutions, $Z_s \gg n_s$, and therefore

$$\frac{n_s}{Z_s - n_s} \approx \frac{n_s}{Z_s} = N_s$$

(mole fraction of the solute), and so Eq. (125) becomes

$$n_m = \frac{Z_m}{1 + (1/N_s)\exp(\overline{\Delta G}^0_{ads}/\kappa T)}. \tag{126}$$

The charge per square centimeter of adsorbed layer is given by

$$\sigma_1 = \sum_i z_i e(n_m)_i. \tag{127}$$

For a binary electrolyte, Eqs. (126) and (127) give

$$\sigma_1 = zeZ_m\left\{\left[1 + \frac{\exp(\overline{\Delta G}^0_{+\text{ads}}/\kappa T)}{N_s}\right]^{-1} - \left[1 + \frac{\exp(\overline{\Delta G}^0_{-\text{ads}}/\kappa T)}{N_s}\right]^{-1}\right\}. \quad (128)$$

Equation (128) is based on the assumption that the adsorption of cations and anions takes place independently. This implies that, on the membrane surface, there are Z_m sites to adsorb cations only and Z_m sites to adsorb anions only. Such a model is valid for a heterogeneous membrane surface that at a given pH has both cationic and anionic groups active. Most membrane systems have only one type of groups active at physiological pH. To overcome this difficulty, 1 in the denominator may be replaced by 2. But for dilute solutions, the term $(\overline{\Delta G}^0_{\text{ads}}/\kappa T) \gg N_s$, so that 1 or 2 in the denominator can be neglected. For most practical membrane systems, Eq. (128) reduces to

$$\sigma_1 = \pm zeZ_m N_s \exp(-\overline{\Delta G}^0_{\pm\text{ads}}/\kappa T), \quad (129)$$

where the positive sign refers to the case of adsorption of cations and negative sign refers to anion adsorption.

The term $\overline{\Delta G}^0_{\pm\text{ads}}$ can be written

$$\overline{\Delta G}^0_{\pm\text{ads}} = ze(\psi_\delta \pm \phi_\pm), \quad (130)$$

where ψ_δ is the surface potential at the Stern layer arising from coulombic interaction and ϕ is the "adsorption potential" arising from interactions, noncoulombic in nature.

Expressing N_s in terms of bulk concentration C [see Eq. (40) of Chapter 2: $N_s \approx CM_1/10^3$, where C is in moles per liter], Eq. (129) for a cation selective membrane can be written

$$\sigma_1 = zeZ_m \frac{M_1 C}{1000} \exp\left[-\frac{ze(\psi_\delta + \phi)}{\kappa T}\right]. \quad (131)$$

σ_2 corresponding to the diffuse double layer is given by Eq. (102) with the sign changed because Eq. (102) has been applied to the membrane surface. Thus the basic equation of the Stern theory may be written [Eq. (116)]

$$\frac{\varepsilon'}{4\pi\delta'}(\psi_0 - \psi'_\delta) = zeZ_m \frac{M_1 C}{1000} \exp\left(-\frac{ze(\psi_\delta + \phi)}{\kappa T}\right)$$

$$- \sqrt{\frac{RT\varepsilon}{2000\pi}} C^{1/2}\left[\exp\left(-\frac{ze\psi_\delta}{\kappa T}\right) - 1\right]^{1/2}, \quad (132)$$

where $|\psi'_\delta| > |\psi_\delta|$ and $\delta' < \delta$. This expression contains a number of unknown quantities and so is not very useful. However, a relation between ψ_δ and C can be derived for some systems.

IV. Distribution of Ions and Potential Differences at Interfaces

For systems where ϕ is small, $\psi_0 \gg \psi'_\delta$ and $\exp(-ze\psi_\delta/\kappa T) \gg 1$, Eq. (132) can be regarded as a quadratic in $[C\exp(-zF\psi_\delta/RT)]^{1/2}$. Thus

$$zeZ_m \frac{M_1}{1000} C \exp\left(-\frac{ze\psi_\delta}{\kappa T}\right) - \sqrt{\frac{\varepsilon RT}{2000\pi}} \sqrt{C \exp\left(-\frac{ze\psi_\delta}{\kappa T}\right)} - \frac{\varepsilon'}{4\pi\delta}\psi_0 = 0. \tag{133}$$

The solution of Eq. (133) has one real value giving

$$[C \exp(-ze\psi_\delta/\kappa T)]^{1/2} = \text{const.}$$

or

$$\psi_\delta = \frac{\kappa T}{ze} \ln C + \text{const.} \tag{134}$$

In the case of membranes selective to cations or anions, a similar equation can be derived from the Stern equation (124) directly by making a few assumptions. Equation (124) can be written

$$n_m = \frac{C(Z_m - n_m)}{1000/M_1} \exp\left(\frac{-ze(\psi_\delta + \phi)}{\kappa T}\right), \tag{135}$$

where n_m is the number of counterions of valence z adsorbed per square centimeter from a bulk solution of concentration C moles per liter and

$$\frac{Z_s - n_s}{n_s} \approx \frac{Z_s}{n_s} = \frac{1000}{M_1 C}.$$

The term $Z_m - n_m$ allows for nonzero size of the counterions placing a limit on n_m. Solving Eq. (135) for n_m gives Eq. (135) in the form

$$n_m = \frac{Z_m}{1 + (1000/CM_1)\exp[ze(\psi_\delta + \phi)/\kappa T]}. \tag{136}$$

For dilute solutions, the second term of the denominator is large compared to unity, and hence Eq. (136) becomes

$$n_m = \frac{Z_m C M_1}{1000} \exp\left(\frac{-ze(\psi_\delta + \phi)}{\kappa T}\right).$$

If the adsorption energy is small, this equation simplifies to

$$n_m = \frac{Z_m C M_1}{1000} \exp\left(\frac{-ze\psi_\delta}{\kappa T}\right). \tag{137}$$

If a value for δ is assumed, the concentration of ions C_m on the membrane surface due to adsorption can be related to bulk concentration. Let $\delta = 3$ Å.

If a counterion occupies an area of 20 Å² of the membrane surface, then $Z_m = 5 \times 10^{14}$ sites/cm² and $C_m = n_m/(18 \times 10^{12})$ (g ions/liter). Substituting these values in Eq. (137) gives the relation

$$C_m = 0.5C \exp(-ze\psi_\delta/\kappa T). \tag{138}$$

Equation (138) differs slightly from the Boltzmann distribution law [see Eq. (3)].

Groups or sites on the membrane surface in general are of the weak-acid or weak-base type, and so their ionization will depend on the pH of the medium in which they exist. The surface potential ψ will alter the pH value at the surface. The surface concentration of hydrogen ions can be expressed as

$$C_{H^+(m)} = C_{H^+(bulk)} \exp(-e\psi/\kappa T). \tag{139}$$

Expressing in terms of pH ($-\log C_{H^+}$), Eq. (139) becomes

$$pH_m = pH_b + (e\psi/2.303\kappa T). \tag{140}$$

When $\psi = 0$, membrane-surface pH will be equal to the pH of the medium. If ψ is negative, $pH_m < pH_b$ because the charge attracts hydrogen ions close to the membrane surface. But the difficulty in using Eq. (140) is that the value of ψ is not known. The appropriate ψ to choose is ψ_0 since the groups whose ionization is affected by the pH exist on the surface of the membrane. In such cases ψ_0 can be calculated by using Eq. (106) or (107).

V. Electrokinetic Phenomena

The potential ψ_δ is sometimes identified with the electrokinetic or zeta potential, which is represented by ζ. Electrokinetic behavior depends on the potential at the surface of shear between the charged membrane surface and the electrolyte solution. The exact location of the shear plane (a region of rapidly changing viscosity) is another unknown characteristic of the electrical double layer. In all probability a certain amount of solvent, in addition to ions, will be bound to the charged surface and form the electrokinetic unit in the Stern layer. Therefore it seems reasonable to suppose that the shear plane will be located a small distance farther away from the surface than the Stern plane. Consequently ζ will be a little smaller than ψ_δ (see Fig. 14 for details).

When there is no specific adsorption, the double layer is entirely diffuse (Gouy). When specific adsorption takes place counterion adsorption generally prevails over coion adsorption. A typical double-layer situation would

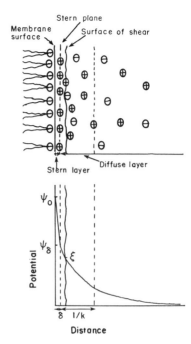

Fig. 14. Schematic representation of an electrical double layer according to Stern theory.

be that given in Fig. 14. It is possible that polyvalent or surface-active counterions would cause reversal of charge within the Stern layer, i.e., ψ_0 and ψ_δ to have opposite signs (see Fig. 15). It is also possible that surface active coions would act to create a situation in which ψ_δ would have the same sign as ψ_0 and be greater in magnitude, as shown in Fig. 16.

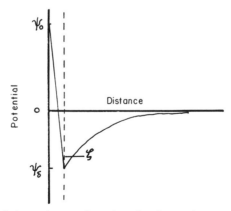

Fig. 15. Reversal of charge due to adsorption of surface active or polyvalent counterions. The potential is reversed in the diffuse double layer. ψ_0 may be little affected by the specific adsorption, but ζ potential has changed sign.

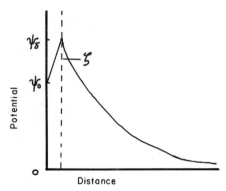

Fig. 16. Adsorption of surface active coions. ψ_δ is made greater (as also ζ) than ψ_0.

In tests of the double-layer theory, identity of ψ_δ and ζ is customarily assumed. Experimental results suggest that errors introduced by that assumption are generally small, especially at hydrophilic surfaces. Any difference between ψ_δ and ζ will be more pronounced at high potentials ($\psi_\delta = 0$ when $\zeta = 0$) and at high electrolyte concentration. Under the conditions of the latter, compression of the diffuse part of the double layer takes place and more of the potential drop from ψ_δ to zero will take place within the shear plane. Measurement of ζ directly cannot be made. However, it can be calculated from electrokinetic data derived from experimental study of one of the electrokinetic phenomena such as electrophoresis or electro-osmosis.

There are a number of electrokinetic phenomena that involve electricity and relative movement between a charged surface and bulk solution. These phenomena may arise from the action of an electric field along the phase boundary, leading to a movement, or they may arise from the motion of two phases along each other, leading to transport of electricity. The former are exemplified by electro-osmosis and electrophoresis, and the latter by streaming potential and sedimentation potential. In each case a plane of slip between the double layer and the medium is involved, and the results of measurements may be interpreted in terms of ζ potential at the shear surface or in terms of charge density at the slip plane.

A. *Electro-Osmosis*

Electro-osmosis is the movement of a liquid with respect to a membrane pore wall as a result of an applied electric field E. The velocity v of the liquid will depend on the ζ measured between the liquid and a point in the liquid far away from the wall surface (Fig. 17). When a steady liquid movement has been set up under the applied electrical potential gradient E V/cm, each layer of liquid of thickness dx (see Fig. 17) will move with a uniform velocity

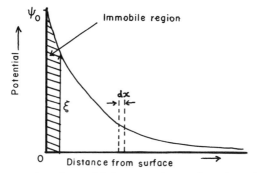

Fig. 17. Representation of potential drop away from a charged membrane surface.

parallel to the wall, the total force on such a layer being zero, i.e., the electrical forces on counterions in this particular liquid layer will be balanced by the frictional drag on neighboring layers. Thus, in the steady state

$$E\rho \, dx = \eta \left(\frac{dv}{dx}\right)_{x+dx} - \eta \left(\frac{dv}{dx}\right)_{x} = \eta \frac{d^2v}{dx^2} \, dx, \tag{141}$$

where ρ is the volume charge density. Substituting the Poisson equation (98) of Chapter 2 into Eq. (141) gives

$$-\frac{E\varepsilon}{4\pi} \frac{d^2\psi}{dx^2} = \eta \frac{d^2v}{dx^2}, \tag{142}$$

which on integration becomes

$$\left(\frac{dv}{dx}\right)_x^a = -\frac{E\varepsilon}{4\pi\eta} \left(\frac{d\psi}{dx}\right)_x^a, \tag{143}$$

where a is the radius of the membrane pore. If the pore width is large compared with $1/\varkappa$, i.e., if $\varkappa a \gg 1$, then the conditions for $x = a$ are $dv/dx = 0$ and $d\psi/dx = 0$. Thus the profile of the flow velocity is given by

$$\frac{dv}{dx} = -\frac{E\varepsilon}{4\pi\eta} \frac{d\psi}{dx}. \tag{144}$$

Since ψ becomes zero at a small distance from the wall of the membrane pore, this corresponds to piston or plug flow (see Fig. 18a, b, c). Further integration of Eq. (144) gives

$$[v]_{x=0}^{x=a} = -\frac{E\varepsilon}{4\pi\eta} [\psi]_{x=0}^{x=a}. \tag{145}$$

At the plane of shear, i.e., $x = 0$, $\psi = \zeta$ and flow velocity $= 0$ (no slippage). Also at $x = a$, $\psi = 0$ and flow velocity $= v$, and therefore the final relation,

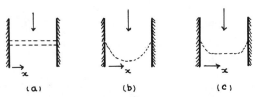

Fig. 18. Different types of flow: (a) piston or plug flow with linear velocity profile; (b) laminar flow, fluid at the pore wall moving very slowly and the rest having a parabolic profile; (c) steep velocity gradient at the wall and the rest having uniform velocity.

assuming constancy of ε and η, becomes

$$v = \frac{E\varepsilon}{4\pi\eta}\zeta. \tag{146}$$

Here ζ is a constant characteristic of the surface and independent of E.

Although direct measurement of electro-osmotic velocity v is possible, usually volume flow rate v' of liquid in the pore is measured. It is given by

$$v' = \pi a^2 v. \tag{147}$$

Ohm's law may be applied to eliminate πa^2. If i is the current and k_{sp} is the specific conductance of the solution of length 1 cm, then

$$i = E\pi a^2 k_{sp}. \tag{148}$$

Combination of Eqs. (147) and (148) eliminates πa^2, giving the relation

$$v' = iv/Ek_{sp}, \tag{149}$$

or if Eqs. (147) and (148) are combined with Eq. (146), the relation becomes

$$v' = i\varepsilon\zeta/4\pi k_{sp}\eta. \tag{150}$$

As the quantities v', i, and k_{sp} can be measured, ζ can be found from this equation.

Validity of Ohm's law for single-capillary systems is questionable because of surface conductance. Equation (148) must be written

$$i = E\pi a^2 k_{sp} + E2\pi a k_{sps}, \tag{151}$$

where k_{sps} is the surface conductance (Ω^{-1}). Thus Eq. (150) in general becomes

$$v' = \frac{i}{4\pi(k_{sp} + 2k_{sps}/a)} \frac{\varepsilon\zeta}{\eta}. \tag{152}$$

To use Eq. (152), surface conductance must be determined or conditions such as high k_{sp} and large a may be employed. Under these conditions, Eq. (152) reduces to Eq. (150).

There are several methods for determining surface conductance of capillaries. Of these, one method consists in using a membrane of known pore size. The conductance of the solution (e.g., KCl) is measured before and after inserting the membrane between the electrodes. Use of a collodion membrane of pore size 40 Å and of pore area 10% in 10^{-3} M KCl solution whose specific conductance is 140×10^{-6} (Ω^{-1} cm^{-1}) gave a value of 0.844 for the ratio (membrane conductance)/(bulk conductance). As the membrane conductance is completely due to the solution in the pores, the conductance of the pore solution is $0.844 \times 9 = 7.6$ times that of the free solution. Thus

$$7.6 = \frac{2\pi a k_{sps} + \pi a^2 k_{sp}}{\pi a^2 k_{sp}} = \frac{2k_{sps}}{k_{sp} a} + 1.$$

As $k_{sp} a = 140 \times 10^{-6} \times 40 \times 10^{-8}$, $k_{sps} = 0.19 \times 10^{-9} \, \Omega^{-1}$.

If an experimental setup shown in Fig. 19a is used, application of an electric field causes electroosmotic flow, and there will be net flow of liquid through the membrane, causing a difference in hydrostatic pressure. The difference in pressure causes counterflow, and in the steady state the electroosmotic pressure developed just balances the flow at the walls of the membrane pores (see Fig. 19b). The relation between the laminar flow (see Fig. 18b) and pressure is derived as follows.

The driving force acting on a pore is $\pi r^2 \Delta P$, where r is the distance from the center of the pore whose radius is a. The viscous force opposing the flow is $-\eta 2\pi r l \, dv/dr$. Here l is the length of the pore and dv/dx is the velocity

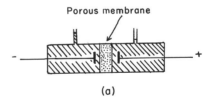

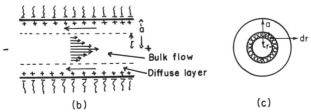

Fig. 19. (a) Apparatus (schematic) for the measurement of electro-osmotic pressure. (b) Enlarged view of one pore. (C) End view of a pore.

gradient. The negative sign indicates that v decreases as r increases. Equating the two forces and integrating gives

$$-\int_v^0 dv = \frac{\Delta P}{\eta 2l} \int_r^a r\, dr, \qquad (153)$$

and so

$$v = \frac{\Delta P}{4\eta l}(a^2 - r^2). \qquad (154)$$

This is the equation of a parabola.

To find the volume of fluid v' crossing any section of the pore in unit time, consider the thin-walled element in Fig. 19c. The volume flow rate of fluid dv' crossing the ends of the element of the pore is $dv' = v\, dA$, where $dA = 2\pi r\, dr$, the shaded area. Substituting for v from Eq. (154) gives

$$dv' = \frac{\Delta P}{4\eta l}(a^2 - r^2) 2\pi r\, dr. \qquad (155)$$

The volume flow rate v' moving across the entire cross section is obtained by integrating over all elements between $r = 0$ and $r = a$, and so

$$v' = \frac{\pi \Delta P}{2\eta l} \int_{r=0}^{r=a} (a^2 - r^2) r\, dr,$$

$$v' = \frac{\pi \Delta P a^4}{8\eta l}. \qquad (156)$$

Equation (156) is called the Poiseuille law.

Combination of Eqs. (148), (150), and (156) gives a relation for the electroosmotic pressure ΔP. Thus

$$\Delta P = \frac{2\zeta \varepsilon E l}{\pi a^2}. \qquad (157)$$

For an aqueous medium at 25°C, Eq. (146) gives

$$v = \frac{E\,(V)\,\zeta\,(V)\,78.3}{4\pi \times 0.89 \times 10^{-2} \times 300^2} = 7.8 \times 10^{-3}\,E\zeta \quad (\text{cm/s}),$$

and Eq. (157) gives

$$\Delta P = \frac{2\zeta\,(V)\,El\,(V)\,78.3}{\pi a^2 \times 300^2}$$

$$= 5.53 \times 10^{-4}\,\zeta El/a^2 \quad (\text{dyn/cm}^2)$$
$$= 4.2 \times 10^{-8}\,\zeta El/a^2 \quad (\text{cm Hg}),$$

where potentials are in volts and lengths in centimeters. If ζ is 100 mV, then a 1-V potential applied gives a pressure of 1 cm Hg if the pore radius is about 1 μm.

B. Streaming Potential and Streaming Current

The potential induced across a membrane when a solution is forced through the membrane pores is called the streaming potential. It arises from the liquid flow carrying with it the mobile part of the electrical double layer near the walls of the membrane pores.

Assuming laminar flow (Fig. 18b), pore radius large compared with $1/\kappa$, and surface conductance negligible, the streaming potential E_{str} is proportional to ζ. The flow profile (see Fig. 18b) is given by Eq. (154), which on differentiation gives the velocity gradient very close to pore wall. Thus

$$\left(\frac{dv}{dr}\right)_{r=a} = -\frac{\Delta P a}{2\eta l}. \tag{158}$$

The velocity v at any distance x ($x = a - r$) from the wall is given by

$$v = \int_0^x \left(\frac{dv}{dx}\right)_{x=0} dx = \frac{\Delta P a x}{2\eta l}. \tag{159}$$

The streaming current is given by

$$I_{str} = 2\pi a \int_0^a \rho v \, dx. \tag{160}$$

Substitution of the Poisson equation for ρ and Eq. (159) for v in Eq. (160) gives

$$I_{str} = -\frac{\Delta P a^2}{4\eta l} \int_0^a \varepsilon \frac{d^2\psi}{dx^2} x \, dx. \tag{161}$$

If ε is independent of x, integration by parts gives

$$I_{str} = -\frac{\Delta P \varepsilon a^2}{4\eta l} \left\{ \left[\frac{d\psi}{dx} x\right]_{x=0}^{x=a} - [\psi]_{x=0}^{x=a} \right\}. \tag{162}$$

As $[d\psi/dx] = 0$ and $\psi = 0$ when $x = a$ and $\psi = \zeta$ when $x = 0$, Eq. (162) reduces to

$$I_{str} = -\frac{\Delta P \varepsilon a^2}{4\eta l} \zeta. \tag{163}$$

This current is responsible for E_{str}. In the steady state, I_{str} must be balanced by a current I conducting the charge back through the liquid and at the surface (in the case of single capillaries) so that $I_{str} + I = 0$. But I is given by

$$I = \frac{E_{str}\pi a^2 k_{sp}}{l} + \frac{E_{str}2\pi a k_{sps}}{l}. \tag{164}$$

Thus, equating Eqs. (163) and (164) gives

$$E_{str} = \frac{\Delta P \varepsilon \zeta a}{4\eta\pi(k_{sp}a + 2k_{sps})} = \frac{\Delta P \varepsilon \zeta}{4\pi\eta[k_{sp} + 2k_{sps}/a]}. \tag{165}$$

When k_{sps} is negligible,

$$E_{str} = \frac{\Delta P \varepsilon \zeta}{4\pi\eta k_{sp}} \quad \text{or} \quad \frac{E_{str}}{\Delta P} = \frac{\varepsilon \zeta}{4\pi\eta k_{sp}}. \tag{166}$$

Streaming-potential measurements, although difficult to make, show that the ratio $E_{str}/\Delta P$ is constant provided the pressure is not too high to cause turbulent flow. If the electrodes at the two surfaces of the membrane are short circuited, the current I_{str} produced by an imposed flow of liquid through the pores of the membrane will go through the short circuit rather than through the liquid. For streamline flow, the current I_{str} is given by Eq. (163). The negative sign shows that when the pore wall and ζ are negative, the flow of positive ions constitutes positive current. This is a useful method for finding values of ζ.

Comparison of Eqs. (150) for electro-osmosis and (166) for streaming potential show that both phenomena can be described by the same parameters. That is,

$$\frac{\text{Volume}}{i} = \frac{E_{str}}{\Delta P} = \frac{\varepsilon \zeta}{4\pi\eta k_{sp}}. \tag{167}$$

Even when the correction for surface conduction is applied [Eqs. (152) and (165)], relation (167) is found to be valid. This relation, known as the Saxen relation, has been known for a long time. This means that an experiment on streaming potential gives the same information as one on electro-osmosis. From these experiments, a value for ζ potential can be derived if either surface conductance is negligible or if the experiment is carried out on a single capillary.

C. Electrophoresis

In electrophoresis the liquid as a whole is at rest while a particle moves through it under the influence of an electric field. In both the phenomena (i.e., electro-osmosis and electrophoresis), the forces acting in the electrical double layer control the relative movements of liquid and solid, and so Eq. (146) will also apply to electrophoresis, v now being the velocity of electrophoretic movement:

$$v = E\varepsilon\zeta/4\pi\eta.$$

For water at 25°C, $\zeta = 129v/E$.

There are complications in more detailed treatment of electrophoretic motion. The effective viscosity in the double layer is affected because of the motion of ions in the mobile part of the double layer. The particle and the ions in the double layer move in opposite directions. This creates a local movement of liquid opposing the motion of the particle, known as electrophoretic retardation. There is also a relaxation effect. Because of the motion of the particle, the double layer lags behind and becomes distorted, because a finite time (relaxation time) is required for the original symmetry to be restored by diffusion and conduction. The resulting asymmetric ion atmosphere exerts an additional retarding force on the particle.

Another difficulty is that the double layer is a source of conductance and so is the surface of the particle itself. As these are difficult to evaluate, nonconducting surfaces or particles must be used for reliable zeta-potential measurements.

VI. Donnan Equilibrium

The system is composed of two solutions separated by a membrane that is impermeable to at least one of the charged species (see Fig. 20). The impermeable species q^{n-} may be some colloidal particle. The presence of a membrane is not essential. In any gel, in an ion-exchange membrane containing charged groups fixed to its matrix, or in a substance, such as clay, ionized into very large and small ions, the equilibria that exist will be of the

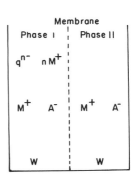

Fig. 20. Donnan equilibrium system.

Donnan type. Three important features are involved in Donnan equilibrium: (1) unequal distribution of ions; (2) osmotic pressure; and (3) potential difference between the two phases.

The elementary theory assumes ideal behavior for both solutions or phases separated by a membrane permeable to solvent w and ions M^+ and A^- but impermeable to ion q of valence n. The condition for equilibrium according to Eq. (128) of Chapter 2 is

$$\mu_i^I + z_i F E^I = \mu_i^{II} + z_i F E^{II}. \tag{168}$$

The temperature and electrochemical potentials of all diffusible species in the two phases are equal; but the osmotic pressures will differ, and so the equality for ideal solutions can be written

$$\mu_i^0 + RT \ln N_i^I + \Pi^I \bar{V}_i + z_i F \psi^I = \mu_i^0 + RT \ln N_i^{II} + \Pi^{II} \bar{V}_i + z_i F \psi^{II}, \tag{169}$$

where N_i is the mole fraction of component i, $\bar{V}_i$ is its partial molar volume, and μ_i^0 is the chemical potential in the standard state. The condition of electroneutrality gives

$$\sum z_i N_i^I = \sum z_i N_i^{II} = 0. \tag{170}$$

Applying these equations to all components of the system gives

$$(\Pi^I - \Pi^{II})\bar{V}_+ + F(\psi^I - \psi^{II}) = RT \ln(N_+^{II}/N_+^I), \tag{171}$$

$$(\Pi^I - \Pi^{II})\bar{V}_- - F(\psi^I - \psi^{II}) = RT \ln(N_-^{II}/N_-^I), \tag{172}$$

$$(\Pi^I - \Pi^{II})\bar{V}_w = RT \ln(N_w^{II}/N_w^I), \tag{173}$$

$$N_+^I = N_-^I + n N_q^I, \tag{174}$$

$$N_+^{II} = N_-^{II} = N^{II}. \tag{175}$$

Adding Eqs. (171) and (172) gives

$$\Delta \Pi = \Pi^I - \Pi^{II} = \frac{RT}{\bar{V}_+ + \bar{V}_-} \ln \frac{N_+^{II} N_-^{II}}{N_+^I N_-^I}. \tag{176}$$

If $\Delta \Pi$ is eliminated between Eqs. (173) and (176), the final condition for the membrane equilibrium of the system of Fig. 20 is

$$\frac{1}{\bar{V}_w} \ln \frac{N_w^{II}}{N_w^I} = \frac{1}{\bar{V}_+ + \bar{V}_-} \ln \frac{N_+^{II} N_-^{II}}{N_+^I N_-^I}. \tag{177}$$

This may be written

$$\frac{N_+^{II} N_-^{II}}{(N_w^{II})^Q} = \frac{N_+^I N_-^I}{(N_w^I)^Q}, \tag{178}$$

VI. Donnan Equilibrium

where Q represents the ratio of the partial molar volumes of electrolyte and solvent water, i.e.,

$$Q = (\bar{V}_+ + \bar{V}_-)/\bar{V}_w. \qquad (179)$$

If the compressibility of both solutions is neglected in the pressure range $\Delta\Pi$, Eq. (178) for dilute solutions (i.e. $N_w^{II} = N_w^{I}$) becomes

$$N_+^{II} N_-^{II} = N_+^{I} N_-^{I}. \qquad (180)$$

Expressed in molar concentrations C, Eq. (180) becomes

$$C_+^{II} C_-^{II} = C_+^{I} C_-^{I},$$
$$C_+^{I} C_-^{I} = (C^{II})^2, \qquad (181)$$
$$C_+^{I} = C_-^{I} + nC_q^{I}.$$

The osmotic pressure $\Delta\Pi$ is given by Eq. (173) as

$$\Delta\Pi = \frac{RT}{\bar{V}_w} \ln \frac{N_w^{II}}{N_w^{I}} = \frac{RT}{\bar{V}_w} \ln \frac{1 - 2N^{II}}{1 - N_+^{I} - N_-^{I} - N_q^{I}}$$

$$\approx \frac{RT}{\bar{V}_w} [N_+^{I} + N_-^{I} - 2N^{II} + N_q^{I}]$$

$$\approx RT(C_+^{I} + C_-^{I} - 2C^{II} + C_q^{I}). \qquad (182)$$

Equation (182) is the van't Hoff equation for osmotic pressure. If $\Delta\Pi$ in Eqs. (171) and (172) is eliminated by the use of Eq. (173), the Donnan membrane potential ψ_{Don} is obtained as

$$\psi_{Don} = \frac{RT}{F} \ln \frac{N_+^{II}}{N_+^{I}} \left(\frac{N_w^{I}}{N_w^{II}}\right)^{Q^+} = \frac{RT}{F} \ln \frac{N_-^{I}}{N_-^{II}} \left(\frac{N_w^{II}}{N_w^{I}}\right)^{Q^-}.$$

Here $Q^+ = (\bar{V}_+/\bar{V}_w)$ and $Q^- = (\bar{V}_-/\bar{V}_w)$. If $N_w^{II} = N_w^{I}$, then the Donnan potential approximates to

$$\psi_{Don} = \frac{RT}{F} \ln \frac{N_+^{II}}{N_+^{I}} = \frac{RT}{F} \ln \frac{N_-^{I}}{N_-^{II}}$$

$$= \frac{RT}{F} \ln \frac{C_+^{II}}{C_+^{I}} = \frac{RT}{F} \ln \frac{C_-^{I}}{C_-^{II}} = \frac{RT}{F} \ln r, \qquad (183)$$

where r is called the Donnan ratio and is given by

$$r = \frac{C_+^{II}}{C_+^{I}} = \frac{C_-^{I}}{C_-^{II}} = \cdots. \qquad (184)$$

From the equation of ion distribution [Eq. (181)], the concentrations of positive and negative ions in phase I can be expressed in terms of salt concentration C of phase II, i.e.,

$$C^I_-(C^I_- + nC^I_q) = (C^{II})^2. \tag{185}$$

Solving this quadratic equation for C^I_- gives

$$C^I_- = [-nC^I_q + \{(nC^I_q)^2 + 4(C^{II})^2\}^{1/2}]/2. \tag{186}$$

When $C^{II} \gg nC^I_q$, Eq. (186) becomes

$$C^I_- = C^{II} - nC^I_q/2 \quad \text{or} \quad C^I_+ = C^{II} + nC^I_q/2. \tag{187}$$

Use of these values in Eq. (182) gives

$$\Delta\Pi = RTC^I_q. \tag{188}$$

The concentration of the impermeable ion alone determines the osmotic pressure, and the Donnan potential (Eq. 183) is reduced to

$$\psi_{\text{Don}} = \frac{RT}{F} \ln\left[C^{II} \bigg/ \left(C^{II} + \frac{nC^I_q}{2}\right)\right]$$

$$\left[\ln x = \frac{x-1}{x} + \cdots\right], \tag{189}$$

$$\psi_{\text{Don}} = -\frac{RT}{2F} \frac{nC^I_q}{C^{II}}.$$

For the condition when $C^{II} \ll nC^I_q$, Eq. (185) yields

$$C^I_-(nC^I_q) = (C^{II})^2,$$

i.e.,

$$C^I_- = (C^{II})^2/nC^I_q \quad \text{or} \quad C^I_+ = [(C^{II})^2/nC^I_q] + nC^I_q.$$

Now the osmotic pressure is given by

$$\Delta\Pi = (n+1)C^I_q RT. \tag{190}$$

Here not only the concentration of the impermeable ion but also the concentration of its counterions determine the osmotic pressure. The Donnan potential is given by

$$\psi_{\text{Don}} = \frac{RT}{F} \ln \frac{C^{II}}{nC^I_q}. \tag{191}$$

Of the three aspects of Donnan equilibrium, viz., distribution of small ions, osmotic pressure, and Donnan potential, the first two are thermodynamically defined equilibrium properties, whereas the last one is ill defined since it cannot be measured directly. However, it can be measured indirectly although approximately by the use of salt bridges.

In the preceding considerations, ideality was assumed. One simple correction for nonideality of solutions is to introduce activity coefficients in Eqs. (169)–(175). Equation (181) can be written

$$(C_+^I f_+^I)(C_-^I f_-^I) = (C^{II})^2 (f_\pm^{II})^2 \quad \text{or} \quad (f_\pm^I)^2 C_+^I C_-^I = (f_\pm^{II} C^{II})^2, \quad (192)$$

i.e., $(a_\pm^I)^2 = (a_\pm^{II})^2$ or $a_\pm^I = a_\pm^{II}$. At equilibrium, mean activity of the electrolyte must be the same in the two phases, although single-ion activities will be different.

The osmotic pressure equation (182) can be corrected by introducing the osmotic coefficient term g. Thus

$$\Delta \Pi = \frac{RT}{\bar{V}_w} \ln \frac{a_w^{II}}{a_w^I} = gRT(C_+^I + C_-^I - 2C^{II} + C_q^I). \quad (193)$$

In general, the Donnan potential is given by

$$\psi_{\text{Don}} = \frac{RT}{z_i F} \ln \frac{a_i^{II}}{a_i^I}, \quad (194)$$

where a_i is the activity of ion i of valence z_i to which the membrane is permeable. These equations, without reliable values for activity coefficients involving single ions, are of little value. However, for dilute solutions, the Debye–Hückel limiting law may be used to estimate values for f_i.

When there are ions of different valences, their distribution under equilibrium conditions [see Eqs. (184) and (192)] can be written

$$\frac{a_+^{II}}{a_+^I} = \frac{a_-^I}{a_-^{II}} = \sqrt{\frac{a_{++}^{II}}{a_{++}^I}} = \sqrt{\frac{a_{--}^I}{a_{--}^{II}}} = \sqrt[3]{\frac{a_{+++}^{II}}{a_{+++}^I}} = \cdots = r. \quad (195)$$

In the preceding considerations, it has been assumed that the impermeable ion was of a strong base or strong acid so that its degree of dissociation is independent of pH changes in the solution. But if they are weak acids, weak bases, or ampholytes such as proteins, dissociation constants have to be taken into account in the consideration of membrane equilibrium. For example take the case of the salt of weak base R whose cation is RH^+. The system at equilibrium is

$$\begin{array}{c|c}
\text{I} & \text{II} \\
RH^+, R & H_3O^+ \ Cl^- \\
H_3O^+ \ H_2O & H_2O \\
Cl^- & \\
\hline
\multicolumn{2}{c}{\text{membrane}}
\end{array} \quad (196)$$

In phase I, the equilibrium is

$$RH^+ \underset{}{\overset{+H_2O}{\rightleftharpoons}} R + H_3O^+$$

Both R and RH^+ are impermeable.

Neglecting activity coefficients, the condition for membrane equilibrium is given by Eq. (192). Thus

$$[C_{H_3O^+}C_{Cl^-}]_I = [C_{H_3O^+}C_{Cl^-}]_{II} = (C_{H_3O^+})_{II}^2. \tag{197}$$

In phase I, $C_R + C_{RH^+} = C_0$, where C_0 is the total concentration of salt and the dissociation constant or the hydrolysis constant of the salt K_h according to the above reaction is given by

$$K_h = \frac{C_R C_{H_3O^+}}{C_{RH^+}} = \frac{C_{H_3O^+}(C_0 - C_{RH^+})}{C_{RH^+}}.$$

Thus

$$C_{RH^+} = [C_0 C_{H_3O^+}]/(K_h + C_{H_3O^+}).$$

But the condition of electroneutrality gives

$$(C_{Cl^-})_I = (C_{H_3O^+})_I + (C_{RH^+})_I = (C_{H_3O^+})_I \left[1 + \frac{C_0}{K_h + C_{H_3O^+}}\right]_I.$$

Substituting this into Eq. (197) gives

$$(C_{H_3O^+})_{II} = \left[C_{H_3O^+}\sqrt{1 + \frac{C_0}{K_h + C_{H_3O^+}}}\right]_I. \tag{198}$$

Thus the Donnan ratio is given by

$$r = \frac{(C_{H_3O^+})_{II}}{(C_{H_3O^+})_I} = \sqrt{1 + \frac{C_0}{K_h + (C_{H_3O^+})_I}} \tag{199}$$

and the Donnan potential by

$$\psi_{Don} = \frac{RT}{2F} \ln\left[1 + \frac{C_0}{K_h + (C_{H_3O^+})_I}\right]. \tag{200}$$

When the acidity of the solution is increased, r approaches unity and ψ_{Don} becomes zero.

Now if the system (196) above contains a weak acid HA, the membrane being impermeable to both acid HA and the anion A^- and the other side of the membrane being KOH solution, analogous considerations give the equation

$$r = \frac{(C_{OH^-})_I}{(C_{OH^-})_{II}} = \left(\sqrt{1 + \frac{C_0}{K_h + (C_{OH^-})_I}}\right)^{-1}, \quad (201)$$

where K_h is the hydrolysis constant of the salt corresponding to the reaction

$$A^- + H_2O \rightleftharpoons HA + OH^-.$$

In this case r is less than unity, and with decreasing acidity of the solution, it approaches unity and ψ_{Don} again tends to zero.

A case of considerable interest is a protein solution near its isoelectric point where the following equilibria exist.

$$H_2O + {}^+H_3NRCOOH \rightleftharpoons H_3O^+ + {}^+H_3NRCOO^-$$

$${}^+H_3NRCOO^- + H_2O \rightleftharpoons H_3O^+ + H_2NRCOO^-$$

The protolytic equilibrium constants in dilute solution are

$$K_1 = \frac{C_{H_3O^+} C_{+R^-}}{C_{+R}} \quad (202)$$

and

$$K_2 = \frac{C_{H_3O^+} C_{R^-}}{C_{+R^-}}. \quad (203)$$

The system at equilibrium may be represented by

I	II
$R^+, {}^+R^-, R^-$	H_3O^+ Cl^-
H_3O^+ Cl^-	
H_2O	H_2O

membrane

The membrane is impermeable to ${}^+R$, ${}^+R^-$, and R^-, and the total concentration of ampholyte is C_0. The following relations are valid:

$$(C_{Cl^-})_I = C_{H_3O^+} + C_{+R} - C_{R^-} \quad (204)$$

(condition of electroneutrality),

$$C_{+R^-} = C_0 - C_{+R} - C_{R^-}. \quad (205)$$

Combining Eqs. (202) and (203) to eliminate C_{+R^-} gives

$$C_{+R} = \frac{(C_{H_3O^+})_I^2 C_{R^-}}{K_1 K_2}. \quad (206)$$

Substitution of Eq. (206) into Eq. (204) gives

$$(C_{Cl^-})_I = (C_{H_3O^+})_I + C_{R^-}\left[\frac{(C_{H_3O^+})_I^2 - K_1K_2}{K_1K_2}\right]. \tag{207}$$

Substitution of Eq. (203) to eliminate C_{+R^-} and Eq. (206) to eliminate C_{+R} in Eq. (205) gives on simplification

$$C_{R^-} = C_0 \bigg/ \left(1 + \frac{(C_{H_3O^+})_I}{K_2} + \frac{(C_{H_3O^+})_I^2}{K_1K_2}\right). \tag{208}$$

Substituting Eq. (208) into Eq. (207) gives

$$(C_{Cl^-})_I = (C_{H_3O^+})_I + \frac{C_0[(C_{H_3O^+})_I^2 - K_1K_2]}{K_1K_2 + K_1(C_{H_3O^+})_I + (C_{H_3O^+})_I^2}. \tag{209}$$

Substituting again Eq. (209) into Eq. (197) gives

$$(C_{H_3O^+})_{II} = \sqrt{(C_{H_3O^+})_I^2 + (C_{H_3O^+})_I \frac{C_0[(C_{H_3O^+})_I^2 - K_1K_2]}{K_1K_2 + K_1(C_{H_3O^+})_I + (C_{H_3O^+})_I^2}}. \tag{210}$$

Thus the Donnan ratio is given by

$$r = \frac{(C_{H_3O^+})_{II}}{(C_{H_3O^+})_I} = \sqrt{1 + \frac{C_0[(C_{H_3O^+})_I^2 - K_1K_2]}{(C_{H_3O^+})_I[K_1K_2 + K_1(C_{H_3O^+})_I + (C_{H_3O^+})_I^2]}}. \tag{211}$$

At the isoelectric point, i.e., $C_{+R} = C_{R^-}$, Eqs. (202) and (203) give $(C_{H_3O^+})^2 = K_1K_2$, and so $r_{\text{isoelectric}} = 1$. The Donnan potential is zero.

For high values of $C_{H_3O^+}$, K_1 and K_2 may be neglected and Eq. (211) simplifies to Eq. (199). As pH decreases, r must decrease and reach unity. This means both r and ψ_{Don} pass through a maximum with change of pH. On the alkaline side of isoelectric point $C_{H_3O^+}^2 < K_1K_2$ and ψ_{Don} becomes negative. So the membrane potential must change sign and increase with increasing pH, pass through a maximum, and reach zero again when the solution is highly alkaline. These predictions have been confirmed by experiments.

VII. Donnan Equilibrium in Charged Membranes

In the equilibria described already, the membrane served as a barrier permeable to small ions and impermeable to large ionized or ionizable molecules. If the ionized or ionizable groups are immobilized by being fixed to the membrane matrix, e.g., ion-exchange and biological membranes, and the membrane remained permeable to other ions, the equilibrium between

VII. Donnan Equilibrium in Charged Membranes

the membrane phase and the aqueous phase surrounding it is governed by the several equations given above to describe the distribution of solute, solvent, and the attendant osmotic pressure and membrane potential.

In general the equilibrium condition for the electrolyte ij is $\bar{\mu}_{ij} = \mu_{ij}$ where the overbar is used to indicate the membrane phase. Equation (176) can be written

$$\ln(a_{ij}/\bar{a}_{ij}) = \Delta \Pi \bar{V}_{ij}/RT. \tag{212}$$

If ij dissociates into v_i and v_j ions, Eq. (212) becomes

$$\ln \frac{m_i^{v_i} m_j^{v_j} \gamma_{\pm}^{v}}{\bar{m}_i^{v_i} \bar{m}_j^{v_j} \bar{\gamma}_{\pm}^{v}} = \frac{\Delta \Pi \bar{V}_{ij}}{RT}. \tag{213}$$

Equation (213) combined with the equation of electroneutrality, i.e., $\sum z_i \bar{m}_i + \omega \bar{X} = 0$, where $\bar{X}$ is the concentration of fixed groups attached to the membrane matrix and ω is the sign of these groups (equal to -1 for negatively charged membrane and $+1$ for positively charged membrane), gives an expression for $\bar{a}_+$ or $\bar{a}_-$. For a selective membrane and (1:1) electrolyte, $\bar{a}_+$ and $\bar{a}_-$ are given by

$$\bar{a}_+ = -\frac{\omega \bar{X} \bar{\gamma}_+}{2} + \left[\frac{(\omega \bar{X} \bar{\gamma}_+)^2}{4} + \frac{a_\pm^2 \gamma_\pm^2}{\bar{\gamma}_\pm^2 \exp(\Pi \bar{V}/RT)} \right]^{1/2},$$

$$\bar{a}_- = \frac{\omega \bar{X} \bar{\gamma}_-}{2} + \left[\frac{(\omega \bar{X} \bar{\gamma}_-)^2}{4} + \frac{a_\pm^2 \gamma_\pm^2}{\bar{\gamma}_\pm^2 \exp(\Pi \bar{V}/RT)} \right]^{1/2}. \tag{214}$$

Under the assumptions that all single-ion activity coefficients in the membrane phase are unity, that the osmotic pressure term is negligible, and that $\omega \bar{X}$ remains constant at all values of external electrolyte concentration, Eq. (214) simplifies to Eq. (186). Thus

$$\bar{m}_+ = -\frac{\omega \bar{X}}{2} + \left[\left(\frac{\omega \bar{X}}{2} \right)^2 + a^2 \right]^{1/2},$$

$$\bar{m}_- = \frac{\omega \bar{X}}{2} + \left[\left(\frac{\omega \bar{X}}{2} \right)^2 + a^2 \right]^{1/2}. \tag{215}$$

The Donnan ratio becomes

$$r = \frac{\bar{m}_+}{a} = -\frac{\omega \bar{X}}{2a} + \left[1 + \left(\frac{\omega \bar{X}}{2a} \right)^2 \right]^{1/2}. \tag{216}$$

The Donnan potential in the case of a negatively charged membrane is given by (ignoring activity coefficients)

$$\psi_{\text{Don}} = \frac{RT}{F} \ln\left[\frac{\bar{X}}{2C} + \sqrt{\frac{\bar{X}^2}{4C^2} + 1}\right], \tag{217}$$

where $\bar{X}$ is in moles per liter. A point of considerable interest is the relation of ψ_{Don} to the surface potential ψ_0.

Serious difficulty in relating ψ_{Don} to ψ_0 lies in defining the term $\bar{X}$. For this purpose, the membrane surface phase must be assigned a particular value for the thickness. This is an approximation in that the diffuse part of the double layer would contain some counterions away from the membrane surface. If the Debye–Hückel length $1/\varkappa$ is taken to represent the thickness of this surface phase, then the counterion concentration calculated from ψ_{Don} would be an average of the values calculated from the potentials of the Gouy theory; here the concentrations are averaged from the surface of the membrane to a distance of $1/\varkappa$ from it (see Fig. 21).

ψ_{Don} calculated from the surface phase of thickness $1/\varkappa$ will be less than the ψ_0 since ψ_0 relates to charges existing on a plane. The exact relation can be demonstrated from Eq. (217), which can be written in an equivalent form [$\sinh^{-1} = \ln(x + \sqrt{x^2 + 1}\,)$]

$$\psi_{\text{Don}} = (RT/F)\sinh^{-1}(\bar{X}/2C). \tag{218}$$

Now $\bar{X}$ can be expressed as $\sigma\varkappa\, 10^3/eN$. Substituting the numerical values for the several constants and expressing σ as charges per angstrom squared,

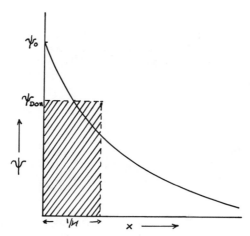

Fig. 21. Typical potential variation with distance from surface with no specific interaction. The double layer is entirely diffuse and the shaded region equal to $1/\varkappa$ (double-layer thickness) is considered to represent the surface phase.

Eq. (218) becomes

$$\psi_{Don} = \frac{RT}{F} \sinh^{-1}\left(\frac{2 \times 136\sigma}{\sqrt{C}}\right). \quad (219)$$

This equation may be compared with Eq. (106). When $(\sqrt{C}/\sigma) = 136$, ψ_{Don} is 81% of ψ_0. The deviation becomes larger or smaller as $1/\varkappa$ becomes large (i.e., dilute solution) or small (i.e., high C).

VIII. Membrane Potential

In the case of charged membranes each face of which is in equilibrium with a solution of the same electrolyte but of different concentration, two Donnan potentials exist, one at each membrane–solution interface. The total electrical potential arising across the membrane is composed of not only these two Donnan potentials ψ'_{Don} and ψ''_{Don} but also a diffusion potential ϕ arising from the presence of a concentration gradient $\Delta C (= C'' - C')$ across the membrane. This is shown in scheme of Fig. 22. Thus the membrane potential according to Teorell, Meyer, and Sievers (TMS theory) is given by

$$E = \psi'_{Don} + \psi''_{Don} + \phi. \quad (220)$$

Teorell, Meyer, and Sievers (see Teorell, 1953) assumed ideal conditions (negligible $\Pi\bar{V}$ term, invariance of $\bar{X}$ and $\bar{\gamma}_i = 1$). ψ_{Don} under these conditions for a negatively charged membrane is given by Eq. (217). Thus

$$\psi'_{Don} = \bar{\psi}' - \psi' = \frac{RT}{F} \ln \frac{\sqrt{\bar{X}^2 + 4a'^2} + \bar{X}}{2a'},$$

$$\psi''_{Don} = \psi'' - \bar{\psi}'' = \frac{RT}{F} \ln \frac{\sqrt{\bar{X}^2 + 4a''^2} + \bar{X}}{2a''}. \quad (221)$$

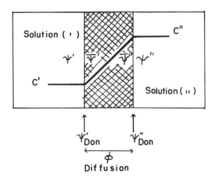

Fig. 22. Schematic representation of several potentials at the membrane–solution interfaces and across the membrane.

The diffusion potential $\phi = \bar{\psi}'' - \bar{\psi}'$ within the membrane may be equated to that of a constrained diffusion junction. Teorell used the expression

$$\phi = \frac{\bar{u}_+ - \bar{u}_-}{\bar{u}_+ + \bar{u}_-} \frac{RT}{F} \ln \frac{\bar{u}_+ \bar{m}'_+ + \bar{u}_- \bar{m}'_-}{\bar{u}_+ \bar{m}''_+ + \bar{u}_- \bar{m}''_-}. \tag{222}$$

Substituting for $\bar{m}_+$ and $\bar{m}_-$ from Eq. (215), Eq. (222) becomes

$$\phi = \bar{U} \frac{RT}{F} \ln \frac{\sqrt{\bar{X}^2 + 4a'^2} + \bar{U}\bar{X}}{\sqrt{\bar{X}^2 + 4a''^2} + \bar{U}\bar{X}}, \tag{223}$$

where $\bar{U} = (\bar{u}_+ - \bar{u}_-)/(\bar{u}_+ + \bar{u}_-)$.

The total membrane potential E is given by the algebraic sum of Eqs. (221) and (223). Thus

$$E = \frac{RT}{F} \left[\bar{U} \ln \frac{\sqrt{\bar{X}^2 + 4a''^2} + \bar{U}\bar{X}}{\sqrt{\bar{X}^2 + 4a'^2} + \bar{U}\bar{X}} + \ln \frac{a''}{a'} \frac{\sqrt{\bar{X}^2 + 4a'^2} + \bar{X}}{\sqrt{\bar{X}^2 + 4a''^2} + \bar{X}} \right]. \tag{224}$$

Three special cases of Eq. (224) are of interest:

(1) When the outside concentration is large compared to the fixed ion concentration, i.e., $a \gg \bar{X}/2$, Eq. (224) reduces to

$$E = (RT/F)(t_+ - t_-)\ln(a''/a'). \tag{225}$$

(2) When $a \ll \bar{X}/2$, Eq. (224) reduces to the Nernst equation

$$E_{\max} = (RT/F)\ln(a''/a'), \tag{226}$$

corresponding to the maximum potential across a permselective membrane.

(3) When $a \cong \bar{X}/2$, a simplification is difficult. But if the approximations made in case 1 are carried through, Eq. (224) reduces to

$$E = (RT/F)(\bar{t}_+ - \bar{t}_-)\ln(a''/a'), \tag{227}$$

where $\bar{t}_+$ and $\bar{t}_-$ are the transport numbers in the membrane phase.

Equation (227) can be written using Eq. (226) in the form

$$E = E_{\max}(2\bar{t}_+ - 1)$$

or

$$\bar{t}_+ = E/2E_{\max} + 0.5. \tag{228}$$

Equation (228) has often been used to calculate transport numbers in the membrane phase from membrane potential measurements. The transport number so calculated can be used to evaluate the permselectivity P_s of the membrane, which is defined as

$$P_s = (\bar{t}_+ - t_+)/(1 - t_+). \tag{229}$$

IX. Some Applications of the Double-Layer Theory

The double-layer theory has been applied to explain several membrane phenomena. It was first applied in 1965 to biological membranes by Chandler, Hodgkin, and Meves to explain the shift in the sodium inactivation curve on the voltage axis following variation in the ionic strength of the internal perfusion solution. Later it was used to explain the decrease in bilayer membrane resistance following treatment with potassium iodide and chloride solutions. If G_0^I is the bilayer membrane conductance in KI solution, then

$$G_0^I = \text{const } \bar{C}_1. \tag{230}$$

$\bar{C}_1$, the concentration of iodide at the membrane surface, is related to the bulk concentration C_1 by the Boltzmann relation, and so Eq. (230) becomes

$$G_0^I = \text{const } C_1 \exp(\psi^I F/RT). \tag{231}$$

If addition of KCl gives conductance G_0^{II}, then

$$G_0^{II} = \text{const } C_1 \exp(\psi^{II} F/RT). \tag{232}$$

Thus

$$G_0^{II}/G_0^I = \exp(\Delta\psi F/RT), \tag{233}$$

where $\Delta\psi = \psi^{II} - \psi^I$, the change in surface potential. This change in $\Delta\psi$ calculated from Eq. (233) agrees with ψ_G predicted by Eq. (106).

Differentiation of Eq. (105) when $z = 1$, σ is constant, and $\psi_0 = \psi_G$ gives

$$C^{1/2}\left(d\psi_G \frac{F}{2RT}\right)\left(\cosh\frac{F\psi_G}{2RT}\right) + \sinh\frac{F\psi_G}{2RT}\frac{1}{2}C^{-1/2}\,dC = 0$$

or

$$\frac{d\psi_G}{d\ln C} = -\left(\frac{RT}{F}\right)\tanh\frac{F\psi_G}{2RT}. \tag{234}$$

For small values of ψ_G and C, Eq. (234) becomes

$$\Delta\psi_G = -\frac{RT}{F}\ln\frac{C_1}{C_2}\left(\tanh\frac{F\psi_G}{2RT}\right). \tag{235}$$

Surface potentials of monolayers of acidic lipids (phosphatidic acid and phosphatidyl serine) measured with the help of an americium air electrode and Vibron 33 B electrometer conformed to Eq. (235) over a wide range of surface charge density and salt concentration. If the charged monolayers are placed back to back to form a bilayer and used to separate two aqueous phases, the membrane potential is given by

$$E = \Delta\psi_G^{II} - \Delta\psi_G^{I}, \tag{236}$$

where II and I indicate the two faces of the bilayer membrane. If the electrolyte concentration C^I in contact with face I is held constant and if C^{II}, the concentration of electrolyte in contact with face II, is varied, then application of Eq. (235) gives

$$E = -\frac{RT}{F} \ln \frac{C^I}{C^{II}} \left(\tanh \frac{\psi_G F}{2RT} \right). \tag{237}$$

The time-dependent behavior of potentials arising across bilayer membranes of phosphatidyl choline, serine, and phosphatidic acid and their mixtures has been found to conform to Eq. (237). Similarly, ζ potentials evaluated from electrophoretic measurements of decane droplets containing phosphatidyl choline and phosphatidic acid as a function of electrolyte concentration showed that the slope $d\zeta/d \ln C$ was equal to $d\psi_G/d \ln C$.

When the surface potential is large $(-\psi_0 \gg RT/F)$, Eq. (103) becomes

$$\exp(-F\psi_0/RT) = 272^2 \sigma^2 / C^+$$

for monovalent ions,

$$\exp(-F\psi_0/RT) = 272\sigma / \sqrt{C^{2+}}$$

for divalent ions, and

$$\exp(-F\psi_0/RT) = \frac{272^{2/3} \sigma^{2/3}}{3\sqrt{C^{3+}}}$$

for trivalent ions. These equations indicate that monovalent and divalent ions will have the same effect on ψ_0 when

$$C^{2+} = (C^+)^2 / 272^2 \sigma^2,$$

and divalent and trivalent ions will have the same effect on ψ_0 when

$$(C^{3+})^2 = (C^{2+})^3 / 272^2 \sigma^2. \tag{238}$$

In crayfish axons, it has been found that divalent cations (Ca, Mg, Sr, Ba) and trivalent cations (La, Y, Eu) had the same effect on the threshold for spike generation provided the cations were of the same valence. A trivalent ion concentration of 225 μM or divalent ion concentration of 13.5 mM was found just optimal in not changing the threshold for excitation. These values can be used in Eq. (238) to derive a value of 1 $e/43$ Å^2 for σ, the charge density on the membrane of crayfish nerve.

IX. Some Applications of the Double-Layer Theory

In the consideration of the Gouy–Chapman double-layer theory, the binding of ions (e.g., cations) to charges (negative sites) on the membrane has been ignored. This binding can be taken into account by considering the relations

$$C + S \rightleftharpoons CS \tag{239}$$

$$K_a = \frac{[CS]}{[C][S]}, \tag{240}$$

where [C] is the concentration of cation at the membrane containing sites of concentration [S] and K_a is the association constant (liters per mole). As the surface charge is due to the presence of free charged sites (S), it follows that

$$\frac{\sigma}{\sigma_{\max}} = \frac{[S]}{[S_{\max}]} = \frac{1}{1 + K_a[C]}, \tag{241}$$

where $[S_{\max}] = [S] + [CS] = [S][1 + K_a(C)]$.

Relating (C) to bulk concentration C_b by the Boltzmann relation gives

$$\sigma = \frac{\sigma_{\max}}{1 + K_a C_b \exp(-zF\psi_0/RT)}. \tag{242}$$

If both monovalent and divalent cations are involved in neutralizing the sites on the membrane, Eq. (242) becomes

$$\sigma = \frac{\sigma_{\max}}{1 + K_{a(1)} C_b^+ \exp(-F\psi_0/RT) + K_{a(2)} C_b^{2+} \exp(-2F\psi_0/RT)}, \tag{243}$$

where $K_{a(1)}$ and $K_{a(2)}$ are the association constants of monovalent and divalent cations respectively.

Equations (104) and (242) or (243) have been used to derive values for both $\sigma_{\max}$ and K_a. This was done by making measurements of shifts along the voltage axis of conductance (g_{Na} and/or g_K) versus voltage curve (i.e., measurements of $\Delta V_{1/2}$, the change in $V_{1/2}$ where $V_{1/2}$ is the potential at which Na or K conductance of the membrane, for example of the squid axon, has reached one-half its maximum value) following changes in the calcium concentration from a standard solution such as 100 mM Ca artificial seawater (ASW) to a test solution for example of 10 mM Ca ASW. This change in $V_{1/2}$ is used as a measure of the change in surface potential. Equations (104) and (242) or (243) are solved for the two Ca concentrations with an iterative procedure by using different values for $\sigma_{\max}$ and K_a until the best fit of the observed change, i.e., change in $V_{1/2}$, is obtained. This type of analysis involving curve fitting ($\Delta V_{1/2}$ plotted as a function of Ca concentration) using Eqs. (104) and (242) or (243) has been used by several investigators to derive values for $\sigma_{\max}$ and K_a, some of which are given in Table II.

TABLE II

Values of Surface Charge Density σ_{max} and K_a Determined by Measurement of Shift in Conductance–Voltage Curve Following Change in External Calcium Concentration

Membrane system	Shift in g_{Na} curve σ_{max} (1 e/Å²)	K_a (liters/mol)	Shift in g_k curve σ_{max} (1 e/Å²)	K_a	Reference
Squid axon	1/120		1/120		Begenisich, 1975, from the data of Frankenhaeuser and Hodgkin, 1957
Myxicola axon	1/100–1/140	0–0.2	1/120–1/280	0.1	Gilbert and Ehrenstein, 1969
	1/77		1/275–1/350	0.1–0.2	Begenisich 1975
			1/77		Schauf, 1975
Crayfish axon		$\sigma_{max} = 1/43^a$			D'Arrigo, 1973
Frog node	1/90	0	small		Begenisich, 1975, from the data of Hille, 1968
	1/225	0.04			Gilbert and Ehrenstein, 1970, from the data of Hille, 1968
	1/111–1/173	0.035–0.83			Hille et al., 1975
Xenopus node	1/160	0	1/417–1/588	5.0	Mozhayeva and Naumov, 1970, 1972
			1/290		Begenisich, 1975, from the data of Brismar, 1973
	1/70				Brismar, 1973
					Vogel, 1973

[a] Value determined by measurement of shift in threshold for excitation; see Eq. (238).

X. Model-System Approach to Evaluation of Surface Charge Density

Equation (224) may be used to estimate $\bar{X}$ and $(\bar{u}_+/\bar{u}_-)$ by the measurement of membrane potential under conditions where C_2/C_1 is maintained constant under varying conditions of concentration. It is a curve-fitting procedure where values of $\bar{X}$ and $\bar{u}_+/\bar{u}_-$ are used to construct several curves by plotting values of E against $\log(1/C_2)$. The experimental curve shifted horizontally and ran parallel to one of the theoretical curves. The shift gave $\log \bar{X}$, and the coinciding curve gave a value for $\bar{u}_+/\bar{u}_-$. A more elegant method is due to Kobatake and Kamo (1973), who derived an equation for the membrane potential similar to Eq. (224), in which the stoichiometric charge density has been replaced by an effective charge density $\overline{\theta X}$ where $\bar{\theta}$ ($0 < \bar{\theta} < 1$) is a constant characteristic of the membrane–electrolyte system and represents the fraction of the counterions that are free. The relevant equations pertaining to evaluation of $\overline{\theta X}$ from measurements of membrane potential are the following:

For a charged membrane separating the same (1:1) electrolyte solutions, the mobilities u_i and the activity coefficients γ_i are considered to be given by

$$\bar{u}_+ = u_+ \frac{\bar{m}_- + \overline{\theta X}}{\bar{m}_- + \bar{X}}, \qquad \bar{u}_- = u_-, \tag{244}$$

$$\bar{\gamma}_+ = \gamma_- \frac{\bar{m}_- + \overline{\theta X}}{\bar{m}_- + \bar{X}}, \qquad \bar{\gamma}_- = \gamma_-, \tag{245}$$

where $\bar{m}_i$ is the molality (moles/1000 g water) of i ($i = +$ for cation and $i = -$ for anion). For a negatively charged membrane, the electroneutrality condition gives

$$\bar{m}_+ = \bar{m}_- + \bar{X}. \tag{246}$$

Substituting Eqs. (245) and (246) in the Donnan relation

$$\bar{m}_+ \bar{\gamma}_+ \bar{m}_- \bar{\gamma}_- = m^2 \gamma_+ \gamma_- \tag{247}$$

gives a quadratic equation whose solution is given by

$$\bar{m}_- = \frac{\overline{\phi X} \pm \sqrt{(\overline{\phi X})^2 + 4m^2}}{2}. \tag{248}$$

When a charged membrane separates two solutions of the same electrolyte of molalities m_1 and m_2, the membrane potential is given by Eq. (227):

$$E = \frac{RT}{F} [\bar{t}_{+(\text{app})} - \bar{t}_{-(\text{app})}] \ln \frac{m_2}{m_1}. \tag{249}$$

The transport numbers referred to the membrane phase in Eq. (249) are called apparent because mass transport across the membrane is ignored or considered too small. Rearrangement of Eq. (249) gives [see Eq. (228)]

$$\bar{t}_{+(\text{app})} = E/2E_{\max} + 0.5. \tag{250}$$

The transport number of anion $\bar{t}_-$ is given by

$$\bar{t}_- = \frac{\bar{u}_-\bar{m}_-}{\bar{u}_+\bar{m}_+ + \bar{u}_-\bar{m}_-}. \tag{251}$$

Substitution of Eqs. (244), (246), and (248) in Eq. (251) gives, on simplification, the relation

$$\bar{t}_- = 1 - t_+ \frac{1 + (1 + 4q^2)^{1/2}}{(1 + 4q^2)^{1/2} + (2t_+ - 1)}, \tag{252}$$

where the transport number of cation in aqueous phase $t_+ = u_+/(u_+ + u_-)$ and $q = (m/\overline{\phi X})$.

Equating $\bar{t}_-$ to $\bar{t}_{-(\text{app})}$ [$\bar{t}_{-(\text{app})}$ calculated from Eq. (250), viz., $\bar{t}_{-(\text{app})} = 1 - \bar{t}_{+(\text{app})}$ agreed to within 2% of the value calculated from Eq. (252)] and averaging m_1 and m_2, that is,

$$q = \frac{m_1 + m_2}{2\overline{\phi X}}, \tag{253}$$

Eq. (252) rearranges to

$$P_s = 1/(1 + 4q^2)^{1/2} \tag{254}$$

or

$$q = \sqrt{1 - P_s^2}/2P_s, \tag{255}$$

where P_s, the permselectivity of the membrane, is given by

$$P_s = \frac{\bar{t}_{+(\text{app})} - t_+}{t_+ - \bar{t}_{+(\text{app})}(2t_+ - 1)}. \tag{256}$$

By measurement of membrane potential as a function of external concentration, $\overline{\phi X}$ on the membrane can be evaluated with the help of Eqs. (250), (256), (255), and (253), in that order.

To evaluate $\bar{X}$ for biomembranes, definite values for both $\overline{\phi X}$ and $\bar{\phi}$ are needed. A value for $\overline{\phi X}$ corresponding to the physiological state of the membrane, i.e., at physiological concentrations of extracellular and intracellular potassium ions (K_0 and K_i), can be obtained, but to derive a value for $\bar{X}$, some procedure to obtain a value for $\bar{\phi}$ must be devised, as its value is unknown for biological membranes. An assumption that can be made is that

TABLE III

Values of Membrane Parameters $\overline{\phi X}$ (mol/kg water), $\overline{\phi}$, $\overline{X}$ (mol/kg water), Surface Charge Density σ (1 $e/\text{Å}^2$) Derived for Several Biological Systems from Resting Membrane Potential Data[a]

Biological system	$\overline{\phi X}$	$\overline{\phi}$ for acid group		$\overline{X}$ for acid group		σ for acid group		Average
		strong	weak	strong	weak	strong	weak	
Squid axon	0.300	0.286	0.364	1.05	0.824	1/158	1/201	1/180
Crayfish axon	0.443	0.220	0.279	2.01	1.59	1/82	1/105	1/94
Myxicola axon	0.444	0.258	0.328	1.72	1.36	1/97	1/122	1/110
Frog muscle (sartorius and semitendinosus)	0.325	0.169	0.214	1.92	1.52	1/87	1/110	1/99
Barnacle muscle	0.170	0.196	0.248	0.868	0.680	1/191	1/242	1/217

[a] From Lakshminarayanaiah (1977).

the values of $\bar{\phi}$ determined for ion-exchange membranes containing strong or weak acid groups are applicable to biomembranes, since such groups are considered to exist in biological membranes although their exact nature and the organic skeleton to which they are attached may be different. The values of $\bar{\phi}$ determined for ion-exchange model membranes are found to fit the following equations:

$$-\log \bar{\phi} = 0.205 - 0.492 \log C$$

(for strong acid ion-exchange membranes),

$$-\log \bar{\phi} = 0.097 - 0.497 \log C$$

(for weak acid ion-exchange membranes), where C is the ambient electrolyte concentration, i.e., $C = (K_0 + K_i)/2$. Thus values for $\bar{\phi}$ corresponding to the ambient physiological state of the biological membrane may be derived. These values of $\bar{\phi}$ together with the values of $\overline{\phi X}$ and $\bar{X}$ derived for various biological membranes are given in Table III.

Conversion of volume charge density $\bar{X}$ (moles/1000 g membrane water) to surface charge density σ requires additional assumptions. These assumptions are (1) that the Davson–Danielli model is applicable to biological membrane; (2) that membrane water is that associated with polar head groups that project into intracellular and extracellular phases of the cell; (3) that the thickness of this water is equivalent to the thickness of the electrical double layer, about 10 Å; and (4) that the charges are uniformly smeared over both the intracellular and extracellular membrane surfaces. The values of σ derived from these considerations are given in Table III. Several of these values (squid and *Myxicola* axons) compare favorably with those given in Table II.

References

Adamson, A. W. (1960). "Physical Chemistry of Surfaces." Wiley (Interscience), New York.
Begenisich, T. (1975). *J. Gen. Physiol.* **66**, 47.
Bockris, J. O'M., and Reddy, A. K. N. (1970). "Modern Electrochemistry," Vols. 1 and 2. Plenum, New York.
Brismar, T. (1973). *Acta Physiol. Scand.* **87**, 474.
Chandler, W. K., Hodgkin, A. L., and Meves, H. (1965). *J. Physiol.* (*London*) **180**, 821.
Conway, B. E. (1965). "Theory and Principles of Electrode Processes." Ronald Press, New York.
D'Arrigo, J. S. (1973). *J. Physiol.* (*London*) **231**, 117.
Davies, J. T., and Rideal, E. K. (1961). "Interfacial Phenomena." Academic Press, New York.
Frankenhaeuser, B., and Hodgkin, A. L. (1957). *J. Physiol.* (*London*) **137**, 218.
Gilbert, D. L., and Ehrenstein, G. (1969). *Biophys. J.* **9**, 447.
Gilbert, D. L., and Ehrenstein, 6. (1970). *J. Gen. Physiol.* **55**, 822.
Hille, B. (1968). *J. Gen. Physiol.* **51**, 221.
Hille, B., Woodhull, A. M., and Shapiro, B. I. (1975). *Philos. Trans. R. Soc. London, Ser. B* **270**, 301.

Khortum, G., and Bockris, J. O'M. (1951). "Text-Book of Electrochemistry," Vols. 1 and 2. Elsevier, Amsterdam.
Kobatake, Y., and Kamo, N. (1973). *Prog. Polym. Sci. Jpn.* **5**, 257.
Kruyt, H. R., ed. (1952). "Colloid Science," Vol. 1. Elsevier, Amsterdam.
Lakshminarayanaiah, N. (1969). "Transport Phenomena in Membranes." Academic Press, New York.
Lakshminarayanaiah, N. (1977). *Bull. Math. Biol.* **39**, 643.
Lakshminarayanaiah, N. (1979). *Subcell. Biochem.* **6**, 401.
MacDonald, R. C., and Bangham, A. D. (1972). *J. Membr. Biol.* **7**, 29.
Mozhayeva, G. N., and Naumov, A. P. (1970). *Nature (London)* **228**, 164.
Mozhayeva, G. N., and Naumov, A. P. (1972). *Biofizika* **17**, 412, 618, 801.
Parsons, R. (1954). *In* "Modern Aspects of Electrochemistry" (J. O'M. Bockris and B. E. Conway, eds.), p. 103. Butterworth, London.
Robinson, R. A., and Stokes, R. H. (1959). "Electrolyte Solutions," 2nd ed. Academic Press, New York.
Schauf, C. L. (1975). *J. Physiol. (London)* **248**, 613.
Teorell, T. (1953). *Prog. Biophys. Biophys. Chem.* **3**, 305.
Vogel, W. (1973). *Experientia* **29**, 1517.

Chapter 4

ELECTRICAL POTENTIALS ACROSS MEMBRANES

The current theories of membrane behavior may be classified roughly into three groups. This is based on the nature of the flux equation used in the treatment. The theories based on Nernst–Planck flux equations or their refinements form one group. The theories based on the principles of irreversible thermodynamics fall into the second group. The theories forming the third group use the concepts of the theory of rate processes.

These several theoretical descriptions supplement one another although, depending on the system under consideration, one description may prove more suitable than the other. In recent years, the concepts of the theory of rate processes are being used a great deal in the description of several membrane phenomena. This, in large measure, is due to the easy accessibility of the computers.

The several membrane theories as applied to considerations of potentials arising across membranes of different types are summarized in this chapter.

Application of Nernst–Planck flux equation [see Eq. (197) of Chapter 2] to the problem of electrical potentials arising across a diffusion zone was considered in Chapter 2. In Chapter 3 the considerations of Teorell, Meyer, and Sievers (TMS theory) to derive an expression for membrane potential across a charged membrane were presented. In that consideration ideality was assumed; that is, activity coefficients in the membrane phase and transport of water through the membrane were ignored. The contribution of water transfer to the total membrane potential across a membrane can be estimated from the equation

$$\psi_w = -\frac{RT}{F} \int_I^{II} \bar{t}_w \, d\ln a_w,$$

where $\bar{t}_w$ is the transport number of water (moles per Faraday) and a_w is the activity of water.

I. Bi- and Multi-Ionic Potentials

The electrical potential across a membrane of a bi-ionic cell

Electrolyte	Membrane	Electrolyte
AX		BX
a'		a''

can be simply written

$$E = (RT/F)\ln(\bar{u}_A/\bar{u}_B) \qquad (1)$$

or as

$$E = (RT/F)\ln(a'_A \bar{u}_A / a''_B \bar{u}_B). \qquad (2)$$

For a general case where the total bi-ionic potential is considered in accordance with the concepts of the TMS theory—two Donnan potentials and one diffusion potential—the total potential is given by

$$E = \frac{RT}{F}\left[\frac{\bar{D}_B - \bar{D}_A}{z_A \bar{D}_A - z_B \bar{D}_B}\ln\frac{z_B \bar{D}_B}{z_A \bar{D}_A} + \frac{1}{z_A z_B}\ln K_B^A \right.$$
$$\left. + \frac{z_A - z_B}{z_A z_B}\ln\frac{\bar{C}}{\bar{C}'_A} + \frac{1}{z_B}\ln\frac{C'_A}{C''_B} + \ln\frac{\gamma_A^{'1/z_A}}{\gamma_B^{''1/z_B}}\right], \qquad (3)$$

where $\bar{D}$s are diffusion coefficients, $\bar{C} = z_A \bar{C}_A + z_B \bar{C}_B = \bar{X}$ is the total counterion concentration, and C'_A and C''_B are the bulk concentrations of solutions. It is assumed that $\bar{D}_A/\bar{D}_B$, $\ln K_B^A$, and $\bar{C}$ are constants. There are no coions and convection in the membrane, and the boundary condition $(\bar{C}''_B/C''_B) = (\bar{C}'_A/C'_A) = \bar{C}/C$ holds.

K_B^A is the corrected molar selectivity coefficient for the exchange reaction

$$\bar{B} + A \rightleftharpoons \bar{A} + B \qquad (4)$$

and is given by the ratio of the activity coefficients in the membrane

$$K_B^A = \frac{\bar{\gamma}_B^{z_A}}{\bar{\gamma}_A^{z_B}} = \left(\frac{\bar{C}_A}{a_A}\right)^{z_B}\left(\frac{a_B}{\bar{C}_B}\right)^{z_A}. \qquad (5)$$

When the valences of ions are equal ($z_A = z_B$), Eq. (3) becomes

$$E = \frac{RT}{zF}\ln\frac{\bar{D}_A \bar{\gamma}_B a'_A}{\bar{D}_B \bar{\gamma}_A a''_B}. \qquad (6)$$

I. Bi- and Multi-Ionic Potentials

Equation (2) is an approximate form of Eq. (6) where $z = 1$ and $\bar{u}_A/\bar{u}_B = (\bar{D}_A\bar{\gamma}_B)/(\bar{D}_B\bar{\gamma}_A)$.

For a multi-ionic membrane system containing two solutions (AX, BX, CX, ...) and (PX, QX, RX, ...) on either side of a membrane the total membrane potential is given by

$$E = E_{\text{diff}} + \psi'_{\text{Don}} - \psi''_{\text{Don}},$$

where

$$E_{\text{diff}} = \frac{RT}{z_i F} \ln\left[\left(\sum_i \bar{D}_i \bar{C}'_i\right) \bigg/ \left(\sum_j \bar{D}_j \bar{C}''_j\right)\right]$$

$$\psi'_{\text{Don}} - \psi''_{\text{Don}} = \frac{RT}{z_i F} \ln\left\{\left[\sum_i \left(\frac{a'_i}{\bar{a}'_i}\right)\right] \bigg/ \left[\sum_j \left(\frac{a''_j}{\bar{a}''_j}\right)\right]\right\} \quad (7)$$

$$E = \frac{RT}{z_i F} \ln\left\{\left[\sum_i \bar{D}_i \bar{C}'_i \left(\frac{a'_i}{\bar{a}'_i}\right)\right] \bigg/ \left[\sum_j \bar{D}_j \bar{C}''_j \left(\frac{a''_j}{\bar{a}''_j}\right)\right]\right\}$$

$$= \frac{RT}{z_i F} \ln\left\{\left[\sum_i \bar{D}_i \left(\frac{a'_i}{\bar{\gamma}_i}\right)\right] \bigg/ \left[\sum_j \bar{D}_j \left(\frac{a''_j}{\bar{\gamma}_j}\right)\right]\right\},$$

where i refers to A, B, C, ... and j to P, Q, R, ... ions.

Equation (7) may be applied to the membrane system (LX)'|membrane| (LX, MX, NX, ...)'', where (LX)' is the reference solution and (LX) on side ('') is the unknown, to estimate the activity of unknown LX in the mixture. If the membrane is ideal and permeable only to L, measurement of the concentration potential would give the activity of a''_L as

$$E = \frac{RT}{z_L F} \ln \frac{a'_L}{a''_L}.$$

But if the membrane is not ideal and permeable to M, N, ... but not to X, Eq. (7) may be written (assuming equivalence of counterions)

$$E = \frac{RT}{z_i F} \ln\left[\left(\frac{\bar{D}_L a'_L}{\bar{\gamma}_L}\right) \bigg/ \left(\sum_i \frac{\bar{D}_i a''_i}{\bar{\gamma}_i}\right)\right].$$

This equation may be rewritten

$$E = \frac{RT}{z_i F} \ln \frac{a'_L}{a''_L + (\bar{D}_M \bar{\gamma}_L/\bar{D}_L \bar{\gamma}_M) a''_M + (\bar{D}_N \bar{\gamma}_L/\bar{D}_L \bar{\gamma}_N) a''_N + \cdots}. \quad (8)$$

But $(\bar{D}_M \bar{\gamma}_L/\bar{D}_L \bar{\gamma}_M)$, etc., may be evaluated by measuring bi-ionic potentials of systems concerned such as (MX|LX), (NX|LX), etc., with equal activities in both solutions.

The general Nernst–Planck flux equation (195) of Chapter 2 has been applied by several investigators to analyze the electrical potentials in selective membranes separating two solutions and permeable only to counterions A and B. In these treatments, the activity coefficients of counterions within the membrane have been included by empirical relations worked out to fit the experimental data. Some of the expressions are very unwieldy. One such treatment of interest and importance, because of its use in the evaluation of ion selectivity coefficients of ion-selective membrane electrodes, is that worked out by Eisenman and Karreman (1962, 1967).

The equilibrium constant K of the reaction (4) is given by

$$K = a_B \bar{a}_A / \bar{a}_B a_A. \tag{9}$$

Rothmund and Kornfeld (1918) showed that the equilibrium constant is given by

$$K = (a_B/a_A)(\bar{x}_A/\bar{x}_B)^n, \tag{10}$$

where the $\bar{x}$s are mole fractions and n depends on the properties of the membrane only.

Equations (9) and (10) give the relations

$$\bar{a}_A = \bar{C}_A \bar{\gamma}_A = p(\bar{x}_A)^n \quad \text{and} \quad \bar{a}_B = \bar{C}_B \bar{\gamma}_B = p(\bar{x}_B)^n, \tag{11}$$

where p is a proportionality constant depending on the properties of the membrane and $n = 1$ for ideal behavior.

The electroneutrality condition gives

$$\bar{C}_A + \bar{C}_B = \bar{X} \tag{12}$$

and

$$\bar{x}_A + \bar{x}_B = 1, \tag{13}$$

where

$$(\bar{x}_A = \bar{C}_A/\bar{X}) \quad \text{and} \quad \bar{x}_B = (\bar{C}_B/\bar{X}). \tag{14}$$

Combining Eqs. (11) and (14) gives

$$\bar{\gamma}_A = \frac{p(\bar{x}_A)^n}{\bar{x}_A \bar{X}} = \left(\frac{p}{\bar{X}}\right)(\bar{x}_A)^{n-1} \quad \text{and} \quad \bar{\gamma}_B = \left(\frac{p}{\bar{X}}\right)(\bar{x}_B)^{n-1}. \tag{15}$$

Total current carried by ions A and B ($I = I_A + I_B$) may be calculated in terms of the $\bar{x}$s [Eq. (14)] by using the Nernst–Planck flux equation. For the condition $I = 0$, the relation

$$d\bar{E} = -\frac{RT}{F} \frac{\bar{u}_A d\bar{x}_A + \bar{u}_B d\bar{x}_B + \bar{u}_A \bar{x}_A d\ln \bar{\gamma}_A + \bar{u}_B \bar{x}_B d\ln \bar{\gamma}_B}{\bar{u}_A \bar{x}_A + \bar{u}_B \bar{x}_B} \tag{16}$$

is obtained. Substitution of the values for $\bar{\gamma}_A$ and $\bar{\gamma}_B$ from Eq. (15) into Eq. (16)

I. Bi- and Multi-Ionic Potentials

gives

$$d\bar{E} = -\frac{nRT}{F} d\ln(\bar{u}_A \bar{x}_A + \bar{u}_B \bar{x}_B). \tag{17}$$

Integration of Eq. (17) between the limits $x = 0$ and $x = d$ (the thickness of the membrane) gives the diffusion potential in the membrane. That is,

$$\bar{E}_{\text{diff}} = \bar{E}_{(d)} - \bar{E}_{(0)} = \frac{nRT}{F} \ln \frac{\bar{u}_A \bar{x}_A(0) + \bar{u}_B \bar{x}_B(0)}{\bar{u}_A \bar{x}_A(d) + \bar{u}_B \bar{x}_B(d)}. \tag{18}$$

Membrane concentrations can be related to the outside concentrations by Eqs. (10) and (13). Thus

$$\bar{x}_A = \frac{K^{1/n} a_A^{1/n}}{a_B^{1/n} + K^{1/n} a_A^{1/n}}, \tag{19}$$

$$\bar{x}_B = \frac{a_B^{1/n}}{a_B^{1/n} + K^{1/n} a_A^{1/n}}. \tag{20}$$

Substitution of these values into Eq. (18) gives the diffusion potential. Thus

$$E_{\text{diff}} = \frac{nRT}{F} \ln \frac{(a_B')^{1/n} + (\bar{u}_A/\bar{u}_B) K^{1/n} (a_A')^{1/n}}{(a_B')^{1/n} + K^{1/n} (a_A')^{1/n}}$$
$$- \frac{nRT}{F} \ln \frac{(a_B'')^{1/n} + (\bar{u}_A/\bar{u}_B) K^{1/n} (a_A'')^{1/n}}{(a_B'')^{1/n} + K^{1/n} (a_A'')^{1/n}}. \tag{21}$$

The two Donnan potentials can be calculated using the nth-power relation of Eq. (11). For the ion-exchange reaction (4), the electrochemical potentials of A and B in the aqueous phase must be equal to the corresponding electrochemical potentials in the membrane phase at equilibrium. This means

$$F\psi_{\text{Don}} = \mu_A - \bar{\mu}_A = \mu_A^0 - \bar{\mu}_A^0 + RT \ln(a_A/\bar{a}_A)$$
$$= \mu_B - \bar{\mu}_B = \mu_B^0 - \bar{\mu}_B^0 + RT \ln(a_B/\bar{a}_B). \tag{22}$$

Equations (11), (19), and (20) substituted into Eq. (22) give on simplification

$$\psi_{\text{Don}} = \text{const} + \frac{nRT}{F} \ln(a_B^{1/n} + K^{1/n} a_A^{1/n}). \tag{23}$$

The total potential is given by

$$E = E_{\text{diff}} + \psi_{\text{Don}}' - \psi_{\text{Don}}''.$$

Thus adding Eqs. (21) and (23) gives for ions $i(B)$ and $j(A)$

$$E = \frac{nRT}{F} \ln \frac{(a_i')^{1/n} + (K_{ij}^{\text{pot}} a_j')^{1/n}}{(a_i'')^{1/n} + (K_{ij}^{\text{pot}} a_j'')^{1/n}}, \tag{24}$$

where $K_{ij}^{pot} = K(\bar{u}_j/\bar{u}_i)^n$ and is called the selectivity coefficient. If the concentrations on side (″) are held constant, as in a membrane electrode unit, Eq. (24) can be written

$$E = \text{const} + \frac{nRT}{F} \ln[a_i^{1/n} + (K_{ij}^{pot}a_j)^{1/n}]. \tag{25}$$

Equation (25) is valid for a monovalent ion (primary) in the presence of another monovalent ion (interfering). The equation applicable to a primary divalent ion in presence of monovalent ions is

$$E = \text{const} + \frac{nRT}{2F} \{(a_i^{2+})^{1/n} + [K_{ij}^{pot}(a_j^+)^2]^{1/n}\}. \tag{26}$$

The equation applicable to a primary divalent ion in presence of other divalent ions is

$$E = \text{const} + \frac{nRT}{2F} \ln[(a_i^{2+})^{1/n} + (K_{ij}^{pot}a_j^{2+})^{1/n}]. \tag{27}$$

These equations for $n = 1$ may be written in a general form as

$$E = E^0 + \frac{RT}{z_iF} \ln\left[a_i + \sum_{i \neq j} K_{ij}^{pot}(a_j)^{z_i/z_j}\right], \tag{28}$$

where E^0 is a constant, i is the primary ion of valence z_i to which the membrane electrode is selective, and j is the interfering ion of valence z_j.

II. Determination of Selectivity Coefficients K_{ij}^{pot}

The preceding equations are useful in the evaluation of responses of ion-selective electrodes, both macro and micro types. This evaluation consists in determining values for the selectivity coefficient K_{ij}^{pot}—the smaller the value, the greater the selectivity of the electrode to the primary ion i than it is to the interfering ion j. There are several methods to determine the values of K_{ij}^{pot}. These may be divided into two categories: one using simple electrolyte solutions and the other using mixed solutions of electrolytes.

Category I: This is also called the separate-solution method. For the primary ion i of valence z_i only in solution (i.e., $a_j = 0$), Eq. (28) becomes

$$E_i = E_i^0 + (RT/z_iF) \ln a_i. \tag{29}$$

If the solution contains only ion j of valence z_j, Eq. (28) becomes

$$E_j = E_i^0 + (RT/z_iF) \ln K_{ij}^{pot} a_j^{z_i/z_j}. \tag{30}$$

Subtracting Eq. (29) from Eq. (30) gives on simplification

$$\log K_{ij}^{\text{pot}} = \frac{(E_j - E_i)z_i F}{2.303RT} + \log a_i - \log a_j^{z_i/z_j}. \tag{31}$$

When $a_i = a_j = a$, Eq. (31) becomes

$$\log K_{ij}^{\text{pot}} = \frac{(E_j - E_i)z_i F}{2.303RT} + \left(1 - \frac{z_i}{z_j}\right) \log a. \tag{32}$$

This method is recommended by the International Union of Pure and Applied Chemistry (IUPAC) only if the electrode exhibits a Nernstian response. This method is not so desirable, because the conditions do not represent the actual conditions under which the electrodes are used, generally mixtures.

When $z_i = z_j = z$, Eq. (32) simplifies to

$$\log K_{ij}^{\text{pot}} = (E_j - E_i)zF/2.303RT. \tag{33}$$

Several studies indicate that Eq. (31) gave the "least constant, most concentration-dependent selectivity coefficients." The equation that generated the most nearly constant selectivity coefficients is

$$\log K_{ij}^{\text{pot}} = \frac{(E_j - E_i)z_j F}{2.303RT} + \log a_i^{z_j/z_i} - \log a_j. \tag{34}$$

This is equivalent to writing Eq. (28) for the bi-ionic case as

$$E = E^0 + \frac{RT}{F} \ln[a_i^{1/z_i} + (K_{ij}^{\text{pot}} a_j)^{1/z_j}]. \tag{35}$$

This suggestion was made by Buck and Stover (1978). Another method to determine K_{ij}^{pot} is to choose the concentrations of ions i and j separately so that they generate the same potential, i.e., $E_i = E_j$. This is a cumbersome procedure, little used in practice; but K_{ij}^{pot} is simply given by [see Eq. (31)]

$$K_{ij}^{\text{pot}} = a_i/a_j^{z_i/z_j}. \tag{36}$$

Category II: This is called the mixture or fixed interference method. The concentration of the interfering ion j is kept constant and that of the primary ion is varied. The measured potentials are plotted against the activity of the primary ion i. The intersection of the extrapolation of the linear portions of the curve will give the value of a_i to be used in Eq. (36) to calculate K_{ij}^{pot}. As the concentration of j to that of i is increased, eventually the electrode shows no response to i, and the plot of potential against $\log a_i$ becomes horizontal. At this point the conditions of Eq. (36) are approximated, and hence it is used to calculate K_{ij}^{pot}.

Another method to use mixed solutions to evaluate K_{ij}^{pot} is to employ Eq. (28) directly. Initially E_i measured in solution containing only the primary ion of activity a_i is given by Eq. (29). Then j ion of activity a_j is added and the E_{ij} potential is measured. This potential is given by Eq. (28). Thus, combining Eqs. (28) and (29) gives

$$E_{ij} - E_i = \frac{RT}{z_i F} \ln\left[\frac{a_i + K_{ij}^{pot} a_j^{z_i/z_j}}{a_i}\right]. \tag{37}$$

Rearrangement of Eq. (37) gives

$$a_j^{z_i/z_j} K_{ij}^{pot} = a_i[\exp\{(E_{ij} - E_i)z_i F/RT\} - 1]. \tag{38}$$

Thus K_{ij}^{pot} can be calculated.

A variant of this method is to add increasing quantities of ion j to solution containing a constant concentration of i and measure E_{ij} after each addition. The right-hand side of Eq. (38) plotted against $a_j^{z_i/z_j}$ should give a straight line whose slope gives the value for K_{ij}^{pot}. But this procedure is found to be insensitive at low values of K_{ij}^{pot}. Under such conditions, the concentration of j is kept high and ion i is added gradually. For these conditions, the following equations may be written:

$$E'_{ij} = E^0 + (RT/z_i F) \ln[a'_i + K_{ij}^{pot} a_j'^{z_i/z_j}], \tag{39}$$

$$E''_{ij} = E^0 + (RT/z_i F) \ln[a''_i + K_{ij}^{pot} a_j''^{z_i/z_j}]. \tag{40}$$

Subtracting Eq. (39) from Eq. (40) gives on rearrangement

$$K_{ij}^{pot}\left[\exp\left\{\frac{(E''_{ij} - E'_{ij})z_i F}{RT}\right\} a_j'^{z_i/z_j} - a_j''^{z_i/z_j}\right] = a''_i - a'_i\left[\exp\left\{\frac{(E''_{ij} - E'_{ij})z_i F}{RT}\right\}\right]. \tag{41}$$

Since all quantities are known, K_{ij}^{pot} can be evaluated graphically or numerically.

In the absence of diffusion potential [no mobility terms; see Eq. (23)], the selectivity of a solid-state membrane (e.g., silver halides, metal sulfides) K_{ij}^{pot} becomes equivalent to the equilibrium constant, which for the general reaction

$$\frac{1}{a}(M_a i_b) + \frac{m}{n} j \rightleftharpoons \frac{1}{n}(M_n j_m) + \frac{b}{a} i$$

becomes

$$K_{ij} = \bar{a}_j^{m/n} a_i^{b/a} / \bar{a}_i^{b/a} a_j^{m/n}. \tag{42}$$

If S_{Mi} is the solubility product of the solid $M_a i_b$, then $S_{Mi} = a_M^a a_i^b$. Similarly, S_{Mj} is given by $S_{Mj} = a_M^n a_j^m$. Thus

$$S_{Mi}^{1/a}/S_{Mj}^{1/n} = a_i^{b/a}/a_j^{m/n}. \tag{43}$$

Substituting Eq. (43) into Eq. (42) gives

$$K_{ij} = \frac{S_{Mi}^{1/a}}{S_{Mj}^{1/n}} (\bar{a}_j^{m/n} \bar{a}_i^{-b/a}). \tag{44}$$

When the membrane is in equilibrium with a solution containing equivalent concentrations of i and j, Eq. (44) may be written as

$$K_{ij} = \frac{S_{Mi}^{1/a}}{S_{Mj}^{1/n}} [\bar{a}_j^{m/n - b/a}]. \tag{45}$$

When the valences of i and j are unity, Eq. (45) simplifies to

$$K_{ij} = S_{Mi}/S_{Mj}. \tag{46}$$

Equation (46) substituted into Eq. (23) gives

$$E = \text{const} + \frac{RT}{F} \ln\left(a_i + \frac{S_{Mi}}{S_{Mj}} a_j\right). \tag{47}$$

In the case of silver halide solid-state electrodes, the selectivity coefficient K_{ij}^{pot} ($= K_{ij}$) calculated by using the solubility products agreed with experimental values.

III. Integration of Nernst–Planck Flux Equation

The integration of Nernst–Planck flux equation carried out by early investigators to evaluate the diffusion potential was given in Chapter 2. Those equations are applicable to any diffusion barrier or membrane that is not of the ionic or charged type. Applications of Nernst–Planck flux equations to charged membranes are general and more useful in that they can be easily reduced to apply to uncharged membranes (e.g., constrained diffusion barrier).

In 1943 Goldman integrated the Nernst–Planck flux equation, viz.,

$$I_i = -z_i u_i \left(RT \frac{dC_i}{dx} + z_i F C_i \frac{dE}{dx} \right),$$

for the case of a charged membrane and obtained

$$I_i = \frac{RT\bar{u}_i(\bar{C}_2 - \bar{C}_1 + \omega \bar{X} EF/RT)(g - z_i)[\bar{C}_{i(2)} \exp(-z_i EF/RT) - \bar{C}_{i(1)}]}{d\bar{C}_2 \exp(-z_i EF/RT) - \bar{C}_1 - z_i \omega \bar{X}[\exp(-z_i EF/RT) - 1]}, \tag{48}$$

where d is the membrane thickness and the constant g is determined by

$$\frac{EF}{RT} = g \ln \frac{\bar{C}_2 - g\omega\bar{X}}{\bar{C}_1 - g\omega\bar{X}}. \tag{49}$$

$\bar{C}_2$ and $\bar{C}_1$ are the total ion concentrations in the membrane at side 1 ($x = 0$) and side 2 ($x = d$), and E is the potential difference across the membrane. $\bar{C}_{i(2)}$ and $\bar{C}_{i(1)}$ are the concentrations of the ion species i in the membrane. It is difficult to determine the membrane concentrations and even harder to relate them to external concentrations without making extra assumptions.

Teorell (1953), among others, also integrated the ideal Nernst–Planck equation for univalent ions and gave a solution equivalent to Eq. (49). The current is given by

$$I_i^+ = -K^+ \bar{u}_i [\bar{C}_{i(2)}^+ \xi - \bar{C}_{i(1)}^+], \tag{50}$$

where

$$K^+ = \frac{RT}{d} \frac{(\bar{C}_2^+ - \bar{C}_1^+) - 0.5\omega\bar{X}\ln\xi}{\bar{C}_2^+\xi - \bar{C}_1^+} \frac{\ln \bar{k}\xi}{\ln \bar{k}} \tag{51}$$

and

$$\bar{k} = \frac{\bar{C}_2^+ + 0.5(\ln \bar{k}\xi)/(\ln \bar{k})\omega\bar{X}}{\bar{C}_1^+ + 0.5(\ln \bar{k}\xi)/(\ln \bar{k})\omega\bar{X}}$$

$$= \frac{\bar{C}_2^- - 0.5(\ln \bar{k}\xi)/(\ln \bar{k})\omega\bar{X}}{\bar{C}_1^- - 0.5(\ln \bar{k}\xi)/(\ln \bar{k})\omega\bar{X}}, \tag{52}$$

and $\xi = \exp(-EF/RT)$.

Teorell (1953) related the membrane interface concentrations to bulk concentrations on either side of the membrane by invoking the Donnan law. Therefore Eq. (50) can be written in terms of the Donnan ratio as

$$I_i^+ = -K^+ \bar{u}_i [C_{i(2)}^+ r_2 \xi - C_{i(1)}^+ r_1]. \tag{53}$$

Equations (48) and (53) are quite general, and when $\omega\bar{X} = 0$ [and $z_i = \pm 1$ in the case of Eq. (48)], they become equivalent to the Behn equation (205) of Chapter 2.

A special case of interest is the one involving biological cell membranes that separate equal concentrations of ions existing in the intracellular and extracellular phases, i.e., $C_1 = C_2$ and $\omega\bar{X} = 0$. For this case, Eq. (51) becomes

$$K^+ = \frac{RT}{d} \frac{\ln \xi}{\xi - 1}$$

III. Integration of Nernst–Planck Flux Equation

and Eq. (53) becomes

$$I^+_{i(C_1=C_2)} = -\frac{RT}{d}\bar{u}_i \frac{\ln\xi}{\xi-1}[C^+_{i(2)}\xi - C^+_{i(1)}]r. \tag{54a}$$

Similarly,

$$I^-_{i(C_1=C_2)} = -\frac{RT}{d}\bar{v}_i \frac{\ln\xi}{\xi-1}[C^-_{i(2)} - C^-_{i(1)}\xi]\frac{1}{r}. \tag{54b}$$

For the condition $\omega\bar{X} = 0$, $r = 1$ (i.e., $\bar{C}_i = C_i$) and substituting for $\ln\xi$, Eq. (54) becomes

$$I_i = \frac{\bar{u}_i FE}{d}\frac{\bar{C}_{i(2)}\exp(-EF/RT) - \bar{C}_{i(1)}}{\exp(-EF/RT) - 1}. \tag{55}$$

This equation was derived by Goldman (1943), who assumed that the membrane material contained a large number of dipoles whose orientations changed to reduce the nonlinearity of the electric field across the membrane. In effect a constant field given by the quotient $-\bar{E}/d$ exists across the membrane. This makes the mathematical treatment of the problem much simpler. Thus when $dE/dx = -\bar{E}/d$, the ideal Nernst–Planck equation becomes

$$I_i = -z_i\bar{u}_i\left[RT\frac{d\bar{C}_i}{dx} - z_iF\bar{C}_i\left(\frac{\bar{E}}{d}\right)\right].$$

Rearrangement gives

$$dx = -z_i\bar{u}_i RT\frac{d\bar{C}_i}{I_i - z_i^2\bar{u}_i F\bar{C}_i(\bar{E}/d)}. \tag{56}$$

Equation (56) can be integrated to give

$$[x]_0^d = -\frac{d}{(z_i\bar{E}F/RT)}\left[\ln(I_i - z_i^2\bar{u}_i F\bar{C}_i\left(\frac{\bar{E}}{d}\right)\right]_{\bar{C}_i(x=0)}^{\bar{C}_i(x=d)}. \tag{57}$$

Substitution of the boundary conditions $\bar{C}_i(x = d) = \bar{C}_{i(1)}$ and $\bar{C}_i(x = 0) = \bar{C}_{i(2)}$ gives

$$I_i = \frac{z_i^2\bar{u}_i F\bar{E}}{d}\frac{\bar{C}_{i(2)} - \bar{C}_{i(1)}\exp(z_i\bar{E}F/RT)}{1 - \exp(z_i\bar{E}F/RT)}. \tag{58}$$

Equation (58) becomes identical to Eq. (55) when $z_i = 1$.

There are several deductions that can be made from Eq. (58) for the case of univalent ions. When $\bar{C}_{i(2)} = \bar{C}_{i(1)} = C$, Eq. (58) becomes

$$I = uFCE/d.$$

This is an equation for electrical migration in a homogeneous solution. When $\bar{E}$ becomes zero, Eq. (58) becomes indeterminate. However, L'Hospital's rule may be used to evaluate Eq. (58) for that condition (i.e., $\bar{E}_1 - \bar{E}_2 = \bar{E}$ and $\bar{E}_1 \to \bar{E}_2$). L'Hospital's rule says that

$$\lim_{x \to a} \frac{f(x)}{\phi(x)} = \lim_{x \to a} \frac{\partial f(x)}{\partial \phi(x)}$$

when $f(a)$ and $\phi(a)$ are both zero or ∞. Thus

$$I = RT\bar{u}(C_2 - C_1)/d,$$

which is the equation for diffusion in a linear concentration gradient.

The equation for the membrane potential $\bar{E}$ can be derived from Eq. (58). For the ith positive univalent ion, Eq. (58) is rewritten

$$I_i^+ = \frac{F\bar{E}}{d} \frac{\bar{u}_i^+ \bar{C}_{i(2)}^+ - \bar{u}_i^+ \bar{C}_{i(1)}^+ \exp(\bar{E}F/RT)}{1 - \exp(\bar{E}F/RT)}. \tag{59}$$

For the ith negative univalent ion, Eq. (58) is rewritten

$$I_i^- = -\frac{F\bar{E}}{d} \frac{\bar{u}_i^- \bar{C}_{i(2)}^- - \bar{u}_i^- \bar{C}_{i(1)}^- \exp(-\bar{E}F/RT)}{1 - \exp(-\bar{E}F/RT)}$$

or

$$I_i^- = \frac{F\bar{E}}{d} \frac{\bar{u}_i^- \bar{C}_{i(2)}^- \exp(\bar{E}F/RT) - \bar{u}_i^- \bar{C}_{i(1)}^-}{1 - \exp(\bar{E}F/RT)}. \tag{60}$$

When no current exists in the membrane,

$$\sum_{i=1}^{n} I_i^+ = \sum_{i=1}^{n} I_i^-.$$

Applying this condition to Eqs. (59) and (60) gives on simplification

$$\sum_{i=1}^{n} \bar{u}_i^+ \bar{C}_{i(2)}^+ + \sum_{i=1}^{n} \bar{u}_i^- \bar{C}_{i(1)}^- = \exp\left(\frac{\bar{E}F}{RT}\right)\left[\sum_{i=1}^{n} \bar{u}_i^+ \bar{C}_{i(1)}^+ + \sum_{i=1}^{n} \bar{u}_i^- \bar{C}_{i(2)}^-\right].$$

Thus

$$\bar{E} = \frac{RT}{F} \ln\left[\left(\sum_{i=1}^{n} \bar{u}_i^+ \bar{C}_{i(2)}^+ + \sum_{i=1}^{n} \bar{u}_i^- \bar{C}_{i(1)}^-\right)\right.$$
$$\left. \times \left(\sum_{i=1}^{n} \bar{u}_i^+ \bar{C}_{i(1)}^+ + \sum_{i=1}^{n} \bar{u}_i^- \bar{C}_{i(2)}^-\right)^{-1}\right]. \tag{61}$$

This is the Goldman equation for the membrane potential. Also, this equation is similar to Eq. (18) applicable to two cations. The form of Eq. (61) is

III. Integration of Nernst–Planck Flux Equation

not very useful since the concentrations of ions in the membrane are unknown. Hodgkin and Katz (1949) recast Eq. (61) into a form in which it is used a great deal in the description of bioelectric potentials existing across resting cell membranes. Hodgkin and Katz assumed that the concentrations of ions at the inner edges of the membrane were directly proportional to those in the aqueous phases in contact with the membrane edges. Thus $\bar{C}_{i(2)} = C_{i(2)}\beta_i$ and $\bar{C}_{i(1)} = C_{i(1)}\beta_i$, where β_i is called the partition or distribution coefficient of i between the membrane and aqueous solution.

In cellular systems, the inside of the cell usually has a high concentration of potassium ions and the outside has a high concentration of sodium ions. There will be on either side of the cell membrane a number of anions, chief among which are the chloride ions. These three principal ions seem to control the electrical activity of most living cells. So, applying Eq. (59) to the currents carried by these three principal ions, the potassium current I_K (dropping the overbar on E and using K for potassium concentration and i and o for intracellular and extracellular, respectively) is given by

$$I_K = \frac{\bar{u}_K F E \beta_K}{d} \frac{(K)_o - (K)_i \exp(EF/RT)}{1 - \exp(EF/RT)}. \tag{62}$$

Substituting P_i for permeability, which is equated to $\bar{u}_i \beta_i (RT/Fd)$, Eq. (62) becomes

$$I_K = P_K F^2 \frac{E}{RT} \frac{(K)_o - (K)_i \exp(EF/RT)}{1 - \exp(EF/RT)}. \tag{63}$$

Similarly,

$$I_{Na} = P_{Na} F^2 \frac{E}{RT} \frac{(Na)_o - (Na)_i \exp(EF/RT)}{1 - \exp(EF/RT)}, \tag{64}$$

$$I_{Cl} = P_{Cl} F^2 \frac{E}{RT} \frac{(Cl)_i - (Cl)_o \exp(EF/RT)}{1 - \exp(EF/RT)}. \tag{65}$$

The total current I ($= I_K + I_{Na} + I_{Cl}$) is therefore given by

$$I = \frac{EF^2}{RT} P_K \frac{L - M \exp(EF/RT)}{1 - \exp(EF/RT)}, \tag{66}$$

where

$$M = (K)_i + \frac{P_{Na}}{P_K}(Na)_i + \frac{P_{Cl}}{P_K}(Cl)_o, \tag{67}$$

$$L = (K)_o + \frac{P_{Na}}{P_K}(Na)_o + \frac{P_{Cl}}{P_K}(Cl)_i. \tag{68}$$

When there is no current across the membrane, i.e., $I = 0$, membrane potential $E_m \cdot (= E)$ is given by

$$E_m = (RT/F)\ln(L/M),$$

which is equivalent to

$$E_m = \frac{RT}{F} \ln \frac{P_K(K)_o + P_{Na}(Na)_o + P_{Cl}(Cl)_i}{P_K(K)_i + P_{Na}(Na)_i + P_{Cl}(Cl)_o}. \tag{69}$$

This equation has been called the Goldman–Hodgkin–Katz or constant-field equation in the biological literature.

In the derivation of Eq. (55), which is equivalent to Eq. (58), constancy of the electric field is not assumed. Similarly, one can also derive the Goldman–Hodgkin–Katz equation without the assumption of constant field from the Behn equation (205) of Chapter 2.

Equation (66) may be simplified if chloride anions are in thermodynamic equilibrium across the membrane. This means $I_{Cl} = 0$, and according to Eq. (65)

$$E_m = (RT/F)\ln[(Cl)_i/(Cl)_o]. \tag{70}$$

Moreover, if no current is flowing across the membrane, $I_{Na} + I_K = 0$. These conditions give

$$E_m = \frac{RT}{F} \ln \frac{P_K(K)_o + P_{Na}(Na)_o}{P_K(K)_i + P_{Na}(Na)_i} = \frac{RT}{F} \ln \frac{(Cl)_i}{(Cl)_o}. \tag{71}$$

In the Goldman–Hodgkin–Katz treatments given above, E_m is due entirely to diffusion based on considerations given to the solubility of ions in the membrane. Since charged species are involved, distribution of ions at the membrane–solution interfaces will depend on the membrane potential. Although this aspect is not considered in the Goldman–Hodgkin–Katz treatment, it has been treated in the TMS theory (see Chapter 3) by introducing the Donnan ratio. In spite of these refinements, the TMS theory has been found to be inapplicable to cell membranes. The effects of potentials ψ_i and ψ_o at the membrane–solution interfaces on the distribution of ions, neglected in the Goldman–Hodgkin–Katz considerations, have been treated by Polissar (1954), who showed that neither $\psi_i = \psi_o = 0$ nor $-\psi_i = \psi_o$ is true. Consequently, the partition coefficient must be multiplied by an exponential factor containing the surface potential. The relation $\bar{C}_j = C_j \beta_j$ for the two interfacial boundaries must be written

$$\bar{C}_{j(1)} = \beta_j C_{j(i)} \exp(\psi_i F/RT), \tag{72a}$$

$$\bar{C}_{j(2)} = \beta_j C_{j(o)} \exp(\psi_o F/RT). \tag{72b}$$

III. Integration of Nernst–Planck Flux Equation

These equations may be written for the three principal ions in the two phases thus:

$$(\bar{K})_i = \beta_K(K)_i \exp(\psi_i F/RT), \qquad (\bar{K})_o = \beta_K(K)_o \exp(-\psi_o F/RT),$$
$$(\overline{Na})_i = \beta_{Na}(Na)_i \exp(\psi_i F/RT), \qquad (\overline{Na})_o = \beta_{Na}(Na)_o \exp(-\psi_o F/RT), \quad (73)$$
$$(\overline{Cl})_i = \beta_{Cl}(Cl)_i \exp(-\psi_i F/RT), \qquad (\overline{Cl})_o = \beta_{Cl}(Cl)_o \exp(\psi_o F/RT).$$

Now substituting these and the relation $P_i = \bar{u}_i \beta_i RT/Fd$ into Eq. (59) gives, for the condition $I = 0$, the relation

$$E = \frac{RT}{F} \ln \frac{P_K(K)_o \exp(-F\psi_o/RT) + P_{Na}(Na)_o \exp(-F\psi_o/RT) + P_{Cl}(Cl)_i \exp(-F\psi_i/RT)}{P_K(K)_i \exp(F\psi_i/RT) + P_{Na}(Na)_i \exp(F\psi_i/RT) + P_{Cl}(Cl)_o \exp(F\psi_o/RT)}. \quad (74)$$

But the total membrane potential E_m across the membrane is composed of ψ_o, E, and ψ_i. Thus

$$E_m = \psi_o + E + \psi_i.$$

To compute E, ψ_o and ψ_i may be written

$$\psi_o + \psi_i = \frac{RT}{F} \ln[\exp(\psi_o F/RT) \exp(\psi_i F/RT)]. \quad (75)$$

Adding Eqs. (74) and (75) gives the total membrane potential as

$$E_m = \frac{RT}{F} \ln \frac{P_K(K)_o + P_{Na}(Na)_o + P_{Cl}(Cl)_i \exp[F(\psi_o - \psi_i)/RT]}{P_K(K)_i + P_{Na}(Na)_i + P_{Cl}(Cl)_o \exp[F(\psi_o - \psi_i)/RT]}. \quad (76)$$

Although ψ_o and ψ_i cannot be measured, Polissar (1954) was able to estimate, by a trial-and-error procedure, values for ψ_o and ψ_i. Expressions for ψ_o and ψ_i may be formulated on the assumption that electroneutrality prevails in the membrane, i.e., $(\overline{Na}) + (\bar{K}) = (\overline{Cl})$. Thus Eq. (72) gives

$$\beta_K(K)_i \exp(\psi_i F/RT) + \beta_{Na}(Na)_i \exp(\psi_i F/RT) = \beta_{Cl}(Cl)_i \exp(-\psi_i F/RT),$$

which on simplification becomes

$$\psi_i = \frac{RT}{2F} \frac{\beta_{Cl}(Cl)_i}{\beta_K(K)_i + \beta_{Na}(Na)_i}. \quad (77)$$

Similarly,

$$\psi_o = \frac{RT}{2F} \frac{\beta_K(K)_o + \beta_{Na}(Na)_o}{\beta_{Cl}(Cl)_o}. \quad (78)$$

The concentrations of ions across a squid axon membrane are known. Thus by assigning suitable values for the β_i, values for ψ_i and ψ_o may be calculated. As the resting membrane potential across the squid axon is known, i.e., E_m is known, values of ψ_i and ψ_o and E may be derived by a trial-and-error procedure. Polissar found that the magnitudes of ψ_i and ψ_o were of the same order as E, and further for E to become equal to E_m so that ψ_i and ψ_o became zero (no surface potentials), β_K has to attain a negative value. Thus the condition $\psi_i = \psi_o = 0$ cannot be true.

Despite the correction introduced to change Eq. (69) to Eq. (76), the simple formulation of Hodgkin and Katz (1949), viz., $\bar{C}_j = \beta_j C_j$, holds well since in several practical cases the chloride permeability term has been found to be negligible.

A healthy cell maintains the ionic K and Na gradients across the cell membrane by extruding Na and absorbing K. These functions are attributed to the actions of a Na–K pump. If the pump is electroneutral, the number of Na ions pumped out of the cell will be equal to the number of K ions pumped into the cell. If it is electrogenic, more Na ions are pumped out for each K ion pumped in. The Goldman–Hodgkin–Katz equation has been modified by several investigators to account for this current generated by the operation of the pump.

In a biological tissue where a Na–K exchange pump operates, the ionic currents are given by

$$I_{\text{Na(net)}} = I_{\text{Na(P)}} + I_{\text{Na(A)}}, \tag{79}$$

$$I_{\text{K(net)}} = I_{\text{K(P)}} + I_{\text{K(A)}}, \tag{80}$$

where P and A represent passive and active processes. When membrane current is zero,

$$I_{\text{Na}} + I_{\text{K}} + I_{\text{Cl}} = I_{\text{Na(P)}} + I_{\text{Na(A)}} + I_{\text{K(P)}} + I_{\text{K(A)}} + I_{\text{Cl}} = 0. \tag{81}$$

The Na–K pump properties are (1) a coupling ratio r indicating the number of Na ions pumped out for every K ion pumped in, and (2) a parameter α linking active Na efflux to passive Na influx. These are equal in the steady state, i.e., $\alpha = 1$, but are different in the nonsteady state when the tissue is loaded with Na. Thus

$$r = -[I_{\text{Na(A)}}/I_{\text{K(A)}}] \tag{82}$$

and

$$\alpha = -[I_{\text{Na(A)}}/I_{\text{Na(P)}}]. \tag{83}$$

Equations (82) and (83) give

$$I_{\text{K(A)}} = -(1/r)I_{\text{Na(A)}} = (\alpha/r)I_{\text{Na(P)}} \tag{84}$$

III. Integration of Nernst–Planck Flux Equation

Substituting Eqs. (83) and (84) into Eq. (81) gives on rearrangement

$$I_{\text{Na(P)}}[1 - \alpha + \alpha/r] + I_{\text{K(P)}} + I_{\text{Cl}} = 0. \tag{85}$$

Substitution for $I_{\text{K(P)}}$, $I_{\text{Na(P)}}$, and I_{Cl} from Eqs. (63)–(65) into Eq. (85) gives on simplification

$$E = \frac{RT}{F} \ln \frac{P_{\text{K}}(\text{K})_o + P_{\text{Na}}(1 - \alpha + \alpha/r)(\text{Na})_o + P_{\text{Cl}}(\text{Cl})_i}{P_{\text{K}}(\text{K})_i + P_{\text{Na}}(1 - \alpha + \alpha/r)(\text{Na})_o + P_{\text{Cl}}(\text{Cl})_o}. \tag{86}$$

This equation was derived by El-Sharkawy and Daniel (1975). A similar equation for the condition $I_{\text{Cl}} = 0$ was derived by Sjödin and Ortiz (1973), who assume $f_{\text{Na}} = 1/\alpha$ and $f_{\text{K}} = -I_{\text{K(P)}}/I_{\text{K(A)}}$. Thus

$$\frac{f_{\text{Na}}}{f_{\text{K}}} = \frac{I_{\text{Na(P)}}}{I_{\text{Na(A)}}} \frac{I_{\text{K(A)}}}{I_{\text{K(P)}}} = -\frac{1}{r} \frac{I_{\text{Na(P)}}}{I_{\text{K(P)}}}$$

or

$$I_{\text{Na(P)}} + I_{\text{K(P)}} r \frac{f_{\text{Na}}}{f_{\text{K}}} = 0. \tag{87}$$

Substituting the values for $I_{\text{K(P)}}$ and $I_{\text{Na(P)}}$ from Eqs. (63) and (64) into Eq. (87) gives on simplification

$$E = \frac{RT}{F} \ln \frac{P_{\text{Na}}(\text{Na})_o + P_{\text{K}}(\text{K})_o r f_{\text{Na}}/f_{\text{K}}}{P_{\text{Na}}(\text{Na})_i + P_{\text{K}}(\text{K})_i r f_{\text{Na}}/f_{\text{K}}}. \tag{88}$$

When $\alpha = 1$ and $I_{\text{Cl}} = 0$ in Eq. (86) and $f_{\text{Na}} = f_{\text{K}}$ in Eq. (88), both the equations reduce to the form first derived by Mullins and Noda (1963). That is,

$$E = \frac{RT}{F} \ln \frac{P_{\text{Na}}(\text{Na})_o + r P_{\text{K}}(\text{K})_o}{P_{\text{Na}}(\text{Na})_i + r P_{\text{K}}(\text{K})_i}. \tag{89}$$

In the steady state, Eqs. (79) and (80) become

$$I_{\text{Na(P)}} = -I_{\text{Na(A)}}, \tag{90}$$

$$-I_{\text{K(P)}} = I_{\text{K(A)}}, \tag{91}$$

Subtracting Eq. (91) from Eq. (90) gives

$$I_{\text{Na(P)}} + I_{\text{K(P)}} = -[I_{\text{Na(A)}} + I_{\text{K(A)}}] = -I_{\text{el}}, \tag{92}$$

where I_{el} is the current produced by the Na–K pump and is negative when $r > 1$.

Substituting from Eqs. (63) and (64) into Eq. (92) gives on rearrangement

$$E = \frac{RT}{F} \ln \frac{P_{\text{K}}(\text{K})_o + P_{\text{Na}}(\text{Na})_o - I_{\text{el}} RT/F^2 E}{P_{\text{K}}(\text{K})_i + P_{\text{Na}}(\text{Na})_i - I_{\text{el}} RT/F^2 E}. \tag{93}$$

This equation was given by Moreton (1969). Two approximations of Eq. (93) are possible: (1) $P_K(K)_i$ is large and $I_{el}(RT/F^2E)$ has little effect on the denominator; (2) $P_{Na}(Na)_i$ is also low. Thus Eq. (93) simplifies to

$$\exp\left(\frac{EF}{RT}\right) = \frac{(K)_o}{(K)_i} + \frac{P_{Na}(Na)_o}{P_K(K)_i} - \frac{I_{el}RT}{F^2 E P_K(K)_i}. \tag{94}$$

If $r > 1$, the last term leads to membrane hyperpolarization. If the pump is inhibited by ouabain, I_{el} is eliminated and the Eq. (94) reduces to

$$\exp\left[\left(\frac{F}{RT}\right)(E - \Delta E)\right] = \frac{(K)_o}{(K)_i} + \frac{P_{Na}(Na)_o}{P_K(K)_i}, \tag{95}$$

where $E - \Delta E$ is the membrane potential measured in presence of pump inhibiting ouabain. ΔE is the depolarization observed on treatment of the tissue with ouabain. Subtraction of Eq. (95) from Eq. (94) gives

$$\exp(EF/RT)_{\text{pump}} - \exp[(E - \Delta E)F/RT]_{\text{no pump}} = -I_{el}RT/F^2 E P_K(K)_i. \tag{96}$$

The electrogenic current or flux can be calculated from Eq. (96) provided P_K and $(K)_i$ values are known. $(K)_i$ can be evaluated from a plot of $\exp[(E - \Delta E)F/RT]$ against $(K)_o$ [see Eq. (95)]. The slope is equal to $1/(K)_i$, and from the intercept on the Y axis P_{Na}/P_K can be evaluated. P_K can be computed from an equation given by Hodgkin and Katz (1949).

Differentiation of Eq. (66), i.e., $(dI/dE)_{I \to 0}$ gives $G(=1/R)$, the membrane conductance. Thus

$$G = \frac{F^3 P_K}{R^2 T^2} E \frac{M \cdot L}{L - M}. \tag{97}$$

Introducing the simplifying assumptions given above gives

$$G = \frac{F^3 P_K}{R^2 T^2} E \frac{[(K)_o + (P_{Na}/P_K)(Na)_o](K)_i}{-(K)_i + (P_{Na}/P_K)(Na)_o + (K)_o}. \tag{98}$$

Thus by measurement of total membrane resistance, G can be determined, and this enables evaluation of P_K from Eq. (98). This in turn enables determination of I_{el} or flux J_{el} with the help of Eq. (96). Also, I_{el} can be evaluated by the use of Ohm's law provided the pump potential contribution ΔE is not large, i.e., less than 8 mV. This involves measurement of membrane resistance and electrogenic potential ΔE_{el},

$$\Delta E_{el} = R_m I_{el} = R_m F J_{el}. \tag{99}$$

In view of Eq. (97), Eqs. (63)–(65) may be rewritten

$$I_K = \frac{RT}{F} G\left[\frac{(K)_i}{M} - \frac{(K)_o}{L}\right], \tag{100}$$

$$I_{Na} = \frac{RT}{F}\frac{P_{Na}}{P_K} G\left[\frac{(Na)_o}{M} - \frac{(Na)_i}{L}\right], \tag{101}$$

$$I_{Cl} = \frac{RT}{F}\frac{P_{Cl}}{P_K} G\left[\frac{(Cl)_i}{M} - \frac{(Cl)_o}{L}\right]. \tag{102}$$

These equations are useful in calculating ion permeabilities.

Another consequence of Eq. (55) or (58) is that one can derive the Ussing equation [flux ratio = (influx)/(outflux)] from it. Equation (62) may be written in the form

$$I_j = P_j \frac{(z_j EF^2)/RT}{1 - \exp(z_j EF/RT)}\left[C_{j(o)} - C_{j(i)}\exp\left(\frac{z_j EF}{RT}\right)\right], \tag{103}$$

where P_j is the permeability of j and E is the membrane potential. The net flux given by Eq. (103) may be considered as a difference between two unidirectional fluxes, the efflux I_{ef} and the influx I_{in}, where

$$I_{ef} = P_j \frac{z_j F^2 E/RT}{1 - \exp(z_j EF/RT)} C_{j(i)}\exp\left(\frac{z_j EF}{RT}\right) \tag{104}$$

and

$$I_{in} = P_j \frac{z_j F^2 E/RT}{1 - \exp(z_j EF/RT)} C_{j(o)}. \tag{105}$$

The flux ratio $f = (I_{in}/I_{ef})$ is then given by

$$f = \frac{I_{in}}{I_{ef}} = \frac{C_{j(o)}}{C_{j(i)}}\exp\left(\frac{-z_j EF}{RT}\right) \tag{106}$$

or

$$RT \ln f = \bar{\mu}_o - \bar{\mu}_i,$$

where the electrochemical potential $\bar{\mu} = \mu_o + RT \ln C + zFV$ and $E = V_i - V_o$. Thus

$$f = \exp[(\bar{\mu}_o - \bar{\mu}_i)/RT]. \tag{107}$$

This relation is valid generally for passive and independent movement of ions. For uncharged solutes following simple diffusion, the flux ratio is reduced to the ratio of concentrations of the solute.

Several factors may affect passive transport of a solute in such a way as to deviate from the predictions of Eq. (106). Some of these factors are active transport (movement of a species against its own gradient), solvent drag (fluxes of solute and solvent proceed by the same pathways), isotope interaction, and single-file diffusion or long pore effect.

If the ion is in flux equilibrium, i.e., $f = 1$, then Eq. (106) becomes

$$C_{j(i)}/C_{j(o)} = \exp(-z_j EF/RT). \tag{108}$$

This equation serves as a valuable criterion for active transport, which is indicated if $C_{j(i)}/C_{j(o)} > \exp(-z_j EF/RT)$. On the other hand, if the membrane potential values lead to $C_{j(i)}/C_{j(o)} < \exp(-z_j EF/RT)$, mediated transport is indicated.

The total membrane current density I is given by

$$I = \sum_j I_j \tag{109}$$

and

$$\begin{aligned} I_j &= z_j F(J_{j(\text{influx})} - J_{j(\text{outflux})}) \\ &= z_j F J_{j(\text{ef})}(J_{j(\text{in})}/J_{j(\text{ef})} - 1). \end{aligned} \tag{110}$$

Partial or ion conductance is given by

$$G_j = \left(\frac{\partial I_j}{\partial E}\right)_{I \to 0} = z_j F J_{j(\text{ef})} \frac{\partial}{\partial E}\left(\frac{J_{j(\text{in})}}{J_{j(\text{ef})}}\right) + z_j F\left[\frac{J_{j(\text{in})}}{J_{j(\text{ef})}} - 1\right]\frac{\partial J_{j(\text{ef})}}{\partial E}. \tag{111}$$

At flux equilibrium $J_{j(\text{in})} = J_{j(\text{ef})} = J_j$, the second term on the right-hand side of Eq. (111) is zero.

Equation (106) may be written

$$J_{j(\text{in})}/J_{j(\text{ef})} = \exp[z_j F(E_j - E)/RT], \tag{112}$$

where E_j is the equilibrium potential of j and is given by

$$E_j = (RT/z_j F)\ln(C_{j(o)}/C_{j(i)}). \tag{113}$$

Thus

$$\frac{\partial}{\partial(E_j - E)}\left[\frac{J_{j(\text{in})}}{J_{j(\text{ef})}}\right] = \frac{J_{j(\text{in})}}{J_{j(\text{ef})}} z_j \frac{F}{RT}. \tag{114}$$

The first term on the right-hand side of Eq. (111) can be obtained from Eq. (114). Thus

$$G_j = z_j^2 F^2 J_j/RT \tag{115}$$

or the total conductance of the membrane due to all ions moving across the membrane is given by

$$G = \sum_j G_j = \frac{F^2}{RT}\sum_j z_j^2 J_j. \tag{116}$$

These equations are useful in computing single-ion and total-membrane conductances from flux measurements.

Another important application of Eq. (69) relates to evaluation of membrane selectivity to different ions. There are two methods commonly used to accomplish this, and they involve voltage clamp measurements. This is a technique used to measure early inward and late outward membrane currents when the membrane potential is held constant at any given voltage. Usually electronic feedback is used to hold the membrane potential constant. Generally in nerve and muscle fibers when the membrane is held by electronic feedback at a holding potential of, say, -70 to -80 mV, imposition of a depolarizing pulse of given strength and duration leads to an early inward current due to flow of Na ions followed by a late outward current due to flow of K ions. These currents can be separated by the use of pharmacological or other agents. Currents obtained at several membrane voltages can be plotted to give the current–voltage $(I-V)$ curve. In the case of the Na-carrying system, the inward current increases with an increase in voltage across the membrane, reaches a maximum, and decreases with further increase in voltage. Thus at a certain membrane voltage, the current due to flow of Na ions becomes zero; that voltage at which this happens is called the reversal or equilibrium potential E_e.

In the first method used to determine ionic selectivity of the membrane, the amplitudes of currents or conductances in a control and test solutions are compared. To be applicable, it is assumed that the number of open channels are the same in each solution and that there is no saturation or block of open channels by the control or test solution. This means that the independence principle of Hodgkin and Huxley (1952) is applicable.

The independence principle assumes that the movement of ions is independent of any other ion. It can predict the net flux of an ion under one set of ionic concentrations from the net flux at the same membrane potential but in another set of ionic concentrations. It follows from Eqs. (110) and (112).

Equation (110) for two ionic environments gives

$$\frac{I'_j}{I_j} = \frac{J'_{j(\text{in})} - J'_{j(\text{ef})}}{J_{j(\text{in})} - J_{j(\text{ef})}}. \tag{117}$$

In these experiments, only the external ionic concentration of j is changed, and the internal concentrations of ions remain unchanged. Thus

$$J_{j(\text{ef})} = J'_{j(\text{ef})} \quad \text{and} \quad J'_{j(\text{in})}/J_{j(\text{in})} = a'_{j(\text{o})}/a_{j(\text{o})},$$

where a_j is the activity of j changed from $a_{j(\text{o})}$ to $a'_{j(\text{o})}$. Dividing both the numerator and the denominator of the right-hand side of Eq. (117) by $J_{j(\text{ef})}$ gives on rearrangement

$$\frac{I'_j}{I_j} = \left[\frac{J'_{j(\text{in})}}{J_{j(\text{in})}}\right]\left[\frac{J_{j(\text{in})}}{J_{j(\text{ef})}} - \frac{J'_{j(\text{ef})}J_{j(\text{in})}}{J'_{j(\text{in})}J_{j(\text{ef})}}\right] \bigg/ \left(\left[\frac{J_{j(\text{in})}}{J_{j(\text{ef})}}\right] - 1\right). \tag{118}$$

As the membrane potential E does not change when the j concentration is changed from one step to another, the last term of the numerator in Eq. (118) is unity because of the applicability of Eq. (112). Thus substituting Eq. (112) and the relation $[J'_{j(\text{in})}/J_{j(\text{in})}] = a'_{j(\text{o})}/a_{j(\text{o})}$, Eq. (118) becomes, when $z_j = 1$,

$$I'_j = \frac{I_j[a'_{j(\text{o})}/a_{j(\text{o})}][\exp(F(E_j - E)/RT) - 1]}{\exp(F(E_j - E)/RT) - 1}, \quad (119)$$

where E_j is the reversal potential for the ion j.

The second method of calculating ionic selectivity of the membrane uses the reversal potential E_e and Eq. (69). In this method, the result does not depend on the number of conducting channels. The change in E_e on changing from a control Na solution to a Na substitute (S) will be given by

$$E_{e,s} - E_{e,\text{Na}} = 2.303 \frac{RT}{F} \log \frac{P_s[S]}{P_{\text{Na}}[\text{Na}]}. \quad (120)$$

The ratio of the permeabilities is a measure of the S–Na selectivity of the membrane. In this way, using Na as the reference ion, selectivity of several inorganic and organic cations to biological membrane have been determined.

At some synapses (e.g., frog neuromuscular junction) Eq. (69) fails to predict the reversal potential; but an extended equation taking into account the permeabilities of Ca and Mg ions has been found to predict the reversal potential well. The extended form follows from Eq. (58), which may now be written as [see also Eq. (63)]

$$I_i = z_i^2 P_i \frac{F^2 E}{RT} \frac{(X)_o - (X)_i \exp(z_i EF/RT)}{1 - \exp(z_i EF/RT)}, \quad (121)$$

where $(X)_i$ and $(X)_o$ are the inside and outside concentrations of X ions. Writing Eq. (121) for the several ions involved, viz., Na, K, Ca, Mg, and Cl, and substituting ξ for $\exp(EF/RT)$, Eq. (121) for the condition $I = I_{\text{Na}} + I_{\text{K}} + I_{\text{Ca}} + I_{\text{Mg}} + I_{\text{Cl}} = 0$ becomes a quadratic in ξ, i.e.,

$$a\xi^2 + b\xi + c = 0, \quad (122)$$

where

$$a = (K)_i + \frac{P_{\text{Na}}}{P_K}(\text{Na})_i + 4\frac{P_{\text{Ca}}}{P_K}(\text{Ca})_i + 4\frac{P_{\text{Mg}}}{P_K}(\text{Mg})_i + \frac{P_{\text{Cl}}}{P_K}(\text{Cl})_i,$$

$$b = [(K)_i - (K)_o] + \frac{P_{\text{Na}}}{P_K}[(\text{Na})_i - (\text{Na})_o] + \frac{P_{\text{Cl}}}{P_K}[(\text{Cl})_o - (\text{Cl})_i], \quad (123)$$

$$c = -(K)_o - \frac{P_{\text{Na}}}{P_K}(\text{Na})_o - 4\frac{P_{\text{Ca}}}{P_K}(\text{Ca})_o - 4\frac{P_{\text{Mg}}}{P_K}(\text{Mg})_o - \frac{P_{\text{Cl}}}{P_K}(\text{Cl})_o.$$

Thus the solution of Eq. (122) gives

$$E = \frac{RT}{F} \ln \frac{-b + \sqrt{b^2 - 4ac}}{2a}, \quad (124)$$

where the values of a, b, and c are given by Eq. (123).

Although Eq. (124) may be used for numerical evaluation of the reversal potential, it does not look as neat as Eq. (69). Even to present Eq. (124) in a tidy form for three cations, it calls for some manipulation.

Equating the current carried by Na, K, and Ca ions to zero gives

$$\frac{P_{Na}[(Na)_o - (Na)_i \exp(EF/RT)] + P_K[(K)_o - (K)_i \exp(EF/RT)]}{1 - \exp(EF/RT)}$$

$$+ \frac{4P_{Ca}[(Ca)_o - (Ca)_i \exp(2EF/RT)]}{1 - \exp(2EF/RT)} = 0.$$

Equating $[P_{Ca}/(1 + \exp(EF/RT)] = P'_{Ca}$ gives

$$P_{Na}(Na)_o + P_K(K)_o + 4P'_{Ca}(Ca)_o$$
$$= \exp(EF/RT)[P_{Na}(Na)_i + P_K(K)_i + 4P'_{Ca}(Ca)_i \exp(EF/RT)].$$

Solving for E gives

$$E = \frac{RT}{F} \ln \frac{P_{Na}(Na)_o + P_K(K)_o + 4P'_{Ca}(Ca)_o}{P_{Na}(Na)_i + P_K(K)_i + 4P'_{Ca}(Ca)_i \exp(EF/RT)}. \quad (125)$$

This equation or other extended forms may be used to predict reversal potentials at the neuromuscular junction.

IV. Other Models

Concentration potentials across a charged membrane may be considered to be made up of two Donnan potentials and a diffusion potential, the latter being given by Eq. (69). In this case the cell potential is given by

$$E = -\frac{RT}{F} \ln \left[\frac{r'}{r''} \frac{P''_+ a''_+ + P'_- a'_-}{P'_+ a'_+ + P''_- a''_-} \right] \quad (126)$$

and

$$r = \frac{\bar{X}}{2a} + \sqrt{\left(\frac{\bar{X}}{2a}\right)^2 + 1}.$$

Permeabilities P_+ and P_- can be evaluated by using tracers, for example, ^{24}NaCl and Na^{36}Cl.

One problem of considerable interest is the existence of asymmetry potentials when two solutions of the same concentration of the same electrolyte are present on either side of the membrane. This is possible if the two faces of the membrane are differently charged. Ohki (1971) showed that the membrane potential is given by

$$E = \frac{4\pi d}{\varepsilon_1} \left(\frac{\sigma - \sigma'}{2 + \varepsilon_0 \varkappa d/\varepsilon_1} \right), \tag{127}$$

where ε_0 is the dielectric constant of the aqueous medium in contact with the membrane whose dielectric constant is ε_1, $\varkappa$ is the Debye length, and σ and σ' are the surface charge densities of the two membrane faces. Equation (127), although of interest theoretically, cannot be used in practice, because of the presence of too many unknown parameters.

When a gradient of charge density exists (e.g., one face of the membrane more charged than the other face) in the membrane and if the electrolyte is trapped in between, the potential arising across such a composite membrane or laminates will decay with time depending on the resistance of the barriers in the membrane. A rough approximation is that the potential can be considered as the sum of two concentration cells in series. Thus E is given by

$$E = \frac{RT}{F} \left[(2\bar{t}_{+\text{(high)}} - 1) \ln \frac{a_m}{a_1} - (2\bar{t}_{+\text{(low)}} - 1) \ln \frac{a_m}{a_2} \right], \tag{128}$$

where a_m is the activity of the electrolyte trapped in the composite membrane. When $a_1 = a_2 = a$, the asymmetry potential is given by

$$E = \frac{2RT}{F} [\bar{t}_{+\text{(high)}} - \bar{t}_{+\text{(low)}}] \ln \frac{C_m}{C}, \tag{129}$$

where activities have been replaced by concentrations. The value of E depends on how stable the value of C_m is. If it is changing with time by diffusion, E will always decay with time depending on how C_m is changing with time. For E to be stable, either decay of C_m should be arrested by pumping diffusing ions back as in biological cells or the counterions should be held by electrostatic forces so that stable surface potentials exist at the two membrane–solution interfaces.

Electrical potentials arising across charged membranes attain saturation when they are measured by increasing the concentration on one side and keeping the concentration on the other side constant. This kind of saturation phenomenon is typical of an adsorption regime, which can be described by Langmuir adsorption equation.

Consider the following scheme of transport of metal ion:

$$M_1 + S \underset{k_{-1}}{\overset{k_1}{\rightleftharpoons}} M_1 S \xrightarrow{k_2} S + P \tag{130}$$

M_1 is the number of cations adsorbed to sites S in the membrane, then transported across it and released as the product P. Following the mathematics of enzyme kinetics, one can write

$$\frac{d(M_1S)}{dt} = k_1(M_1)(S) - k_{-1}(M_1S) - k_2(M_1S). \tag{131}$$

In the steady state $d(M_1S)/dt$ is zero, and so Eq. (131) becomes

$$[M_1S] = \frac{k_1(M_1)(S)}{k_{-1} + k_2}. \tag{132}$$

The rate of formation of the product (dP/dt) is given by

$$v = \frac{dp}{dt} = k_3(M_1S) = \frac{k_2 k_1(M_1)(S)}{k_{-1} + k_2}. \tag{133}$$

If the maximum number of sites on the membrane is S_0, then $(S_0) = (S) + (M_1S)$, and so substituting this in Eq. (132) gives

$$(S_0) = (S)\left[1 + \frac{k_1(M_1)}{k_{-1} + k_2}\right]. \tag{134}$$

Thus substituting for (S) from Eq. (134) into Eq. (133) gives

$$v = \frac{k_2 k_1(M_1)(S_0)}{k_{-1} + k_2 + k_1(M_1)}. \tag{135}$$

When the cation concentration M_1 is large, k_{-1} and k_2 will be negligible compared to $k_1(M_1)$ in the denominator of Eq. (135), and so

$$v_{max} = k_2(S_0). \tag{136}$$

Dividing Eq. (135) by Eq. (136) and rearranging gives

$$v = \frac{v_{max}(M_1)}{(M_1) + K_1} \tag{137}$$

where $K_1 = (k_{-1} + k_2)/k_1$, the dissociation constant of the site in the membrane.

The rate of formation of the product is equivalent to the rate of transfer of metal ions across the membrane, i.e., $v = dp/dt = -d(M_1)/dt$. This is equivalent to current, and so Eq. (137) may be written

$$I = \frac{I_{max}(M_1)}{(M_1) + K_1}. \tag{138}$$

But current at constant resistance is proportional to E, and so Eq. (138) in terms of potential becomes

$$E = \frac{E_{max}(M_1)}{(M_1) + K_1}. \tag{139}$$

Inverting Eq. (139) gives

$$\frac{1}{E} = \frac{1}{E_{max}} + \frac{1}{(M_1)} \frac{K_1}{E_{max}}. \tag{140}$$

A double-reciprocal plot of E against (M_1) will give a straight line whose slope is equal to (K_1/E_{max}) and the intercept on the ordinate equal to $1/E_{max}$. Alternatively, when $1/E = 0$, the X-axis intercept gives $K_1 = -(M_1)$. Thus values of E_{max} and K_1 can be derived.

The effect of the presence of another metal ion M_2 that competes with M_1 for the occupancy of the site S may also be considered. Here the additional reaction due to M_2 is

$$M_2 + S \underset{k_{-3}}{\overset{k_3}{\rightleftharpoons}} M_2S \tag{141}$$

At equilibrium

$$k_{-3}(M_2S) = k_3(M_2)(S). \tag{142}$$

The value for (S_0) is now given by

$$(S_0) = (S) + (M_1S) + (M_2S). \tag{143}$$

Substituting for (M_1S) from Eq. (132) and for (M_2S) from Eq. (142) into Eq. (143) gives

$$(S_0) = (S)\left[1 + \frac{k_1(M_1)}{k_{-1} + k_2} + \frac{k_3}{k_{-3}}(M_2)\right]. \tag{144}$$

As the rate of formation of the product dP/dt is still given by Eq. (133), substitution of Eq. (144) into Eq. (133) gives on rearrangement

$$v = \frac{v_{max}(M_1)}{K_1 + (M_1) + (K_1/K_2)(M_2)}. \tag{145}$$

where $K_2 = k_{-3}/k_3$ and v_{max} is given by Eq. (136). In terms of voltage, Eq. (145) becomes on inversion

$$\frac{1}{E} = \frac{1}{E_{max}} + \frac{K_1[1 + (M_2)/K_2]}{E_{max}} \frac{1}{(M_1)}. \tag{146}$$

Again, the plot of $1/E$ against $1/(M_1)$ should give a straight line, or the X-axis intercept when $1/E = 0$ gives

$$-(M_1) = K_1 + (K_1/K_2)(M_2). \tag{147}$$

Thus when values of $-(M_1)$ determined at $1/E = 0$ for several concentrations of M_2 are plotted against (M_2), one would expect a straight line with slope equal to K_1/K_2 and intercept equal to K_1 enabling evaluation of both K_1 and K_2.

V. Liquid Membranes

A liquid ion-exchange membrane is usually formed by dissolving a liquid ion exchanger in a water-immiscible solvent. Unlike solid ion exchangers, which have their ionogenic groups fixed to the membrane matrix, ionogenic groups (sites) of liquid ion exchangers are mobile. Depending on the solvent used to form the membrane, the sites would be completely dissociated (dielectric constant of the solvent is high) or highly associated (dielectric constant of solvent is low) into ion pairs. The behavior of these membranes has been described from the standpoint of electrodiffusion by Conti, Eisenman, Sandblom and Walker (see Eisenman, 1969).

In the case of liquid membranes with electrically charged ligands (see Fig. 1 for the conditions existing in the membrane system), it has been shown that the electrical potential E across the membrane under conditions of zero applied potential is given by

$$E = -\frac{RT}{z_i F}\left\{\ln\left(\sum_i \bar{u}_i a_i'' k_i\right)\bigg/\left(\sum_i \bar{u}_i a_i' k_i\right)\right] + \int 1 + \int 2\right\}, \qquad (148)$$

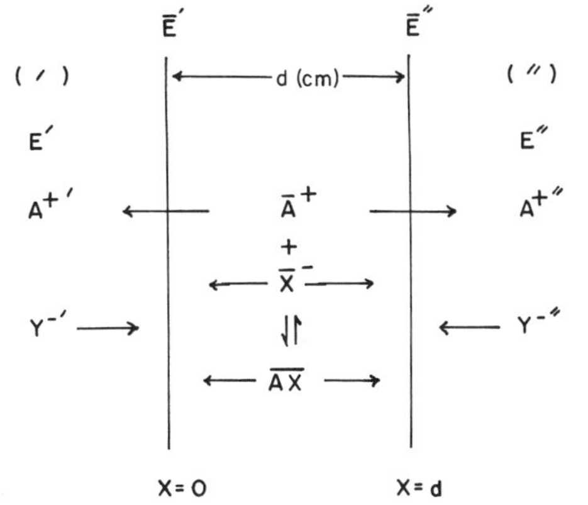

Fig. 1. Schematic diagram of liquid membrane system containing a charged ligand. $\bar{A}^+, \bar{X}^-$, Y^- refer to counterions, sites, and coions, respectively. $\bar{A}\bar{X}$ is the mobile ion pair. A^+ freely crosses the membrane–solution interfaces, whereas $\bar{X}^-$ is confined to the membrane phase. resulting in the exclusion of Y^- from the membrane phase. On the x coordinate, $x = 0$ and $x = d$ indicate the two interfaces and d is the thickness of the liquid membrane. Es are the electrical potentials and overbars refer to the membrane phase. (') and (") are the two aqueous phases on either side of the membrane.

where

$$\int 1 = \int_{l'}^{l''} t \, d \ln \left[\left(\sum_i \bar{u}_{iX} K_{iX} \bar{C}_i \right) \Big/ \left(\sum_i \bar{u}_i \bar{C}_i \right) \right],$$

$$t = \frac{\bar{u}_X \bar{C}_X}{\left[\left(\bar{u}_X \bar{C}_X \Big/ \sum_i \bar{u}_{iX} \bar{C}_{iX} \right) + 1 \right] \sum_i \bar{u}_i \bar{C}_i + \bar{u}_X \bar{C}_X}. \tag{149}$$

The $\bar{u}$s are mobilities in the membrane phase, $k_i = \exp[(\mu_i^0 - \bar{\mu}_i^0)/RT]$. K_{iX} is the association constant related to the reaction

$$\bar{i}^+ + \bar{X}^- \rightleftharpoons \overline{iX},$$

$$\int 2 = \int_{l'}^{l''} \frac{(\bar{u}_X J_X^t / RT) \, dx}{\left(\bar{u}_X + \sum_i \bar{u}_{iX} K_{iX} \bar{C}_i \right) \sum_i \bar{u}_i \bar{C}_i + \bar{u}_X \bar{C}_X \sum_i \bar{u}_{iX} K_{iX} \bar{C}_i}. \tag{150}$$

J_X^t is the total flux $J_X + \sum_i J_{iX}$.

Equation (148) is applicable to both steady-state and non-steady-state conditions. In the steady state $J_X^t = 0$, and so integral 2 is zero. Here two cases are of importance.

(1) When there is complete dissociation, $t = 0$ [Eq. (149)], and so integral $\int 1 = 0$. Therefore Eq. (148) reduces to

$$E = -\frac{RT}{z_i F} \ln \left[\left(\sum_i \bar{u}_i k_i a_i'' \right) \Big/ \left(\sum_i \bar{u}_i k_i a_i' \right) \right]. \tag{151}$$

(2) When there is strong association involving two counterions A and B such that $\bar{C}_X = \bar{C}_A + \bar{C}_B$, $\int 1$ can be written

$$\int 1 = \int_{l'}^{l''} \frac{\bar{u}_X [\bar{C}_A / \bar{C}_B) + 1]}{(\bar{u}_A + \bar{u}_X)(\bar{C}_A / \bar{C}_B) + \bar{u}_B + \bar{u}_X}$$

$$\times d \ln \frac{(\bar{u}_{AX} \bar{C}_A K_{AX} / \bar{C}_B) + \bar{u}_{BX} K_{BX}}{\bar{u}_A (\bar{C}_A / \bar{C}_B) + \bar{u}_B}. \tag{152}$$

Equation (152) integrated and substituted into Eq. (148) gives

$$E = \frac{RT}{z_i F} \left[(1 - \tau) \ln \frac{(\bar{u}_A + \bar{u}_X) k_A a_A' + (\bar{u}_B + \bar{u}_X) k_B a_B'}{(\bar{u}_A + \bar{u}_X) k_A a_A'' + (\bar{u}_B + \bar{u}_X) k_B a_B''} \right.$$

$$\left. + \tau \ln \frac{\bar{u}_{AX} K_{AX} k_A a_A' + \bar{u}_{BX} K_{BX} k_B a_B'}{\bar{u}_{AX} K_{AX} k_A a_A'' + \bar{u}_{BX} K_{BX} k_B a_B''} \right], \tag{153}$$

where

$$\tau = \frac{\bar{u}_X (\bar{u}_{BX} K_{BX} - \bar{u}_{AX} K_{AX})}{(\bar{u}_A + \bar{u}_X) \bar{u}_{BX} K_{BX} - (\bar{u}_B + \bar{u}_X) \bar{u}_{AX} K_{AX}}. \tag{154}$$

V. Liquid Membranes

In the case of liquid membrane with electrically neutral ligand X (see Fig. 2 for the conditions existing in the membrane system), the ligand forms a complex with ions determining the membrane potential.

The following reactions occur in both the aqueous and the membrane phases

$$A^+ + X \rightleftharpoons AX^+, \qquad K_{AX}^+ = \frac{C_{AX}}{a_A C_X}, \qquad (155)$$

$$AX^+ + Y^- \rightleftharpoons AXY, \qquad K_{AXY} = \frac{C_{AXY}}{C_{AX} a_Y}. \qquad (156)$$

For the neutral species (e.g., X, AXY) the partition equilibria are given by

$$X \rightleftharpoons \bar{X}, \qquad k_X = \bar{C}_X / C_X. \qquad (157)$$

The equilibria for the charged species ($R^\pm = A^+, AX^+, Y^-$) are

$$R^\pm \rightleftharpoons \bar{R}^\pm, \qquad k_r = \frac{\bar{a}_r \exp(z_r \bar{E} F / RT)}{a_r \exp(z_r E F / RT)}. \qquad (158)$$

For this liquid–membrane system subject to (a) equal total concentrations of the carrier on both sides of the membrane, (b) presence of only two monovalent ions i and j, and (c) negligible concentration of neutral complexes iXY and jXY, it has been shown that the potential is given by

$$E = \frac{RT}{F} \ln \frac{a_i' + [\bar{u}_{jX} k_{jX} K_{jX}^+ / \bar{u}_{iX} k_{iX} K_{iX}^+] a_j'}{a_i'' + [\bar{u}_{jX} k_{jX} K_{jX}^+ / \bar{u}_{iX} k_{iX} K_{iX}^+] a_j''}$$
$$+ \frac{RT}{F} \ln \frac{1 + K_{iX}^+ a_i'' + K_{jX}^+ a_j''}{1 + K_{iX}^+ a_i' + K_{jX}^+ a_j'}. \qquad (159)$$

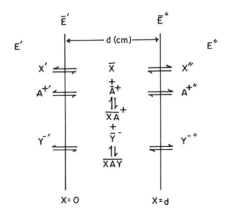

Fig. 2. Schematic diagram of liquid membrane system containing a neutral ligand. Three independent types of species, A^+ (cation), Y^- (anion), and X (electroneutral ligand) are present in all the plases. (') and ('') are the two aqueous phases on either side of the liquid membrane. The Es are the electrical potentials, and overbars refer to the membrane phase. $\bar{X}\bar{A}^+$ and $\bar{X}\bar{A}\bar{Y}$ are complexed cation and the electroneutral complex.

When dilute solutions are used, little complex ion formation takes place and the second term in Eq. (159) becomes negligible. Thus Eq. (159) simplifies to

$$E = \frac{RT}{F} \ln \frac{a'_i + [\bar{u}_{jX}k_{jX}K^+_{jX}/\bar{u}_{iX}k_{iX}K^+_{iX}]a'_j}{a''_i + [\bar{u}_{jX}k_{jX}K^+_{jX}/\bar{u}_{iX}k_{iX}K^+_{iX}]a''_j}. \tag{160}$$

TABLE I
Several Relations for Selectivity between a Primary Ion i and an Interfering Ion j for Different Types of Membranes[a]

Membrane type	Species within the membrane	$K^{pot}_{ij} = 1/K^{pot}_{ji}$	Equation
Solid membrane			
Ion exchanger	X^-, i^{z+}, j^{z+}	$\dfrac{\bar{u}_j}{\bar{u}_i} K_{ij}$	(24)
Silver (halides) (solid state)	M^+, i^{z-}, j^{z-}	$\dfrac{S_{Mi}}{S_{Mj}}$	(46)
Liquid membrane			
Ion exchanger: dissociated	X^-, i^{z+}, j^{z+}	$\dfrac{\bar{u}_j k_j}{\bar{u}_i k_i}$	(151)
associated			
(1) $\tau = 0$ (poorly mobile site)	iX_z, jX_z	$\dfrac{(\bar{u}_j + \bar{u}_X)k_j}{(\bar{u}_i + \bar{u}_X)k_i}$	(153)
when $\bar{u}_X \ll \bar{u}_i, \bar{u}_j$		$\dfrac{\bar{u}_j k_j}{\bar{u}_i k_i}$	
(2) $\tau = 1$ (high mobile site)		$\dfrac{\bar{u}_{jX}k_jK_{jX}}{\bar{u}_{iX}k_iK_{iX}}$	
		$= \dfrac{\bar{u}_{jX}}{\bar{u}_{iX}} K_{ij}$	(153)
With electrically neutral ligand	X, i^{z+}, j^{z+}	$\dfrac{\bar{u}_{jX}k_{jX}K^+_{jX}}{\bar{u}_{iX}k_{iX}K^+_{iX}}$	(160, 162)
	$iX_n^{z+}, jX_n^{z+}, Y^-$	$= \dfrac{\bar{u}_{jX}K_j}{\bar{u}_{iX}K_i}$	
when $\bar{u}_{jX} = \bar{u}_{iX}$		$\dfrac{K_j}{K_i}$	(162)

[a] k_i, k_j: partition coefficients; K_{iX}, K_{jX}: association constants; K_i, K_j: bulk partition coefficients; S_{Mi}, S_{Mj}: solubility products of sparingly soluble precipitates of metal M.

Expressing $k_{iX}K_{iX}^+$ in terms of a bulk partition coefficient K_i for the reaction

$$i^+ + Y^- + X^* \rightleftharpoons iX^{+*} + Y^{-*}$$

where * indicates the organic phase

$$K_i = \frac{a_{iX}^* a_Y^*}{a_i a_Y a_X^*}. \qquad (161)$$

Substituting from Eqs. (155), (157), and (158), Eq. (161) becomes

$$K_i = k_{iX}K_{iX}^+ k_Y/k_X.$$

The potential difference disappears because oppositely charged ions (iX^{+*} and Y^{-*}) are involved. A similar equation may be writen for K_j and as the species X and Y are common for the system, the relation

$$K_j/K_i = k_{jX}K_{jX}^+/k_{iX}K_{iX}^+ \qquad (162)$$

is obtained.

The above equations have been verified by constructing appropriate membrane systems. In Table I are given the several expressions for the selectivity between a primary ion i and an interfering ion j for different types of liquid membranes.

VI. Thermodynamic Approach to Isothermal Membrane Potential

Unlike the theories of membrane potential based on electrodiffusion, which depend on the details of internal structure and properties of the membrane, the thermodynamic approach does not need this information. Staverman derived the relation

$$-F\,dE = \sum \frac{t_i}{z_i} d\mu_i \qquad (163)$$

using the principles of irreversible thermodynamics (see Chapter 6). In a similar way, Scatchard expressed membrane potential as

$$E = -\frac{RT}{F} \int_{a_i'}^{a_i''} \sum_i t_i \, d\ln a_i. \qquad (164)$$

The solutions of activity a_i' and a_i'' extend up to each membrane–solution interface. Equation (164) may be applied to all components moving across the membrane. For (1:1) electrolyte, the mobile components are counterion, coion, water, and the fixed charges of the membrane. If movements of counterion, coion, and water are considered relative to the polymer network to which the charges are fixed, then the summation in Eq. (164) refers only to counterion (+), coion (−), and water (w). Thus Eq. (164) becomes

$$E = -\frac{RT}{F} \int_I^{II} (\bar{t}_+ \, d\ln a_+ - \bar{t}_- \, d\ln a_- + \bar{t}_w \, d\ln a_w). \quad (165)$$

If anion reversible electrodes are used in the membrane cell to measure the cell emf E, then the electrode potential E_{ref} between two such electrodes is given by

$$E_{ref} = (RT/F)\ln(a_-^I/a_-^{II}). \quad (166)$$

Substitution of Eq. (166) and use of the relations $\bar{t}_+ + \bar{t}_- = 1$ and $d\ln a_w = -2 \times 10^{-3} \, mM_1 \, d\ln a_\pm$ [see Eq. (25) of Chapter 2] in Eq. (165) gives

$$E = -\frac{2RT}{F} \int_I^{II} (\bar{t}_+ - 10^{-3} \, mM_1 \bar{t}_w) \, d\ln a_\pm. \quad (167)$$

This equation satisfactorily describes the electrical potentials arising across model membranes when they separate solutions of the same (1:1) electrolyte but of different concentrations.

VII. Kinetic Approach to Membrane Potentials

The theory of rate processes has been applied to the consideration of membrane potentials. Nagasawa and Kobatake (1952) derived the expression

$$E = \frac{RT}{F}\left[\alpha \ln \frac{C_1}{C_2} - \alpha \ln \frac{C_1 + \beta}{C_2 + \beta} + \frac{\bar{u} - \bar{v}}{\bar{u} + \bar{v}} \ln \frac{C_1 + \beta}{C_2 + \beta}\right], \quad (168)$$

where

$$\alpha = \frac{k_2(\bar{u} - \bar{v}) + k_3 \bar{u}}{k_2(\bar{u} + \bar{v}) + k_3 \bar{u}} \quad \text{and} \quad \beta = \frac{k_2(\bar{u} + \bar{v}) + k_3 \bar{u}}{k_1(\bar{u} + \bar{v})}.$$

k_1, k_2, and k_3 are complex functions involving the radius of the pores in the membrane, electrical potential due to fixed charge, and the magnitude of fixed charge present in the membrane. Although it is difficult to express in precise terms what these constants mean, α is considered to indicate the membrane selectivity to the ions and β the extent of ion adsorption on the membrane material. When the potential due to fixed charges is zero, $\alpha = 1$, and β is very small (e.g., porous membrane), Eq. (168) reduces to

$$E = \frac{RT}{F}\left[\ln \frac{C_1}{C_2} - \frac{2\bar{v}}{\bar{u} + \bar{v}} \ln \frac{C_1 + \beta}{C_2 + \beta}\right]. \quad (169)$$

VII. Kinetic Approach to Membrane Potentials

In order to account for osmotic flow, a correction factor k'_3 (osmotic pressure coefficient) has been introduced into Eq. (169). Thus

$$E = \frac{RT}{F}\left[\ln\frac{C_1}{C_2} - \frac{2\bar{v} - k'_3}{\bar{u}+\bar{v}}\ln\frac{C_1+\beta}{C_2+\beta}\right]. \tag{170}$$

These equations have been used to explain membrane potentials observed across both anion and cation selective membranes.

According to the theory of rate processes, ion transport across a membrane may be considered to occur by the ion jumping over energy barriers present in the pores of the membrane. The rate constant for the forward jump is given by

$$k = \nu\exp(-\Delta G/RT),$$

where ν is the vibration frequency and is given by $\nu = \kappa T/h$ ($\approx 6.2 \times 10^{12}$ s^{-1} at $T = 298$ K).

The outflow J_{out} of ion j is given by

$$J_{\text{out}} = C_{j(\text{in})}\nu\exp(-\Delta G_j/RT). \tag{171}$$

In the presence of an electric field (assuming the barrier in the membrane to be located δ cm from the outside surface of the membrane) the outflux is given by

$$J_{\text{out}} = C_{j(\text{in})}\nu\exp(-\Delta G_j/RT)\exp[z_j(1-\delta)EF/RT]. \tag{172}$$

Assuming that the rate constant for the backward jump is the same as the forward rate constant, the influx is given by

$$J_{\text{in}} = C_{j(\text{out})}\nu\exp(-\Delta G_j/RT)\exp(-z_j\delta EF/RT). \tag{173}$$

Thus the net current I_j carried by the ion j is given by $I_j = (J_{\text{out}} - J_{\text{in}})z_j F$. Substituting for J_{out} and J_{in} from Eqs. (172) and (173) gives

$$I_j = z_j F\nu\exp(-\Delta G_j/RT)\exp[z_j(1-\delta)EF/RT] \\ \times [C_{j(\text{in})} - C_{j(\text{out})}\exp(-z_j EF/RT)]. \tag{174}$$

If Na, K, and Ca ions are involved in carrying the current, as is the case at neuromuscular junctions, the reversal potential can be worked out by using Eq. (174) to calculate the total current and then equation $\sum_i I_i = 0$. Thus

$$\exp[(1-\delta)EF/RT][(\text{Na})_i + P_1(\text{K})_i + 2P_2(\text{Ca})_i\exp[(1-\delta)EF/RT]] \\ = \exp(-\delta EF/RT)[(\text{Na})_o + P_1(\text{K})_o + 2P_2(\text{Ca})_o\exp(-\delta EF/RT)].$$

Rearrangement gives

$$E_{\text{rev}} = \frac{RT}{F}\ln\frac{(\text{Na})_o + P_1(\text{K})_o + 2P_2(\text{Ca})_o\exp(-\delta E_{\text{rev}}F/RT)}{(\text{Na})_i + P_1(\text{K})_i + 2P_2(\text{Ca})_o\exp[(1-\delta)E_{\text{rev}}F/RT]}. \tag{175}$$

where

$$P_1 = \frac{\exp(-\Delta G_K/RT)}{\exp(-\Delta G_{Na}/RT)} \quad \text{and} \quad P_2 = \frac{\exp(-\Delta G_{Ca}/RT)}{\exp(-\Delta G_{Na}/RT)}.$$

If changes in surface potential due to binding of ions to surface charges have to be taken into account, then E in Eq. (175) must be replaced by $E - \psi$, where ψ can be evaluated by using the Gouy–Chapman equation.

Equation (175) corrected for surface potential was tested by Lewis (1979) and found inadequate to explain the ion-concentration dependence of reversal potential measured at the frog neuromuscular junction. This discrepancy between the simple single-barrier theory and experimental results is probably due to the assumption that the two rate constants (forward and backward) were equal. A more complex theory based on a binding site regulating the ion flow across the membrane and situated in the valley of two energy barriers in the membrane has been worked out by Lewis and Stevens (1979). In this theory the forward and backward rate constants are different. The additional assumption is that the binding site is vacant before it can be occupied by an ion. This means that the probability of displacing one ion by another ion from the binding site is zero.

The reaction scheme of ion j with the site is shown in Fig. 3. The rate constants are given by

$$\begin{aligned} k_{1j} &= v_1 C_{j(1)} \exp(-\Delta G_1/RT) \exp(-z_j \delta EF/2RT) \\ &= k'_{1j} C_{j(1)} \exp(-z_j \delta EF/2RT), \end{aligned} \quad (176)$$

$$\begin{aligned} k_{-1j} &= v_2 \exp(\Delta G_2/RT) \exp(z_j \delta EF/2RT) \\ &= k'_{-1j} \exp(z_j \delta EF/2RT), \end{aligned} \quad (177)$$

$$\begin{aligned} k_{2j} &= v_3 \exp(-\Delta G_3/RT) \exp[-z_j(1-\delta)EF/2RT] \\ &= k'_{2j} \exp[-z_j(1-\delta)EF/2RT], \end{aligned} \quad (178)$$

$$\begin{aligned} k_{-2j} &= v_4 C_{j(2)} \exp(\Delta G_4/RT) \exp[z_j(1-\delta)EF/2RT] \\ &= k'_{-2j} C_{j(2)} \exp[z_j(1-\delta)EF/2RT]. \end{aligned} \quad (179)$$

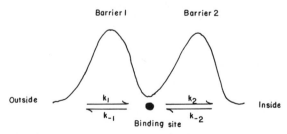

Fig. 3. Two-barrier model of the channel with a binding site situated in the valley of the energy barriers. ks are the rate constants for binding and unbinding.

The binding site can be in one of four states—empty or occupied by ion 1, 2, or 3 (e.g., Na, K, and Ca ions)—the probability of each state being P_0, P_1, P_2, P_3. Thus

$$P_0 + P_1 + P_2 + P_3 = 1 \tag{180}$$

The flux of j in going to the binding site is $P_0(k_{1j} + k_{-2j})$. The flux of j leaving the binding site is $P_j = (k_{2j} + k_{-1j})$. Thus for the three ions, in the steady state, the three relations are given by

$$P_0(k_{1\text{Na}} + k_{-2\text{Na}}) = P_{\text{Na}}(k_{2\text{Na}} + k_{-1\text{Na}}), \tag{181}$$

$$P_0(k_{1\text{K}} + k_{-2\text{K}}) = P_{\text{K}}(k_{2\text{K}} + k_{-1\text{K}}), \tag{182}$$

$$P_0(k_{1\text{Ca}} + k_{-2\text{Ca}}) = P_{\text{Ca}}(k_{2\text{Ca}} + k_{-1\text{Ca}}). \tag{183}$$

Substituting Eqs. (181)–(183) into Eq. (180) gives

$$P_0 = \left(1 + \frac{k_{1\text{Na}} + k_{-2\text{Na}}}{k_{2\text{Na}} + k_{-1\text{Na}}} + \frac{k_{1\text{K}} + k_{-2\text{K}}}{k_{2\text{K}} + k_{-1\text{K}}} + \frac{k_{1\text{Ca}} + k_{-2\text{Ca}}}{k_{2\text{Ca}} + k_{-1\text{Ca}}}\right)^{-1}. \tag{184}$$

One-way flux of j is given by $P_j k_{-1j} - P_0 k_{1j}$. Therefore the current is $I_j = z_j F[P_j k_{-1j} - P_0 k_{1j}]$. The total current is given by

$$I = \sum_j I_j = FP_0 \left[\sum_{j=1}^{n} z_j k_{-1j} \frac{k_{1j} + k_{-2j}}{k_{2j} + k_{-1j}} - \sum z_j k_{1j}\right]. \tag{185}$$

Equation (185) may be written in full for the three ions, Na, K, and Ca. The total current thus obtained on equating to zero gives the reversal potential E_{rev}. The resulting expression is very complicated and the original paper by Lewis and Stevens (1979) should be consulted for details.

References

Buck, R. P., and Stover, F. S. (1978). *Anal. Chim. Acta* **101**, 231.
Eisenman, G. (1962). *Biophys. J., Suppl.* **2**, 314.
Eisenman, G. (1967). *In* "Glass Electrodes for Hydrogen and Other Cations, Principles and Practice" (G. Eisenman, ed.), p. 133. Dekker, New York.
Eisenman, G. (1969). *NBS Spec. Publ. (U.S.)* **314**, 1.
El-Sharkawy, T. Y., and Daniel, E. E. (1975). *Am. J. Physiol.* **229**, 1283.
Eyring, H., Lumbry, R., and Woodbury, J. W. (1949). *Rec. Chem. Prog.* **10**, 100.
Glasstone, S., Laidler, K. J., and Eyring, H. (1941). "The Theory of Rate Processes." McGraw-Hill, New York.
Goldman, D. E. (1943). *J. Gen. Physiol.* **27**, 37.
Helfferich, F. (1962). "Ion-Exchange." McGraw-Hill, New York.
Hille, B. (1971). *J. Gen. Physiol.* **58**, 599.
Hodgkin, A. L. (1951). *Biol. Rev. Cambridge Philos. Soc.* **26**, 339.
Hodgkin, A. L., and Huxley, A. F. (1952). *J. Physiol. (London)* **116**, 449.
Hodgkin, A. L., and Katz, B. (1949). *J. Physiol. (London)* **108**, 37.

Hodgkin, A. L., Huxley, A. F., and Katz, B. (1952). *J. Physiol.* (*London*) **116**, 424.
Karreman, G., and Eisenman, G. (1962). *Bull. Math. Biophys.* **24**, 413.
Kobatake, Y. (1958). *J. Chem. Phys.* **28**, 146, 442.
Kotyk, A., and Janacek, K. (1975). "Cell Membrane Transport," 2nd ed. Plenum, New York.
Lakshminarayanaiah, N. (1969). "Transport Phenomena in Membranes." Academic Press, New York.
Lakshminarayanaiah, N. (1976). "Membrane Electrodes." Academic Press, New York.
Lakshminarayanaiah, N. (1979). *Subcell. Biochem.* **6**, 401.
Lakshminarayanaiah, N. and Siddiqi, F. A. (1971). *Biophys. J.* **11**, 617.
Lewis, C. A. (1979). *J. Physiol.* (*London*) **286**, 417.
Lewis, C. A., and Stevens, C. F. (1979). *In* "Membrane Transport Processes" (C. F. Stevens and R. W. Tsien, eds.), Vol. 3, p. 133. Raven Press, New York.
Lorimer, J. W., Boterenbrood, E. I., and Hermans, J. J. (1956). *Dissuss. Faraday Soc.* **21**, 141.
Moreton, R. B. (1969). *J. Exp. Biol.* **51**, 181.
Mullins, L. J., and Noda, K. (1963). *J. Gen. Physiol.* **47**, 117.
Nagasawa, M., and Kobatake, Y. (1952). *J. Phys. Chem.* **56**, 1017.
Ohki, S. (1971). *J. Colloid Interface Sci.* **37**, 318.
Polissar, M. J. (1954). *In* "The Kinetic Basis of Molecular Biology" (F. H. Johnson, H. Eyring, and M. J. Polissar, eds.), Chapter 11. Wiley, New York.
Rothmund, V., and Kornfeld, G. (1918). *Z. Anorg. Allg. Chem.* **103**, 129.
Scatchard, G. (1953). *J. Am. Chem. Soc.* **48**, 176.
Sjödin, R. A., and Ortiz, O. (1973). *J. Gen. Physiol.* **66**, 283.
Staverman, A. J. (1952). *Trans. Faraday Soc.* **48**, 176.
Teorell, T. (1953). *Prog. Biophys. Biophys. Chem.* **3**, 305.
Ussing, H. H. (1949). *Acta Physiol. Scand.* **19**, 43.
Ussing, H. H. (1952). *Adv. Enzymol.* **13**, 21.

Chapter **5**

KINETIC MODELS OF MEMBRANE TRANSPORT

As a preliminary to the consideration of kinetic equations given later in this chapter for the description of transport of lipophilic and other ions across lipid bilayer and biological membranes by carrier or pore mechanism, some of the equations used to follow enzyme reactions are first outlined.

I. Equations of Enzyme Kinetics

In Chapter 4 the simple kinetic equations (130)–(147) related to metal ion reactions with the charged sites in the membrane giving rise to electrical potentials across membranes are described. Those equations were obtained by following the mathematical logic of Michaelis and Menten, whose equation for enzyme (E) and substrate (A) reaction

$$\text{E} + \text{A} \underset{k_{-1}}{\overset{k_1}{\rightleftarrows}} \text{EA} \xrightarrow{k_2} \text{E} + \text{P} \tag{1}$$

is written in the standard form

$$v = \frac{V(\text{A})}{K_\text{A} + (\text{A})} \tag{2}$$

[compare Eq. (2) with Eq. (137) of Chapter 4], where V is maximum velocity, (A) is substrate concentration, and K_A is the Michaelis–Menten constant. From Eq. (1) it is apparent that $V = k_2(\text{E}_0)$, where (E_0) is the total enzyme concentration and $K_\text{A} = (k_{-1} + k_2)/k_1$.

The reaction scheme (1) is incomplete since significant amounts of both substrate and product exist in equilibrium mixture. If allowance is made for this, reaction (1) should be written

$$E + A \underset{k_{-1}}{\overset{k_1}{\rightleftarrows}} EA \underset{k_{-2}}{\overset{k_2}{\rightleftarrows}} E + P \qquad (3)$$

Since the total concentration of enzyme is $(E_0) = (E) + (EA)$, one can write for the steady state

$$\frac{d(EA)}{dt} = k_1[(E_0) - (EA)](A) + k_{-2}[(E_0) - (EA)](P) - (k_{-1} + k_2)(EA) = 0.$$

Solving for (EA) gives

$$(EA) = \frac{k_1(E_0)(A) + k_{-2}(E_0)(P)}{k_{-1} + k_2 + k_1(A) + k_{-2}(P)}. \qquad (4)$$

The net rate of release of P is given by

$$v = k_2(EA) - k_{-2}[(E_0) - (EA)](P). \qquad (5)$$

Substitution for (EA) from Eq. (4) into Eq. (5) gives on simplification

$$v = \frac{k_1 k_2 (E_0)(A) - k_{-1} k_{-2} (E_0)(P)}{k_{-1} + k_2 + k_1(A) + k_{-2}(P)}. \qquad (6)$$

Two special cases are of interest: (1) when $(P) = 0$, the initial rate of the forward reaction v^f is obtained, and (2) when $(A) = 0$, the initial rate of reverse reaction v^r is obtained. In either case the resulting equations are of the form of the Michaelis–Menten equation (2) with the maximum velocity and the constant given by

$$V^f = k_2(E_0) \quad \text{and} \quad K_A = (k_{-1} + k_2)/k_1$$

for the first case and

$$V^r = k_{-1}(E_0) \quad \text{and} \quad K_P = (k_{-1} + k_2)/k_{-2}$$

for the second case. Substituting these values, Eq. (6) can be written

$$v = \frac{V^f(A)/K_A - V^r(P)/K_P}{1 + (A)/K_A + (P)/K_P}. \qquad (7)$$

This equation is the general reversible form of the Michaelis–Menten equation. Many complicated reactions may be described by Eq. (7). For example, conversion by the enzyme of A into P by the formation of an intermediate EP as shown in the reaction scheme

$$E + A \underset{k_{-1}}{\overset{k_1}{\rightleftarrows}} EA \underset{k_{-2}}{\overset{k_2}{\rightleftarrows}} EP \underset{k_{-3}}{\overset{k_3}{\rightleftarrows}} E + P \qquad (8)$$

In principle the rate equation for scheme (8) may be derived as described above. Here there are two intermediates whose rate of change $d(EA)/dt$ and $d(EP)/dt$ must be equated to zero. Thus

$$\frac{d(EA)}{dt} = k_1(E)(A) + k_{-2}(EP) - (k_{-1} + k_2)(EA), \tag{9}$$

$$\frac{d(EP)}{dt} = k_{-3}(E)(P) + k_2(EA) - (k_3 + k_{-2})(EP). \tag{10}$$

As the total enzyme concentration is $(E_0) = (E) + (EA) + (EP)$, Eqs. (9) and (10) become

$$k_1(A)[(E_0) - (EA) - (EP)] + k_{-2}(EP) - (k_{-1} + k_2)(EA) = 0, \tag{11}$$

$$k_{-3}(P)[(E_0) - (EA) - (EP)] + k_2(EA) - (k_3 + k_{-2})(EP) = 0. \tag{12}$$

Equations (11) and (12) are two simultaneous equations, which can be solved for (EA) and (EP). The solution is

$$\frac{(EA)}{(E_0)} = \frac{k_1(k_{-2} + k_3)(A) + k_{-2}k_{-3}(P)}{\Sigma}, \tag{13}$$

$$\frac{(EP)}{(E_0)} = \frac{k_1 k_2(A) + k_{-3}(k_{-1} + k_2)(P)}{\Sigma}, \tag{14}$$

where

$$\Sigma = k_{-1}k_{-2} + k_{-1}k_3 + k_2 k_3 + k_1(k_2 + k_{-2} + k_3)(A) + k_{-3}(k_{-1} + k_2 + k_{-2})(P).$$

The net reaction rate from left to right is

$$v = k_2(EA) - k_{-2}(EP). \tag{15}$$

Thus substituting for (EA) and (EP) from Eqs. (13) and (14) gives

$$v = \frac{[k_1 k_2 k_3(A) - k_{-1}k_{-2}k_{-3}(P)](E_0)}{\Sigma}. \tag{16}$$

This is similar to Eq. (7), and the definitions of the parameters are

$$V^f = \frac{k_2 k_3(E_0)}{k_{-2} + k_2 + k_3}, \qquad V^r = \frac{k_{-1}k_{-2}(E_0)}{k_{-1} + k_{-2} + k_2},$$

$$K_A = \frac{k_{-1}k_{-2} + k_{-1}k_3 + k_2 k_3}{k_1(k_{-2} + k_2 + k_3)},$$

$$K_P = \frac{k_{-1}k_{-2} + k_{-1}k_3 + k_2 k_3}{k_{-3}(k_{-1} + k_{-2} + k_2)}.$$

Although these equations look complex, K_A and K_P are the dissociation constants of EA and EP, respectively, when the second step of the reaction is rate limiting in either direction. Hence, when k_2 is small (i.e., the time constant is large) compared with $(k_{-2} + k_3)$, $K_A = (k_{-1})/k_1$. Similarly, when k_{-2} is small compared with $(k_{-1} + k_2)$, $K_P = k_3/k_{-3}$. Both will be true when the interconversion of EA and EP becomes rate limiting in both directions; that is, if $k_3 \gg k_2$ and $k_{-1} \gg k_{-2}$.

The other interesting possibility is the release of A and P from their complexes with the same rate constant, i.e., $k_{-1} = k_3$. In this case also $K_A = k_{-1}/k_1$ and $K_P = k_3/k_{-3}$.

The net velocity of a reaction at equilibrium is zero. For this condition, Eq. (7) gives

$$\frac{V^f(A)_\infty}{K_A} = \frac{V^r(P)_\infty}{K_P},$$

and so

$$\frac{V^f K_P}{V^r K_A} = \frac{(P)_\infty}{(A)_\infty} = K, \tag{17}$$

where K is the equilibrium constant of the reaction. Equation (17) is known as the Haldane equation.

In several cases it is possible that the product could act as an inhibitor in which case if Eq. (7) applies, the rate will decrease with accumulation of the product. In the reaction scheme (8) product inhibition would occur in an irreversible way if accumulation of the product caused the enzyme complex EP to be removed or sequestered. Now Eq. (7) becomes

$$v = \frac{V^f(A)/K_A}{1 + (A)/K_A + (P)/K_P} = \frac{V^f(A)}{K_A[1 + (P)/K_P] + (A)}. \tag{18}$$

As the inhibition caused by the accumulated product would be the same as that caused by the added product, one could follow initial rates (little product produced) with different concentrations of added product. At each product concentration, the initial rate of enzyme reaction when different substrate concentrations are used would follow the Michaelis–Menten equation. But one would be getting apparent values for V and K_m. Thus comparing Eqs. (18) and (2) gives

$$V^{app} = V^f \quad \text{and} \quad K_m^{app} = K_m[1 + (P)/K_P].$$

K_m^{app} is larger than K_m and increases with increase in the concentration of P. There are many compounds that act as inhibitors that may be classified as irreversible inhibitors and reversible inhibitors. Irreversible inhibitors combine with the enzyme and decrease its activity to zero. Reversible inhibitors form complexes with the enzyme, and these complexes have different properties from those of the free enzyme. Different names are given to indicate several altered properties. Thus (1) an increase in the value of K_m indicates competitive inhibition; (2) a decreased V value indicates pure noncompetitive inhibition; (3) a decrease in V and K_m values but the ratio of the decreases being constant indicates uncompetitive inhibition; and (4) some combination of (1)–(3) is called mixed inhibition.

When the inhibitor-bound enzyme has no activity, this type of inhibition is called complete or linear inhibition since plots of apparent values of K_m/V and $1/V$ versus inhibitor concentration give straight lines. When inhibitor-bound enzyme retains some activity, it is called partial or hyperbolic inhibition since the shapes of the plots are hyperbolic.

In the case of competitive inhibition both the substrate and the inhibitor compete for the same site thus

$$\text{EI} \xrightleftharpoons{K_I} \text{E} + \text{S} \rightleftharpoons \text{ES} \longrightarrow \text{E} + \text{P}$$

The defining equation is

$$v = \frac{V(S)}{K_m[1 + (I)/K_I] + (S)}. \tag{19}$$

This can be written to conform to Michaelis–Menten form as

$$v = \frac{V^{\text{app}}(S)}{K_m^{\text{app}} + (S)}, \tag{20}$$

where $V^{\text{app}} = V$, $K_m^{\text{app}} = K_m[1 + (I)/K_I]$ and

$$\frac{V^{\text{app}}}{K_m^{\text{app}}} = \frac{V/K_m}{1 + (I)/K_I}. \tag{21}$$

The effect of competitive inhibitor is to increase the apparent value of K_m by a factor of $[1 + (I)/K_I]$, decrease the value of V/K_m by the same factor, and allow V to remain unchanged.

In the case of mixed inhibition, simplest mechanism is

$$\begin{array}{ccc} \text{E} + \text{S} & \rightleftharpoons \text{ES} & \longrightarrow \text{E} + \text{P} \\ K_I \updownarrow & \updownarrow K_I' & \\ \text{EI} & \text{EIS} & \end{array}$$

Linear mixed inhibition occurs if both V^{app} and $V^{\text{app}}/K_m^{\text{app}}$ vary with the inhibitor concentration following the equations

$$V^{\text{app}} = \frac{V}{1 + (I)/K_I'}, \tag{22}$$

$$K_m^{app} = \frac{K_m[1 + (I)/K_I]}{1 + (I)/K_I'}, \quad (23)$$

and

$$\frac{V^{app}}{K_m^{app}} = \frac{V/K_m}{1 + (I)/K_I}. \quad (24)$$

Uncompetitive inhibition is a limiting case of mixed inhibition when $K_I \to \infty$ (i.e. $(I)/K_I \to 0$). Thus Eqs. (22)–(24) become

$$V^{app} = \frac{V}{1 + (I)/K_I'}, \quad (22)$$

$$K_m^{app} = \frac{K_m}{1 + (I)/K_I'}, \quad (25)$$

and

$$\frac{V^{app}}{K_m^{app}} = \frac{V}{K_m}. \quad (26)$$

II. Schematic Method of Deriving Rate Equations

Solving simultaneous equations, for example, Eqs. (11) and (12), is rather complicated. So simplified techniques have been introduced. A useful procedure devised by King and Altman, (1956) is illustrated for the two intermediate mechanism represented by

$$E + A \underset{k_{-1}}{\overset{k_1}{\rightleftarrows}} EA \xrightarrow{k_2} EA' \xrightarrow{k_3} E + Y$$
$$\downarrow X$$

where EA' is the second intermediate and X and Y are the products.

First, the mechanism is written in cyclic form

$$E \underset{k_{-1}}{\overset{k_1 A}{\rightleftarrows}} EA$$
$$k_3 \nwarrow \swarrow k_2$$
$$EA'$$

The concentrations of E, EA, and EA' are proportional to the sums of terms that lead individually or in sequence to the species under consideration. Thus for E the terms are

E: $\underset{k_3}{\overset{k_{-1}}{\leftarrow}} \quad \underset{k_3 \quad k_2}{\nwarrow \swarrow}$

The concentration of E is proportional to $k_{-1}k_3 + k_2k_3$. In each term the number of rate constants is one less than the number of species in the cycle. Similarly, for EA and EA′

$$\text{EA:} \quad \underset{k_3}{\overset{k_1 A}{\rightleftarrows}} \qquad \text{EA′:} \quad \overset{k_1 A}{\underset{k_2}{\rightleftarrows}}$$

Hence

$$(\text{E}) \propto k_{-1}k_3 + k_2k_3,$$

$$(\text{EA}) \propto k_1 k_3 (\text{A}),$$

$$(\text{EA′}) \propto k_1 k_2 (\text{A}).$$

But $(\text{E}_0) = (\text{E}) + (\text{EA}) + (\text{EA′})$, and so

$$\frac{(\text{EA})}{(\text{E}_0)} = \frac{k_1 k_3 (\text{A})}{k_{-1}k_3 + k_2 k_3 + k_1 k_3 (\text{A}) + k_1 k_2 (\text{A})}. \tag{27}$$

The rate of the reaction is

$$v = k_2 (\text{EA}).$$

Substituting for (EA) from Eq. (27) gives

$$v = \frac{k_1 k_2 k_3 (\text{E}_0)(\text{A})}{k_3 (k_{-1} + k_2) + k_1 (k_2 + k_3)(\text{A})}. \tag{28}$$

Comparison of this equation with the Michaelis–Menten form shows that

$$V = \frac{k_2 k_3 (\text{E}_0)}{k_2 + k_3} \quad \text{and} \quad K_\text{A} = \frac{k_{-1} + k_2}{k_1} \frac{k_3}{k_2 + k_3}. \tag{29}$$

Two cases are of interest:

(1) When $k_3 \gg k_2$, Eqs. (29) simplify to

$$V = k_2(\text{E}_0) \quad \text{and} \quad K_\text{A} = (k_{-1} + k_2)/k_1.$$

This result corresponds to that of a single intermediate. The second intermediate is unimportant because of its large rate constant for disappearance.

(2) When $k_2 \gg k_3$, Eqs. (29) reduce to

$$V = k_3(\text{E}_0) \quad \text{and} \quad K_\text{A} = \frac{k_{-1} + k_2}{k_1} \frac{k_3}{k_2}.$$

The Michaelis constant now is complicated and is less than $(k_{-1} + k_2)/k_1$ since (k_3/k_2) is a small fraction.

Another illustration of the schematic method is given for the mechanism

$$E + A \underset{k_{-1}}{\overset{k_1}{\rightleftarrows}} EA + B \underset{k_{-2}}{\overset{k_2}{\rightleftarrows}} EAB \rightleftarrows EXY \underset{k_{-3}}{\overset{k_3}{\rightleftarrows}} EY + X \underset{k_{-4}}{\overset{k_4}{\rightarrow}} E + Y$$

The steady-state measurements give no information about the isomerization $EAB \rightleftharpoons EXY$. The cyclic form of the mechanism can be written as

$$\begin{array}{ccc} E & \underset{k_{-1}}{\overset{k_1 a}{\rightleftarrows}} & EA \\ k_4 \updownarrow k_{-4}Y & & k_{-2} \updownarrow k_2 b \\ EY & \underset{k_3}{\overset{k_{-3}X}{\rightleftarrows}} & EAB \\ & & EXY \end{array}$$

where a, b, X, and Y represent concentrations of A, B, X, and Y, respectively.

The terms for E are

[schematic diagrams]

$$k_{-1}k_{-2}k_{-3}X + k_{-1}k_{-2}k_4 + k_{-1}k_3k_4 + k_2bk_3k_4. \tag{30}$$

For EA, the terms are

$$k_1k_{-2}k_{-3}aX + k_1k_{-2}k_4a + k_1k_3k_4a + k_{-2}k_{-3}k_{-4}XY. \tag{31}$$

For EAB + EXY, they are

$$k_1k_2k_{-3}abX + k_1k_2k_4ab + k_{-1}k_{-3}k_{-4}XY + k_2k_{-3}k_{-4}bXY. \tag{32}$$

For EY, they are

$$k_1k_2k_3ab + k_{-1}k_{-2}k_{-4}Y + k_{-1}k_3k_{-4}Y + k_2k_3k_{-4}bY. \tag{33}$$

If the sum of Eqs. (30)–(33) is represented by Σ, then

$$(E)/(E_0) = \text{Eq. (30)}/\Sigma, \quad (EA)/(E_0) = \text{Eq. (31)}/\Sigma,$$
$$(EAB + EXY)/(E_0) = \text{Eq. (32)}/\Sigma, \quad \text{and} \quad (EY)/(E_0) = \text{Eq. (33)}/\Sigma. \tag{34}$$

The rate of the reaction is given by the sum of the rates of the steps that produce the particular product minus the sum of the rates of the steps that consume the same product. Thus

$$v = \frac{dP}{dt} = k_3(EAB + EXY) - k_{-3}(EY)(X). \tag{35}$$

Substituting for these from Eq. (34) gives on simplification

$$v = (k_1 k_2 k_3 k_4 E_0 ab - k_{-1} k_{-2} k_{-3} k_{-4} E_0 XY)/\Sigma.$$

In coefficient form, this equation is written

$$v = \frac{E_0(C_1 ab - C_2 XY)}{C_3 + C_4 a + C_5 b + C_6 X + C_7 Y + C_8 ab + C_9 aX + C_{10} bX + C_{11} XY + C_{12} abX + C_{13} bXY}, \quad (36)$$

where the coefficients are given by

$$C_1 = k_1 k_2 k_3 k_4, \qquad C_2 = k_{-1} k_{-2} k_{-3} k_{-4},$$
$$C_3 = k_{-1}(k_{-2} + k_3)k_4, \qquad C_4 = k_1(k_{-2} + k_3)k_4,$$
$$C_5 = k_2 k_3 k_4, \qquad C_6 = k_{-1} k_{-2} k_{-3},$$
$$C_7 = k_{-1}(k_{-2} + k_3)k_{-4}, \qquad C_8 = k_1 k_2(k_3 + k_4),$$
$$C_9 = k_1 k_{-2} k_{-3}, \qquad C_{10} = k_2 k_3 k_{-4},$$
$$C_{11} = k_{-3} k_{-4}(k_{-1} + k_{-2}), \qquad C_{12} = k_1 k_2 k_{-3},$$
$$C_{13} = k_2 k_{-3} k_{-4}.$$

For simple mechanisms, the procedure of King and Altman (1956) is convenient. For complex mechanisms, the procedure has been simplified by several investigators. For some of these procedures consult Cornish-Bowden (1979).

III. Enzyme Kinetics of Mediated Transport

Many cell membranes have mechanisms for the specific transport of many substances, both charged and uncharged. The rate of transport increases with increase in concentration of the substance and attain saturation. The explanation is that the solute binds to a mobile carrier, which moves in the membrane from one side to the other, releases the bound molecule and returns back as shown below.

$$\begin{array}{ccc}
\text{Side (')} \ (CS)' & \underset{k_{-2}}{\overset{k_2}{\rightleftarrows}} & (CS)'' \ \text{Side ('')} \\
k_1 S' \updownarrow k_{-1} & & k_{-3} S'' \updownarrow k_3 \\
C' & \underset{k_4}{\overset{k_{-4}}{\rightleftarrows}} & C''
\end{array}$$

C' is the carrier on side (') and C'' is the same carrier on side (''). The complex CS formed on side (') moves to side ('').

The flux J_S of the substrate S on application of the King-Altman technique to the scheme given above yields

$$\begin{aligned}J_S = {} & C_t(k_1k_2k_3k_4S' - k_{-1}k_{-2}k_{-3}k_{-4}S'') \\ & \times \{(k_4 + k_{-4})(k_{-1}k_{-2} + k_{-1}k_3 + k_2k_3) \\ & + k_1[k_3(k_2 + k_4) + k_4(k_2 + k_{-2})]S' \\ & + k_{-3}[k_{-2}(k_{-1} + k_{-4}) + k_{-4}(k_{-1} + k_2)]S'' \\ & + k_1k_{-3}(k_2 + k_{-2})S'S''\}^{-1}, \end{aligned} \qquad (37)$$

where C_t is the total carrier concentration.

When the initial rate is considered, i.e., $S'' = 0$, Eq. (37) simplifies to

$$J_{S(0)} = C_t \frac{k_1k_2k_3k_4S'}{(k_4 + k_{-4})(k_{-1}k_{-2} + k_{-1}k_3 + k_2k_3) + k_1[k_3(k_2 + k_4) + k_4(k_2 + k_{-2})]S'}. \qquad (38)$$

Equation (38) is formally identical to Eq. (2). Thus

$$J_{\max} = C_t \frac{k_2k_3k_4}{k_3(k_2 + k_4) + k_4(k_2 + k_{-2})}$$

and

$$K_m = \frac{(k_4 + k_{-4})(k_{-1}k_{-2} + k_{-1}k_3 + k_2k_3)}{k_1[k_3(k_2 + k_4) + k_4(k_2 + k_{-2})]}.$$

These equations are very general and can be simplified for special cases. See Kotyk and Janacek (1977) for several simplifications.

IV. Eyring Model for Membrane Permeation

With artificial membranes of the polymeric type, the membrane may be considered as a series of potential-energy barriers existing one behind the other, across which material must pass to cross the membrane. To do this, the permeating species must have sufficient energy. A typical potential energy diagram is shown in Fig. 1 for the system solution | membrane | solution. The figure indicates the pathway the species should follow to cross the membrane, and λs indicate the mean jump distance for the species. If C_i is the concentration of the substance (mole/cm^3) at the ith position in the membrane, then the amount of material in 1 cm^2 cross section and length λ_i (the distance between equilibrium maxima in Fig. 1) is $C_i\lambda_i$. The velocity of the forward movement is

$$v_f = k_i C_i \lambda_i, \qquad (39)$$

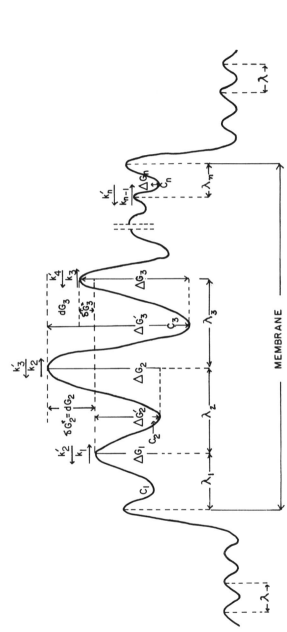

Fig. 1. General representation of the potential-energy diagram showing permeation in a membrane. ks are the rate constants. ΔGs are free-energy terms. λs indicate the jump distances.

where k_i is the specific rate constant for crossing the barrier i, and according to the theory of rate processes is given by

$$k_i = \dot{k}\frac{\kappa T}{h}\exp\left(-\frac{\Delta G_i}{RT}\right), \qquad (40)$$

where $\dot{k}$ is the transmission coefficient generally assumed to be unity. The velocity for backward movement over the ith barrier is

$$v_b = k_{-(i+1)}C_{i+1}\lambda_{i+1}. \qquad (41)$$

The net flux is

$$J = k_i C_i \lambda_i - k_{-(i+1)}C_{i+1}\lambda_{i+1}. \qquad (42)$$

Assumption that $k_i = k_{-(i+1)} = k$ and $\lambda_i = \lambda_{i+1} = \lambda$ gives

$$J = k\lambda(C_i - C_{i+1}) = -k\lambda^2\left(\frac{C_{i+1} - C_i}{\lambda}\right) = -k\lambda^2\frac{dC}{dx}, \qquad (43)$$

where $D = k\lambda^2$.

In the steady state, the flux is the same over every barrier. Therefore for the first barrier

$$J = k_1 C_1 \lambda_1 - k_{-2}C_2\lambda_2, \qquad (44)$$

for the second barrier

$$J = k_2 C_2 \lambda_2 - k_{-3}C_3\lambda_3, \qquad (45)$$

for the third barrier

$$J = k_3 C_3 \lambda_3 - k_{-4}C_4\lambda_4, \qquad (46)$$

for the $(n-1)$th barrier

$$J = k_{n-1}C_{n-1}\lambda_{n-1} - k_{-n}C_n\lambda_n. \qquad (47)$$

Eliminating $C_2\lambda_2$ from Eqs. (44) and (45) gives

$$J = k_1 C_1 \lambda_1 - (k_{-2}/k_2)(J + k_{-3}C_3\lambda_3). \qquad (48)$$

Similarly eliminating $C_3\lambda_3$ from Eqs. (46) and (48) gives

$$J = k_1 C_1 \lambda_1 - (k_{-2}/k_2)[J + (k_{-3}/k_3)(J + k_{-4}C_4\lambda_4)]. \qquad (49)$$

Completion of this process for all the $(n-1)$ steps gives on rearrangement

$$J = \frac{k_1 C_1 \lambda_1 - \dfrac{k_{-2}k_{-3}\cdots k_{-(n-1)}}{k_2 k_3 \cdots k_{n-1}}k_{-n}C_n\lambda_n}{1 + \dfrac{k_{-2}}{k_2} + \dfrac{k_{-2}k_{-3}}{k_2 k_3} + \cdots + \dfrac{k_{-2}\cdots k_{-(n-1)}}{k_2 \cdots k_{n-1}}}. \qquad (50)$$

IV. Eyring Model for Membrane Permeation

According to Eq. (40),

$$\frac{k_{-2}}{k_2} = \exp\left(\frac{\Delta G_2 - \Delta G'_2}{RT}\right) = \exp\left(\frac{\delta G_2}{RT}\right), \quad (51)$$

where δG_2 is the difference in energy of barriers 1 and 2;

$$\frac{k_{-2}k_{-3}}{k_2 k_3} = \exp\left[\frac{(\Delta G_2 - \Delta G'_2) + (\Delta G_3 - \Delta G'_3)}{RT}\right] = \exp\left(\frac{\delta G_3}{RT}\right), \quad (52)$$

where δG_3 is the algebraic sum of dG_2 and dG_3 and is the difference in the barrier heights of 1 and 3. Similarly, one has

$$\frac{k_{-2}k_{-3}\cdots k_{-(n-1)}}{k_2 k_3 \cdots k_{n-1}} = \exp\left(\frac{\delta G_{n-1}}{RT}\right) \quad (53)$$

and

$$\frac{k_{-2}k_{-3}\cdots k_{(n-1)}k_{-n}}{k_1 k_2 k_3 \cdots k_{n-1}} = \exp\left[\frac{(\Delta G_1 + \delta G_{n-1} - \Delta G'_n)}{RT}\right]. \quad (54)$$

But $\Delta G_1 + \delta G_{n-1} - \Delta G'_n = \Delta G_n$, the difference in free energy between positions 1 and n. Thus substituting from Eqs. (51)–(54) in Eq. (50) gives

$$J = \frac{\lambda_1 k_1 [C_1 - (\lambda_n/\lambda_1)\{\exp(\Delta G_n/RT)\} C_n]}{1 + \exp(\delta G_2/RT) + \exp(\delta G_3/RT) + \cdots + \exp(\delta G_{n-1}/RT)}. \quad (55)$$

When the barriers are all of equal height, the denominator of Eq. (55) becomes n, the total number of barriers, since $\delta G_i = 0$. Thus

$$J = \frac{\lambda_1^2 k_1}{\lambda_1 n}\left[C_1 - \frac{\lambda_n}{\lambda_1}\left[\exp\left(\frac{\Delta G_n}{RT}\right) C_n\right]\right]. \quad (56)$$

But $\lambda_1^2 k_1 = D_1$, the diffusion coefficient and $\lambda_1 n = d$, the thickness of the membrane. Thus Eq. (56) can be written

$$J = P(C_1 - C'_n),$$

where $P = D_1/d$ is the permeability.

The general scheme contained in Eq. (50) may be made applicable to a simple membrane transport system in which s number of jumps in solution and m number of jumps in membrane are made by a permeating species. The total number of jumps made by the species along the pathway is $2s + m + 2$ where $2s$ is for the two solutions and the term 2 is for the two membrane–solution interfaces. Thus Eq. (50) reduces to

$$J = \frac{k_s \lambda (C'_1 - C'_n)}{2s + 2k_s/k_{sm} + mk_s k_{ms}/k_{sm} k_m}. \quad (57)$$

As $J = P(C_1' - C_n')$, the permeability P is given by

$$P = \frac{k_s\lambda}{2s + 2k_s/k_{sm} + mk_sk_{ms}/k_{sm}k_m}. \tag{58}$$

Diffusion in solution is generally several times larger than the constants associated with the membrane, and so Eq. (58) becomes

$$P = \frac{\lambda}{2/k_{sm} + mk_{ms}/k_{sm}k_m} = \frac{\lambda k_{sm}k_m}{2k_m + mk_{ms}}. \tag{59}$$

Thus

$$\frac{1}{P} = \frac{2k_m}{\lambda k_{sm}k_m} + \frac{mk_{ms}}{\lambda k_{sm}k_m}. \tag{60}$$

As k_{sm}/k_{ms} may be equated to the partition coefficient β, $k\lambda^2 = D$ and $m\lambda = d$, Eq. (60) may be written

$$\frac{1}{P} = \frac{2\lambda}{D_{sm}} + \frac{d}{D_m\beta}. \tag{61}$$

If the rate determining step is the membrane, then

$$P = D_m\beta/d. \tag{62}$$

If the solution–membrane interface becomes rate limiting, Eq. (61) gives

$$P = D_{sm}/2\lambda, \tag{63}$$

and the permeability becomes independent of the distribution coefficient and the thickness of the membrane.

V. Eyring Model and Biological Membranes

The electrodiffusion model (Nernst–Planck flux equations) dominated the description of ion permeation through membranes. Because of the accumulation of refined data, this approach is proving less satisfactory. The trend now is to apply the Eyring model considering the membrane or membrane channel as a series of energy wells and barriers. The permeating species jump from well to well at a rate that decreases exponentially with the height of the energy barrier. Selectivity of the membrane or channel arises from the different well depths and barrier heights the species have to cross. These views applied to nerve cell membranes by several investigators are summarized in this section.

V. Eyring Model and Biological Membranes

As the biological membrane is thin, there may be only a few barriers to ion transport across it. Depolarization increases membrane permeability to ions. The temperature coefficient for Na and K ion permeation is low (~ 1.3), corresponding to activation energies of about 5 kcal/mol. This indicates that permeation occurs through pores or channels, and barriers to ion passage must exist in these water-filled pores.

A. Potential-Energy Barriers in the Membrane with Independence

Woodbury and colleagues (1970) have considered the experimental data on the giant axon of the squid from the standpoint of the rate theory assuming a model of four potential energy barriers (see Fig. 2). The flux according to Eq. (50) becomes

$$J = \frac{\lambda(k_0 k_1 k_2 k_3 C_0 - k_{-1} k_{-2} k_{-3} k_{-4} C_4)}{k_1 k_2 k_3 + k_{-1} k_2 k_3 + k_{-1} k_{-2} k_3 + k_{-1} k_{-2} k_{-3}}, \quad (64)$$

where C_0 and C_4 are the internal and external ion concentrations of equal ionic strength. The free energy difference between the two solutions is zero. When $C_0 = C_4$, $E = 0$, and $J = 0$. When $E \neq 0$, half transmembrane potential acts in the forward direction and adds to the activation energy and half subtracts in the backward direction. Thus

$$\Delta G_i = \Delta G_{i0} + \tfrac{1}{2} z E_i F$$

and

$$\Delta G_i' = \Delta G_{i0} - \tfrac{1}{2} z E_i F,$$

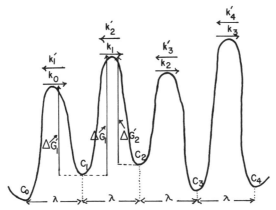

Fig. 2. Four-potential-energy barrier model of an ion channel in a membrane. C_0 and C_4 are the concentrations of ions in internal and external solutions. C_1, C_2, C_3 are concentrations in the potential energy minima. λ is jump distance and ΔGs are free-energy terms.

where E_i is the voltage across the ith barrier. The rate constants thus become

$$k_i = b_i \exp(zE_iF/2RT), \qquad (65)$$

$$k_{-i} = b_{-i}\exp(-zE_iF/2RT), \qquad (66)$$

where $b_i = \dot{k}_i(\kappa T/h)\exp(-\Delta G_{i0}/RT)$, $b_{-i} = \dot{k}_{-i}(\kappa T/h)\exp(-\Delta G'_{i0}/RT)$, and the b_i and b_{-i} are voltage independent.

When $E = 0$ and $C_0 = C_4$, J must be equal to zero. This leads to

$$b_0 b_1 b_2 b_3 = b_{-1} b_{-2} b_{-3} b_{-4}. \qquad (67)$$

The assumption that the transmission coefficients $\dot{k}_i$ and $\dot{k}_{-i}$ are equal gives the relation

$$\Delta G_{00} + \Delta G_{10} + \Delta G_{20} + \Delta G_{30} = \Delta G'_{10} + \Delta G'_{20} + \Delta G'_{30} + \Delta G'_{40}. \qquad (68)$$

If the field in the membrane E is constant, then $E_i = E/4$. Using this relation and Eqs. (65)–(68) in Eq. (64) gives on simplification

$$J = \frac{\lambda(C_0 e^{V/2} - C_4 e^{-V/2})}{e^{3V/8}/b_0 + e^{-3V/8}/b_{-4} + b_{-1}e^{V/8}/b_0 b_1 + b_3 e^{-V/8}/b_{-3}b_{-4}}, \qquad (69)$$

where $V = zEF/RT$.

The ion equilibrium potential E_e is given by $E_e = (RT/zF)\ln(C_4/C_0)$ or $C_4/C_0 = \exp(V_e)$, where $V_e = (zE_eF/RT)$. Other defining terms are $b_0/b_{-4} = \exp(3V_1/4)$ and $b_{-2}/b_2 = b_0 b_1 b_3/b_{-1}b_{-3}b_{-4} = \exp(V_2/4)$ [see Eq. (67)]. Substitution of these into Eq. (69) gives

$$J = \lambda C_0 \exp\left(\frac{V_e}{2}\right)\left[\exp\left(\frac{V}{2}\right)\exp\left(-\frac{V_e}{2}\right) - \exp\left(-\frac{V}{2}\right)\exp\left(\frac{V_e}{2}\right)\right]$$

$$\times \left\{\frac{\exp(-3V_1/8)}{b_{-4}}\left[\exp\left(\frac{3V}{8}\right)\exp\left(-\frac{3V_1}{8}\right) + \exp\left(-\frac{3V}{8}\right)\exp\left(\frac{3V_1}{8}\right)\right]\right.$$

$$\left. + \frac{b_3 \exp(-V_2/8)}{b_{-3}b_{-4}}\left[\exp\left(\frac{V}{8}\right)\exp\left(-\frac{V_2}{8}\right) + \exp\left(-\frac{V}{8}\right)\exp\left(\frac{V_2}{8}\right)\right]\right\}^{-1}.$$

Further simplification yields the relation

$$J = \frac{\lambda(C_0 C_4 b_0 b_{-4})^{1/2} \sinh[(V - V_e)/2]}{\cosh[3(V - V_1)/8] + (b_{-1}b_3/b_1 b_{-3})^{1/2}\cosh[(V - V_2)/8]}. \qquad (70)$$

This equation may be compared with the Goldman–Hodgkin–Katz equation [see Eq. (63) of Chapter 4], which is now written

$$J_j = P_j V \frac{C_{j(0)} - C_{j(i)}\exp V}{1 - \exp V}, \qquad (71)$$

where $V = (z_j EF/RT)$. If E_e is the equilibrium potential for the j ion, then
$$V_e = z_j E_e F/RT = \ln(C_{j(0)}/C_{j(1)}).$$
Equation (71) converted to hyperbolic functions becomes
$$J_j = \frac{P_j V [C_{j(0)} C_{j(i)}]^{1/2} \sinh[(V - V_e)/2]}{\sinh(V/2)}. \tag{72}$$

Both Eqs. (70) and (72) show that the current–voltage relations should be nonlinear. But in the case of Eq. (70) there are several variable parameters (V_1, V_2, and bs) that can be forced to take values to make the current–voltage relation nearly linear over a given voltage range, for example -150 mV $< (V - V_e) < 150$ mV. In the case of squid giant axon in normal Na saline, instantaneous Na current I_{Na} varies linearly with voltage V_m. According to Woodbury and colleagues (1970), for the conditions when $V_1 = V_2 = 52\ F/RT$ (mV) [where $52 = (RT/F)\ln(\text{Na})_o/(\text{Na})_i$, $(\text{Na})_o = 480$ and $(\text{Na})_i = 60\ \mu M/cm^3$ and $(\text{Na})_o/(\text{Na})_i = C_4/C_0$], $(b_0/b_{-4}) = \exp(1.56) = 4.75$, $b_{-2}/b_2 = \exp(2.08/4) = 1.68$, linearized Eq. (70) deviates from linearity by less than 2% if $(b_{-1}b_3)/b_1 b_{-3} = a^2 = 0.36$. Under these constraints, the barrier height differences can be evaluated. The linearized conditions are

$$\ln\frac{b_0}{b_{-4}} = \frac{\Delta G'_{40} - \Delta G_{00}}{RT} = \frac{3}{4}\ln\frac{C_4}{C_0},$$

$$\ln\frac{b_{-2}}{b_2} = \frac{\Delta G_{20} - \Delta G'_{20}}{RT} = \frac{1}{4}\ln\frac{C_4}{C_0}, \tag{73}$$

$$\ln\frac{b_{-1}b_3}{b_1 b_{-3}} = \frac{\Delta G_{10} - \Delta G'_{10} + \Delta G'_{30} - \Delta G_{30}}{RT} = \ln a^2.$$

According to Fig. 2, a barrier $i + 1$ is $\Delta G_{10} - \Delta G'_{10} = RT\delta_{i+1,i}$ higher than the barrier i, and so Eq. (73) becomes

$$\Delta G_{00} - \Delta G'_{40} = -\delta_{14} = \frac{3}{4}\ln\frac{C_4}{C_0} = 1.56,$$

$$\delta_{32} = \frac{1}{4}\ln\frac{C_4}{C_0} = 0.52,$$

$$\delta_{21} - \delta_{43} = 2\ln a = -1.02,$$

and Eq. (68) gives $\delta_{21} + \delta_{32} + \delta_{43} + \delta_{14} = 0$. The δs increase linearly (each increment roughly 0.5) and $\delta_{41} = -\delta_{14} = 1.56\ RT$ corresponds to 930 cal/mole. Thus to generate a linear current–voltage relation, barrier heights should increase progressively in going from the low-concentration to the high-concentration side.

The absolute height of the potential-energy barrier in the membrane corresponds to the variation of maximum I_{Na} with temperature and the Q_{10} of this is about 1.3. Hence the potential energy corresponds to

$$\Delta G = \frac{2.303R(T+10)(T)}{10} \log 1.3,$$

and at 20°C it is $\Delta G = 4700$ cal/mole. According to Eq. (70) the temperature dependence of the flux is contained in the term $(b_0 b_{-4})^{1/2}$.

The current–voltage relation (I_{Na} versus E) is linear only at normal $(Na)_o$. From the independence principle, one can find I'_{Na} at any other $(Na)_o$. Thus, from Eq. (70) it follows that

$$\frac{I'_{Na}}{I_{Na}} = \frac{J'_{Na}}{J_{Na}} = \left[\frac{(Na)'_o}{(Na)_o}\right]^{1/2} \frac{\sinh[F(E - E'_{Na})/2RT]}{\sinh[F(E - E_{Na})/2RT]}.$$

This equation is slightly different from that given by Hodgkin and Huxley [see Eq. (119) of Chapter 4].

In the preceding considerations the potential-energy wells were ignored since in the steady state they play no role in determining the fluxes.

B. Potential-Energy Barrier in the Membrane without Independence

Here no linearity in the current–voltage relation exists. If an ion binding site S exists inside the channel, an ion X can reach the site (saturable) from either inside or outside the membrane. The steps involved in the permeation of X are

$$(X)_{outside} + S \underset{k_{-1}}{\overset{k_1}{\rightleftarrows}} XS \underset{k_{-2}}{\overset{k_2}{\rightleftarrows}} (X)_{inside} + S \quad (74)$$

This is similar to reaction represented by Eq. (3), and the net current in the outward direction is given by

$$I_X = zF \frac{k_{-1}k_{-2}(X)_i - k_1 k_2 (X)_o}{k_{-1} + k_2 + k_{-2}(X)_i + k_1(X)_o}. \quad (75)$$

This for internal ions only becomes

$$I_X(E) = \frac{I_{max}(E)}{1 + K_m(E)/(X)_i},$$

where the voltage-dependent rate coefficients (see Fig. 3) determine $I_{max}(E)$. Thus

$$I_{max}(E) = zFk_{-1} = zFb_{-1} \exp(z\delta EF/2RT) \quad (76)$$

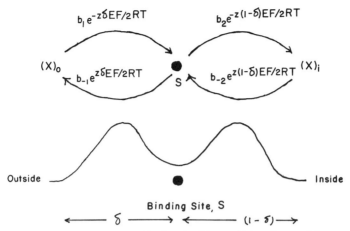

Fig. 3. A two-barrier model with one binding site S (saturable) at distance δ from outside the membrane. δE is the field sensed by the site. bs are the voltage-independent terms.

and

$$K_m(E) = (k_{-1} + k_2)/k_{-2},$$

$$= \frac{b_{-1}\exp(z\delta EF/2RT) + b_2\exp[-z(1-\delta)EF/2RT]}{b_{-2}\exp[z(1-\delta)EF/2RT]} \quad (77)$$

If a permeant ion A is inside and a permeant ion B is outside, one can act as a competitive inhibitor of the other (i.e., without independence), and thus the currents I_A and I_B can be written following Eq. (18) as

$$I_A = \frac{I_{A(max)}(A)}{K_A[1 + (B)/K_B] + (A)}, \quad (78)$$

$$I_B = \frac{-I_{B(max)}(B)}{K_B[1 + (A)/K_A] + (B)}. \quad (79)$$

The total current is given by

$$I_A + I_B = \frac{I_{A(max)}(A)/K_A - I_{B(max)}(B)/K_B}{1 + (B)/K_B + (A)/K_A}. \quad (80)$$

At a certain potential the current may become zero. This will happen when

$$I_{A(max)}(A)/K_A = I_{B(max)}(B)/K_B. \quad (81)$$

Now if (A) = (B), the voltage-independent permeability ratio becomes

$$\frac{P_A}{P_B} = \frac{I_{A(max)}/K_A}{I_{B(max)}/K_B}. \quad (82)$$

In terms of voltage-independent rate coefficients (i.e., the b_i), Eq. (82) as shown by Hille becomes

$$\frac{P_A}{P_B} = \frac{b_{1A}}{b_{1B}} = \frac{b_{-2A}}{b_{-2B}}. \tag{83}$$

The permeability ratio becomes dependent on the ratio of the rates of association of the ions with the site in the membrane.

This kinetic approach of treating ions jumping between vacant sites in a channel has been used in the construction of several models of transport for sodium and potassium ions in nerve and muscle membranes. A kinetic model of four energy barriers and three sites in the three wells and one ion jumping from an empty site to another empty site (one-ion single-file channel) has been described by Hille (1975b). Similarly Hille and Schwarz (1978) have presented a model of multi-ion single-file channel for the potassium channel. This kind of informative kinetic work involves several manipulations, computer time, and other help from computer personnel to carry out. The original papers referred to above must be consulted for details.

The blocking effect (or inhibition) of a channel by an ion can be described by the mechanism of Eq. (74) in which X is the blocking ion, which can reach the site S in the channel from either inside or outside of the membrane.

In the steady state the probability P of the site not being occupied is constant and is given by

$$P = \frac{k_{-1} + k_2}{k_1 + k_2 + k_{-1} + k_{-2}}. \tag{84}$$

The rate constants are shown in Fig. 3. The probability that a channel is not blocked becomes

$$P = \frac{(b_{-1}/b_1)\exp(z\delta EF/RT) + (b_2/b_1)\exp[z(2\delta - 1)FR/2RT]}{(X)_0 + (X)_i(b_{-2}/b_1)\exp(zEF/2RT) + (b_{-1}/b_1)\exp(z\delta EF/RT)} \tag{85}$$
$$+ (b_2/b_1)\exp[z(2\delta - 1)EF/2RT]$$

In the steady state when $E = 0$ [i.e., $I_X = 0$ in Eq. (75)],

$$b_{-2}/b_2 = b_1/b_{-1}. \tag{86}$$

Equation (86) may be used to eliminate any one of the b_i from Eq. (85), which may be used to explain the depression of I_{Na} by hydrogen ions. If I_{Na} is measured at pH 7 and at pH 5, then

$$[I_{Na(pH\ 5)}/I_{Na(pH\ 7)}] = [P_{(pH\ 5)}/P_{(pH\ 7)}], \tag{87}$$

where P is the fraction of channels conducting. Substituting from Eq. (85) gives

$$\frac{I_{\text{Na(pH 5)}}}{I_{\text{Na(pH 7)}}} = \frac{10^{-7} + Q(E)}{10^{-5} + Q(E)}, \tag{88}$$

where

$$Q(E) = (X)_i \frac{b_{-2}}{b_1} \exp\left(\frac{zEF}{2RT}\right) + \frac{b_{-1}}{b_1} \exp\left(\frac{z\delta EF}{RT}\right) + \frac{b_2}{b_1} \exp\left(\frac{z(2\delta - 1)EF}{2RT}\right),$$

b_{-1}/b_1, b_2/b_1, and δ are the three unknowns since (b_{-2}/b_1) can be evaluated by using Eq. (86). Woodhull (1973) measured I_{Na} at pH 5 and 7 as a function of voltage and tried fitting the experimental curve by using Eq. (88) and several values for the three unknowns. Only when b_2/b_1 and b_{-2}/b_1 were zero was a good fit obtained. This means that when (X) is applied from the outside, b_2 and b_{-2} can be ignored. For this condition, Eq. (85) simplifies to

$$P = \frac{Q(E)}{(X)_o + Q(E)}, \tag{89}$$

where

$$Q(E) = (b_{-1}/b_1) \exp(z\delta EF/RT). \tag{90}$$

A plot of $Q(E)$ evaluated from experimental data by using Eq. (88) against voltage on semilogarithmic paper gave a straight line conforming to Eq. (90). Such plots gave values of 0.26 for δ and 3.9 μM for b_{-1}/b_1. This means that the site with a pK_a of 5.4 (3.9×10^{-6} M) is situated one quarter of the way across the thickness of the membrane from the outside.

C. Kinetic View of Selectivity

The selectivity coefficient as defined in Eqs. (9) and (24) of Chapter 4 is an equilibrium selectivity and depends on the selective binding of a cation to the anionic site in the membrane. The high anionic field strength determined the selectivity. This may be so for a carrier but not for a channel. Equation (83) developed for a channel shows that the rate constant of association determines the selective permeation through the channel. Armstrong (1975) has explored this view in depth on a kinetic basis for the potassium channel. His analysis is as follows.

In the scheme of Armstrong (1975) the site in the channel is empty and there is no displacement of one ion by another ion (no "knock-on" mechanism). Na^+ and K^+ ions compete for the same site S. The reactions are

$$K^+ + S \underset{k_{-1}}{\overset{k_1}{\rightleftharpoons}} KS, \tag{91}$$

$$\text{Na}^+ + \text{S} \underset{k_{-2}}{\overset{k_2}{\rightleftarrows}} \text{NaS} \tag{92}$$

$$k_1(\text{K})(\text{S}) = k_{-1}(\text{KS}), \tag{93}$$

$$k_2(\text{Na})(\text{S}) = k_{-2}(\text{NaS}). \tag{94}$$

The ratio of K^+ to Na^+ occupied sites is given by

$$\frac{(\text{KS})}{(\text{NaS})} = \frac{(\text{K}^+)}{(\text{Na}^+)} \frac{k_1}{k_2} \frac{k_{-2}}{k_{-1}}. \tag{95}$$

There are two limiting cases. In case (i) when $k_1 = k_2$, the channel becomes selective to K^+ if $k_{-2} > k_{-1}$. Here selectivity arises from selective binding and is true of carriers such as nonactin, whose complex with K^+ ion dissociates more slowly than its Na^+ complex. In this case of selective binding if $k_1 = k_2$ and $(\text{K}^+) = (\text{Na}^+)$ on both sides of the membrane containing channels, then

$$(\text{KS})/(\text{NaS}) = k_{-2}/k_{-1}. \tag{96}$$

The flux of K^+ through such a membrane from side (') to side ('') would be

$$J_K = (\text{KS})k_{-1}\delta_t \tag{97}$$

and that of Na^+ would be

$$J_{\text{Na}} = (\text{NaS})k_{-2}\delta_t, \tag{98}$$

where δ_t is the fraction of time that either ion in the site is released to the side ('') after dissociation from the site.

If binding is the primary factor, then as $k_1 = k_2$, the time constant τ would be the same for both K^+ and Na^+. Substituting for (KS) from Eq. (96) into Eq. (97) shows that $J_K = J_{\text{Na}}$. This happens in view of the fact that slower dissociation of KS just compensates for the greater occupancy of the site by K^+ ions. So selective binding is not a viable mechanism for a channel.

In case (ii) when $k_{-1} = k_{-2}$, the channel becomes selective to K^+ if $k_1 > k_2$ in Eq. (95). Here selectivity arises from selective exclusion. When $(\text{Na}^+) = (\text{K}^+)$, Eq. (95) becomes $(\text{KS})/(\text{NaS}) = k_1/k_2$, and Eqs. (97) and (98) now give

$$\frac{J_K}{J_{\text{Na}}} = \frac{(\text{KS})k_{-1}\delta_t}{(\text{NaS})k_{-2}\delta_t} = \frac{k_1}{k_2}.$$

Thus the ratio of the association rate constants determine the selectivity or the relative permeability of one over the other.

In cases where the "knock-on" mechanism operates (i.e., site in the channel is always occupied by an ion that leaves only when displaced by another ion), the reactions are

$$K' + KS \underset{k_{-1}}{\overset{k_1}{\rightleftharpoons}} KS + K'' \tag{99}$$

$$K' + NaS \underset{k_{-2}}{\overset{k_2}{\rightleftharpoons}} KS + Na'' \tag{100}$$

$$Na' + KS \underset{k_{-3}}{\overset{k_3}{\rightleftharpoons}} NaS + K'' \tag{101}$$

$$Na' + NaS \underset{k_{-4}}{\overset{k_4}{\rightleftharpoons}} NaS + Na'' \tag{102}$$

where (') and ('') indicate inside and outside of the membrane and KS and NaS represent K^+ and Na^+ occupied sites in the channel. In the steady state

$$\frac{d(KS)}{dt} = k_2(NaS)(K)' - k_{-2}(Na)''(KS) - k_3(Na)'(KS) + k_{-3}(NaS)(K)'' = 0, \tag{103}$$

$$\frac{d(NaS)}{dt} = -k_2(NaS)(K)' + k_{-2}(Na)''(KS) + k_3(Na)'(KS) - k_{-3}(NaS)(K)'' = 0. \tag{104}$$

When $(K)' = (Na)'$ and $(Na)'' = (K)'' = 0$, both Eqs. (103) and (104) give

$$(KS)/(NaS) = k_2/k_3. \tag{105}$$

The unidirectional fluxes of K^+ and Na^+ from side (') to side ('') are given by

$$J_K = k_1(K)'(KS) + k_3(Na)'(KS),$$

$$J_{Na} = k_2(K)'(NaS) + k_4(Na)'(NaS).$$

Thus

$$\frac{J_K}{J_{Na}} = \frac{(KS)}{(NaS)} \left[\frac{k_1 + k_3}{k_2 + k_4} \right] = \frac{k_2}{k_3} \left[\frac{k_1 + k_3}{k_2 + k_4} \right]. \tag{106}$$

If the rates of reaction of K^+ and Na^+ with the K-occupied site are equal, then $k_1 = k_3$ and similarly if the same occurs with the Na-occupied site, then $k_2 = k_4$.

In the case of an ion reacting with an occupied site, two factors seem to operate: the rate of association and the degree of binding. Armstrong (1975) has called them selective access factor α and selective binding factor β, respectively. Operation of the access factor gives $k_1 = \alpha k_3$ and $k_2 = \alpha k_4$. If K^+ or Na^+ ion can displace Na^+ or K^+ from a Na^+ or K^+ occupied site as easily as K^+ or Na^+ from a K^+ or Na^+ occupied site, then $k_1 = k_2$ and $k_3 = k_4$. If K^+ ion is strongly bound to the site by a factor β, then $k_2 = \beta k_1$ and $k_4 = \beta k_3$. If K^+ ion reacts readily and binds strongly with the site, then $k_1 = \alpha k_3$, $k_2 = \alpha k_4$, $k_2 = \beta k_1$ and $k_4 = \beta k_3$. Of these relations, substituting those that are appropriate to eliminate all k_i from Eq. (106) gives

$$\frac{J_K}{J_{Na}} = \frac{(\alpha k_4)}{(k_4/\beta)} \frac{k_1 + (k_4/\beta)}{\beta k_1 + k_4} = \alpha.$$

This means that selectivity is governed by the ratio of the association rate constants and not on dissociation rate constants. According to this concept, selectivity arises by the selective exclusion of the less permeant ion. In this way, despite strong binding, which decreases the flux of the selective ion, the selectivity of the channel to a particular ion is asserted.

VI. Model for Lipid-Soluble Ions

Kinetic treatment of permeation through membranes has been applied in recent years to thin lipid membranes. Several interesting phenomena have been noted, and a variety of sophisticated techniques, both experimental and theoretical to analyze them, have been developed. The several important equations are due, among others, to three main groups, led by Läuger (1981) in West Germany, Haydon (1972; also Hladky, 1979) in England, and Eisenman (Eisenman et al., 1978) in the United States.

Thin lipid bilayer membranes have a very high electrical resistance. They are not permeable to small ions, but are permeable to lipophilic ions, such as tetraphenyl borate, tetraphenyl phosphonium ions, dipicrylamine, which have charges shielded by hydrophillic groups. The energy of interaction of these ions with the lipid membrane contains four terms: (1) U_e, the electrostatic interaction of the ion with the membrane; (2) U_h, the hydrophobic interaction of the organic ion with the membrane; (3) the electrostatic interaction due to interfacial potential ΔE at the membrane–solution interface; and (4) U_s, other short-range interactions such as hydrogen bonding between ion and molecules in the membrane. The potential-energy barrier for ion movement through the lipid membrane is thus given by

$$\Delta U_P = U_e + U_h + zF\Delta E + U_s. \tag{107}$$

Of these, U_e and U_h are the important ones. The shapes of these potential energy functions are shown in Fig. 4.

Electrostatic interaction energy U_e will be independent of position if the membrane is considered as a macro phase, and the potential energy of the

VI. Model for Lipid-Soluble Ions

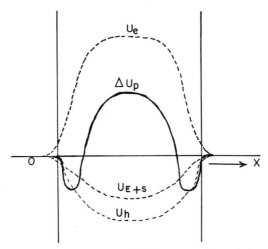

Fig. 4. Total potential energy ΔU_P in a lipid bilayer membrane (schematic). U_{E+S} is the combination of energy due to interfacial potential and other short-range interactions for a negative ion. (After Läuger et al., 1981.)

ion in the membrane may be equated to Born energy for an ion in media of different dielectric constants. Thus

$$U_B = \frac{z^2 e^2}{2r}\left(\frac{1}{\varepsilon_m} - \frac{1}{\varepsilon}\right), \qquad (108)$$

where z is the valence (-1), e the electronic charge, r the radius of the ion considered equivalent to a sphere, ε_m the dielectric constant of the membrane, and ε that of water. The partition coefficient is given by

$$\beta = \bar{C}/C = \exp(-U_B). \qquad (109)$$

Equation (108) is not realistic since the membrane, 50-Å thick, cannot be equated to a macro phase. An ion in an aqueous phase ($\varepsilon \approx 80$) entering the membrane ($\varepsilon \approx 2$) will be subject to image forces. Consequently an ion in the aqueous phase will be repelled from the membrane boundary. On the other hand, the ion in the membrane will be attracted toward the boundary. This force is given by (aqueous phase perfect conductor and ion nonpolarizable)

$$\mathscr{F}_i(x) = -\frac{e^2}{\varepsilon_m (2x)^2}. \qquad (110)$$

As an ion in the membrane "looks out" at two phase boundaries, an infinite set of image charges has to be introduced (see Neumcke and Läuger, 1969) to satisfy the boundary conditions of the electric potential. So the total force $\mathscr{F}(x)$ on the ion is given according to Andersen and Fuchs (1975) by

$$\mathscr{F}(x) = \frac{-e^2}{4\varepsilon_m} \left\{ \frac{1}{x^2} + \frac{1}{d^2} \sum_{n=1}^{\infty} \frac{1}{(n + x/d)^2} - \frac{1}{(n - x/d)^2} \right\}, \quad (111)$$

where d is the thickness of the membrane. The potential energy U_e in the interior of the membrane ($r < x < d - r$) is

$$U_e = U_B - \int_r^x \mathscr{F}(x) \, dx$$

$$= U_B + \frac{e^2}{4\varepsilon_m d} \left[\frac{d}{r} - \frac{d}{x} + \frac{x - r}{d} \sum_{n=1}^{\infty} \frac{1}{(n + x/d)(n + r/d)} \right.$$

$$\left. - \frac{1}{(n - x/d)(n - r/d)} \right]. \quad (112)$$

Although U_B is given by Eq. (108), Andersen and Fuchs (1975) modified it as

$$U_B = \frac{e^2}{2r} \left(\frac{1}{2\varepsilon_m} - \frac{1}{\varepsilon} \right). \quad (113)$$

The results of numerical evaluation of U_e using Eqs. (112) and (113) for $r < x < d - r$ and a third-degree polynomial for $-r < x < r$ and $d - r < x < d + r$ are shown in Fig. 5. Also in Fig. 6 is given a plot of $\exp(U_e/\kappa T)$ to show that although U_e is fairly flat in the middle of the membrane, $\exp(U_e/\kappa T)$ has a clear peak in the middle.

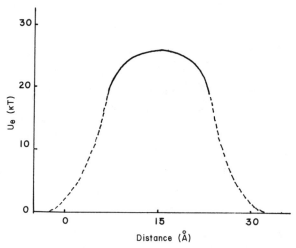

Fig. 5. Image force potential energy barrier for an ion of radius 4.2 Å in a membrane of thickness 30 Å and dielectric constant 2.0. Energy is in units of κT. (After Anderson and Fuchs, 1975.)

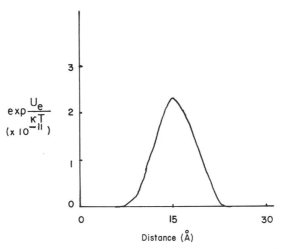

Fig. 6. Image force potential energy barrier same as in Fig. 5 but given as a plot of $\exp(U_e/\kappa T)$ versus distance. (After Andersen and Fuchs, 1975.)

Entry of large organic ions into the membrane is favored by strong hydrophobic interactions. The potential energy of the discharged ion U_h is determined by short-range interactions. The change in U_h at the interface is very steep. It arises from the abrupt changes in the solvation state of the ion. When U_e and U_h are added (see Fig. 4) to give the total potential energy, the net function U_p (other short-range interactions being negligible) shows two minima at the interfaces separated by a broad barrier at the middle of the membrane. So the hydrophobic ions reside at the two minima near the membrane–solution interfaces.

Transport of lipophilic ions across a bilayer membrane can occur by surmounting the energy barrier in the middle of the membrane. If the barrier is broad, it cannot be crossed by the ion in a single jump. Consequently the barrier is assumed to be made up of a number of smaller energy barriers, so that the ion can hop across. The overall transport of the ion is assumed to occur as shown in Fig. 7. N' and N'' (moles/cm^2) are the concentrations of X in the potential minima and $N = N' = N''$ at equilibrium (i.e., in the absence of voltage) and $\beta = N/C$, is the partition coefficient and has the dimensions of length, which is equal to the thickness of a solution layer containing the same quantity of ions as half of the membrane.

The free energy of adsorption is ignored since the surface potential is considered to be low. As the permeable ion is located at the membrane surfaces, there is little space charge in the membrane and the electric field becomes independent of position. If E is the voltage drop across the central barrier,

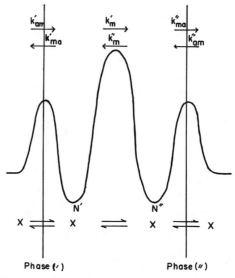

Fig. 7. Transport of lipophilic ion across a lipid bilayer membrane.

the effective voltage is δE (δ has a value between 0 and 1). Then

$$k'_m = k_m \exp(z\delta EF/2RT), \quad (114)$$

$$k''_m = k_m \exp(-z\delta EF/2RT), \quad (115)$$

where k_m is the rate constant in the absence of externally applied voltage E. It is assumed that there is no voltage drop across the interfaces.

The rate of change of N' and N'' is governed by the absorption rate at solution–membrane interface ($'$), desorption rate at membrane–solution interface ($''$) and translocation rate across the central barrier. Thus

$$\frac{dN'}{dt} = k'_{am}C - k'_{ma}N' - k'_m N' + k''_m N'', \quad (116)$$

$$\frac{dN''}{dt} = k''_{am}C - k''_{ma}N'' - k''_m N'' + k'_m N'. \quad (117)$$

In the steady state $dN'/dt = dN''/dt = 0$ and $k'_{am} = \beta k'_{ma}$ and $k''_{am} = \beta k''_{ma}$. Thus

$$\beta = \frac{k'_{am}}{k'_{ma}} = \frac{k''_{am}}{k''_{ma}} = \frac{k_{am}}{k_{ma}}.$$

VI. Model for Lipid-Soluble Ions

Also N' and N'' shift toward steady-state values N'_{ss} and N''_{ss}. Under these conditions, solving Eqs. (116) and (117) for N'_{ss} and N''_{ss} gives

$$N'_{ss} = N \frac{k_{ma} + 2k''_m}{k_{ma} + k'_m + k''_m} \tag{118}$$

and

$$N''_{ss} = N \frac{k_{ma} + 2k'_m}{k_{ma} + k'_m + k''_m}. \tag{119}$$

Equation (116) can be written

$$\frac{dN'}{dt} = k_{ma}N - k_{ma}N' - k'_m N' + k''_m(2N - N')$$

$$= (k_{ma} + 2k''_m)N - N'(k_{ma} + k'_m + k''_m). \tag{120}$$

Substituting for N from Eq. (118) gives on rearrangement

$$\frac{dN'}{N'_{ss} - N'} = dt(k_{ma} + k'_m + k''_m),$$

which on integration becomes

$$[\ln(N'_{ss} - N')]_{N'=N}^{N'=N'(t)} = -(k_{ma} + k'_m + k''_m)[t]_{t=0}^{t}. \tag{121}$$

Simplification of Eq. (121) gives

$$N'(t) = N'_{ss} + (N - N'_{ss})\exp(-t/\tau), \tag{122}$$

where $\tau = 1/(k_{ma} + k'_m + k''_m)$. Similarly $N''(t)$ is given by

$$N''(t) = N''_{ss} + (N - N''_{ss})\exp(-t/\tau). \tag{123}$$

Electrical current density at any given time is given by

$$I(t) = zF(k'_m N' - k''_m N''). \tag{124}$$

Equations (114), (115), (118), (119), (122), and (123) substituted into Eq. (124) give on simplification

$$I(t) = 2zFk_m\beta C \sinh\left(\frac{z\delta EF}{2RT}\right) \frac{k_{ma} + 2k_m \cosh(z\delta EF/2RT)\exp(-t/\tau)}{k_{ma} + 2k_m \cosh(z\delta EF/2RT)}. \tag{125}$$

For small voltages, i.e., $z\delta EF/2RT \ll 1$, Eq. (125) becomes

$$I(t) = z^2 F k_m \beta C \frac{\delta EF}{RT} \frac{k_{ma} + 2k_m \exp(-t/\tau)}{k_{ma} + 2k_m}, \tag{126}$$

and now $\tau = 1/(k_{ma} + 2k_m)$.

At $t = 0$, $I(t) = I_0$, and at $t = \infty$, $I(t) = I_\infty$. Equation (126) gives

$$I_0 = \frac{z^2 F^2}{RT} k_m \beta C \delta E \tag{127}$$

and

$$I_\infty = \frac{z^2 F^2}{RT} k_m \beta C \delta E \frac{k_{ma}}{k_{ma} + 2k_m}. \tag{128}$$

Equation (126) can now be written

$$I(t) = \frac{z^2 F^2}{RT} k_m \beta C \delta E \frac{k_{ma}}{k_{ma} + 2k_m}$$

$$+ \frac{z^2 F^2}{RT} k_m \beta C \delta E \left[1 - \frac{k_{ma}}{k_{ma} + 2k_m} \right] \exp(-t/\tau). \tag{129}$$

Substituting Eqs. (127) and (128) into Eq. (129) gives

$$I(t) = I_\infty + (I_0 - I_\infty) \exp(-t/\tau). \tag{130}$$

The preceding equations for current may now be expressed in terms of membrane conductance G_0, which is given by

$$G_0 = [I(t)/E]_{E \to 0}.$$

Thus Eqs. (127) and (128) become

$$(G_0)_{t \to 0} = G_{00} = \frac{z^2 F^2}{RT} k_m \beta C, \tag{131}$$

$$(G_0)_{t \to 0} = G_{0\infty} = \frac{z^2 F^2}{RT} \beta C \frac{k_m k_{ma}}{k_{ma} + 2k_m}. \tag{132}$$

Voltage dependence of initial conductance $(G)_{t=0}$ can be obtained from Eq. (125). Thus

$$(G)_{t=0} = \left(\frac{I}{\delta E} \right)_{t=0} = 2zF\beta C \frac{k_m}{\delta E} \sinh\left(\frac{z\delta EF}{2RT} \right).$$

Substituting from Eq. (131) gives

$$(G)_{t=0} = (G_0)_{t=0} \frac{2RT}{z\delta EF} \sinh\left(\frac{z\delta EF}{2RT} \right). \tag{133}$$

The kinetic parameters of lipophilic ions taken up by the planar lipid bilayer membranes can be determined by the use of the voltage jump method. Sudden application of voltage to the lipid membrane doped with a lipophilic ion gives rise to a relaxation current, which declines exponentially with a single time constant. Ideally the current should reach a stationary value

$I(\infty)$ for $t \gg \tau$. The current decreases slowly because of the concentration polarization in the aqueous phases.

Equation (128) in terms of Eq. (127) becomes

$$I_\infty = \frac{k_{ma}}{k_{ma} + 2k_m} I_0. \tag{134}$$

If the energy barrier in the middle of the membrane is low, i.e., if $k_m \gg k_{ma}$, then $I_0 \gg I_\infty$ or $G_{00} \gg G_{0\infty}$. This has been found to be true experimentally. Further, the initial exponential decay is given by

$$I(t) = I_0 \exp(-t/\tau) = I_0 \exp(-2k_m t) \tag{135}$$

and

$$I_0 = (F^2/RT) k_m \beta CE \tag{136}$$

for $z = 1$ and $\delta \approx 1$.

A semilogarithmic plot of $I(t)$ against t for small values of t should give a straight line. From the slope, a value for k_m can be derived. The intercept on the current axis gives a value for I_0 and thus a value for β can be derived with the help of Eq. (136). Evaluation of k_{ma} is not possible without an evaluation of the parameters controlling concentration polarization. However, upper and lower limits for the values of k_{ma} have been given (see Table I).

With several of the lipophilic ions (e.g., picrate, tetraphenylborate, dipicrylamine), membrane conductance increases with increase in concentration C, reaches a maximum, and then decreases. This is explained on the basis that the number of ions adsorbed to the interface is limited and that an ion jump across the membrane is possible only if there is a free site on the other surface for the ion to hop into. The probability of such a process is

TABLE I

Dioleoyllecithin Bilayer Membrane Parameters[a]

Ion	Dipicryl amine	Tetraphenyl borate
τ_0 (s)	1.3×10^{-3}	55×10^{-3}
λ_{00}/C (Ω^{-1} cm^{-2} M^{-1})	3×10^4	1×10^3
$C_{0,max}$ (M)	3×10^{-7}	3×10^{-7}
$\lambda_{00,max}$ (Ω^{-1} cm^{-2})	3×10^{-3}	7×10^{-5}
k_m (s^{-1})	380	9
β (cm)	2×10^{-2}	3×10^{-2}
N_s (cm^{-2})	4×10^{12}	5×10^{12}

[a] From Ketterer et al. (1971).

given by $[1 - N''/N_s]$, where N_s is the maximum number of ions which may be adsorbed per unit area. On the basis of considerations already presented, the following equations have been derived by Ketterer et al. (1971):

$$I(t) = zF\left[k'_m N'\left(1 - \frac{N''}{N_s}\right) - k''_m N''\left(1 - \frac{N'}{N_s}\right)\right], \quad (137)$$

$$N = \frac{\beta C}{1 + \beta C/N_s}, \quad (138)$$

$$N' = N\left[1 - \frac{k_m zEF/2RT}{k_{ma}(1 + \beta C/N_s)^2 + 2k_m(1 + \beta C/N_s)}\right], \quad (139)$$

$$N'' = N\left[1 + \frac{k_m zEF/2RT}{k_{ma}(1 + \beta C/N_s)^2 + 2k_m(1 + \beta C/N_s)}\right], \quad (140)$$

$$\lambda_{00} = \frac{z^2 F^2}{RT} \frac{\beta C k_m}{(1 + \beta C/N_s)^2}, \quad (141)$$

$$\lambda_{0\infty} = \frac{z^2 F^2}{RT} \beta C \frac{k_{ma} k_m}{k_{ma}(1 + \beta C/N_s)^2 + 2k_m(1 + \beta C/N_s)}. \quad (142)$$

The maximum of λ_{00} is attained at the concentration

$$C_{0,\max} = N_s/\beta. \quad (143)$$

From this equation N_s may be evaluated.

The maximum conductance is given by

$$\lambda_{00,\max} = (z^2 F^2/RT) k_m N_s. \quad (144)$$

Similarly,

$$C_{\infty,\max} = (N_s/\beta)\sqrt{1 + 2k_m/k_{ma}}, \quad (145)$$

$$\lambda_{0,\infty,\max} = \frac{z^2 F^2}{2RT} \frac{k_{ma} k_m N_s \sqrt{1 + 2k_m/k_{ma}}}{(k_{ma} + k_m)(1 + \sqrt{1 + 2k_m/k_{ma}}) + k_m}. \quad (146)$$

The several parameters evaluated for the different membrane systems are given in Table I.

Bruner (1975) studied the effect of high field, i.e., $z\delta EF/2RT \gg 1$, on charge transfer from one interface to the other (deplete the adsorbed charge and transfer it to the other interface) and found the density of surface charge transfer to be equal to the integral over time of the observed transient current density. The high-amplitude relaxation currents had a nonexponential decay. Charge density transported increased with increase in the intensity of the applied field and attained saturation. The saturating charge density observed in the membrane system—dierucoyl lecithin + 10^{-6} M dipicrylamine—was 1.1×10^{13} cm^{-2} (see also Table I).

VI. Model for Lipid-Soluble Ions

From the rate constant k_m, the energy of activation of the ion for overcoming the central energy barrier in the membrane may be calculated from the relation

$$k_m = v_m \exp(-\Delta G_m/RT). \quad (147)$$

The frequency factor is on of the order of 6×10^{12} sec^{-1}. Using the values of k_m given in Table I gives values of 14 (dipicrylamine) and 16.2 (tetraphenylborate) kcal/mol for ΔG_m. These values are roughly equal to the dielectric energy of the ion in the middle of the membrane. Image force calculations for the tetraphenylborate ion of spherical shape (radius 4.2 Å) gave a value of 17.4 kcal/mol ($\varepsilon_m = 2$ and $d = 50$ Å for the membrane dielectric constant and its thickness, respectively). Similarly, temperature-dependent studies of $\tau\, (= 1/2k_m)$ by Bruner (1975) gave a value of 14 kcal/mol (10^{-7}–10^{-8} M dipicrylamine) for activation enthalpy.

ΔG_m in Eq. (147) may be separated into enthalpy and entropy components. Thus

$$k_m = v_m \exp(\Delta S_m/R) \exp(-\Delta H_m/RT). \quad (148)$$

As the values of ΔG_m are close to experimentally evaluated enthalpies the activation entropy ΔS_m for ion translocation is relatively small.

Potential energy barriers (their heights and shapes) to ion transport have been studied by several investigators. Andersen and Fuchs (1975) considered several models, and their approach based on electrodiffusion regime is outlined.

The number of ions moved through the membrane during a current transient attains an upper limit N_s with increase in potential, and can be written

$$[N_s + \Delta N(E)] - [N_s - \Delta N(E)] = (N'' - N')/2 = \Delta N \quad (149)$$

and

$$N'' + N' = 2N_s. \quad (150)$$

Assuming Boltzmann distribution of ions at the interfaces, one can write

$$C' \exp(-z\phi'F/RT) = C'' \exp(-z\phi''F/RT), \quad (151)$$

$$N = \beta C, \quad (152)$$

and

$$\phi'' - \phi' = \delta E, \quad (153)$$

where ϕs are the electrostatic potentials at the minima due to the applied field. Combining Eqs. (149)–(153) gives on simplification for a negative ion ($z = -1$)

$$(N_s - \Delta N)/(N_s + \Delta N) = \exp(\delta EF/RT). \quad (154)$$

Rearrangement of Eq. (154) gives

$$\Delta N = N_s \tanh(\delta EF/2RT). \tag{155}$$

Inclusion of the potential-energy barrier term into the Nernst–Planck flux equation when the current transients are measured gives for the flux of a negative ion

$$J = -u\left[RT\frac{dC}{dx} - C(x,t)F\frac{d\phi}{dx} + C(x,t)\frac{dU}{dx}\right]. \tag{156}$$

Integration of Eq. (156) gives

$$J = \frac{-uRT[C'' \exp(-\delta EF/2RT) - C' \exp(\delta EF/2RT)]}{\exp(EF/2RT)\int_r^{d-r} \exp\{[U_1(x) - F\phi(x)]/RT\}\,dx}, \tag{157}$$

where r is the distance of energy minima from the interfaces and $U_1(x) = U(x) - U(r)$.

Membrane conductance G_m and barrier conductance G_b are defined according to Andersen and Fuchs (1975) as

$$G_m(E,t) = I(E,t)/[E - E_m(t)], \tag{158}$$

where

$$E_m = (RT/F)\ln[C''(t)/C'(t)], \tag{159}$$

and

$$G_b(E,t) = I(E,t)/[\delta E - E_m(t)]. \tag{160}$$

Substituting Eq. (157) into Eq. (160) gives

$$G_b(E,t)$$

$$= \frac{-uRTF[C'' \exp(-\delta EF/2RT) - C' \exp(\delta EF/2RT)]}{[\delta E - (RT/F)\ln(C''/C')]\exp(EF/2RT)\int_r^{d-r} \exp\{[U_1(x) - F\phi(x)]/RT\}\,dx}. \tag{161}$$

Andersen and Fuchs (1975), while emphasising that $G_m(E,t) \neq G_b(E,t)$ for $\delta < 1$, showed that $G_b(E,t)$ is equal to the classic Nernst–Planck expression for membrane conductance.

The initial current ($t = 0$) when $C' = C'' = C$ and $E_m = 0$, is

$$I(E,0) = \frac{2uRTFC \sinh(\delta EF/2RT)}{\exp(EF/2RT)\int_r^{d-r} \exp\{[U_1(x) - F\phi(x)]/RT\}\,dx}. \tag{162}$$

VI. Model for Lipid-Soluble Ions

Thus the initial membrane conductance $G_m(E,0)$ is given as

$$G_m(E, 0) = \frac{I(E, 0)}{E} = \frac{2uRTFC \sinh(\delta EF/2RT)}{E \exp(EF/2RT) \int_r^{d-r} \exp\{[U_1(x) - F\phi(x)]/RT\} dx}. \quad (163)$$

But normalizing the initial conductance gives

$$\frac{G_m(E, 0)}{G_m(0, 0)} = \frac{I(E, 0)/E}{\lim_{E \to 0} I(E, 0)/E} = \frac{I(E, 0)/\delta E}{\lim_{E \to 0} I(E, 0)/\delta E} = \frac{G_b(E, 0)}{G_b(0, 0)}, \quad (164)$$

where $G_m(0,0)$ and $G_b(0,0)$ are the initial membrane and barrier conductances, respectively. To obtain information about ion movement in the interior of the membrane by the study of the current–voltage characteristics, it is necessary according to Eq. (164) to use normalized initial conductance.

Andersen and Fuchs (1975) have shown that

$$I(E, t) = I(E, 0) \exp[-t/\tau(E)], \quad (165)$$

where

$$\tau(E) = \frac{\beta \exp(EF/2RT) \int_r^{d-r} \exp\{[U_1(x) - F\phi(x)]/RT\} dx}{2uRT \cosh(\delta EF/2RT)}. \quad (166)$$

The important function $f(x)$ governing the characteristics of variation of current with voltage and of time constant with voltage is given by

$$f(x) = \exp\left(\frac{EF}{2RT}\right) \int_r^{d-r} \exp\left(\frac{U_1(x) - F\phi(x)}{RT}\right) dx. \quad (167)$$

Integration of Eq. (167) requires information about the variation of $\phi(x)$ through the membrane. It is assumed that $\phi(x)$ varies linearly through the membrane. $U_1(x)$ is approximated to the electrostatic charging energy $U_e(x)$, which is given by Eq. (112). Andersen and Fuchs (1975) have shown that

$$f(E) = \exp\left(\frac{EF}{2RT}\right) \int_r^{d-r} \exp\left(\frac{U_1(x) - F\phi(x)}{RT}\right) dx$$

$$= \exp\left[\omega\left(\frac{EF}{RT}\right)^2\right] \int_r^{d-r} \exp\left(\frac{U_e(x)}{RT}\right) dx, \quad (168)$$

where ω is a parameter dependent on membrane thickness and has a value of 0.005 for a membrane 30 Å thick. Equation (168) for the ratio $f(E)/f(0)$ becomes

$$f(E)/f(0) = \exp[\omega(EF/RT)^2]. \quad (169)$$

Equations (162) and (165) show that imposition of a potential difference across the membrane an initial current $I(E, 0)$ will flow, and this is due to redistribution of ions between the two interfacial regions. The current decays exponentially and ideally to zero. The decay is due to back diffusion of ions causing decrease in net current. At the end of the stimulus, a new current transient appears in the opposite direction. This is due to concentration difference since $C'' \neq C'$ although $E = 0$, [see Eq. (157)]. From Eq. (157), $I_{E,t}(0,0)$, the initial current when the potential that was on for t msec returns from E to zero, may be obtained as

$$I_{E,t}(0,0) = -uRTF(C'' - C') \bigg/ \int_r^{d-r} \exp\left(\frac{U_1(x)}{RT}\right) dx. \qquad (170)$$

Similarly, $G_{E,t}(0,0)$, the initial barrier conductance, is given by

$$G_{E,t}(0,0) = \frac{uRTF(C'' - C')}{E_{m(E,t)}(0,0) \int_r^{d-r} \exp[U_1(x)/RT] \, dx}. \qquad (171)$$

The "off" current relaxation time constant $\tau(0)$ is given by

$$\tau(0) = \left[\beta \int_r^{d-r} \exp\left(\frac{U_1(x)}{RT}\right) dx\right] \bigg/ 2uRT. \qquad (172)$$

Both the initial relaxation currents when the stimulus is turned "on" for a few milliseconds and then turned "off" can be followed.

The number of coulombs of charge moved through a bilayer during a single current transient is given by $\int_0^\infty I(t) \, dt$, which can be approximated by [see Eq. (165)]

$$\int_0^\infty I(0) \exp\left(\frac{-t}{\tau}\right) dt = I(0) \cdot \tau. \qquad (173)$$

$I(0)$ and τ can be evaluated by plotting $\log I(t)$ versus time. Extrapolating the straight line to $t = 0$ gives $I(0)$, and from the slope of the line τ can be obtained. The total charge moved during "on" response was found equal to the charge moved during "off" response. Charge moved as a function of potential increased with increase in applied potential and attained saturation. The several experimental points can be fitted to Eq. (155) by choosing values for δ and N_s systematically. In the case of bacterial phosphtidyl ethanolamine membranes, best fit was obtained for $\delta = 0.77$, while $\delta = 0.92$ produced a good fit for dioleoyl phosphatidyl ethanolamine membranes.

Tetraphenyl borate ion distribution into the membrane boundary region when the outside concentration was about $10^{-7} M$ was estimated to be about 10^5. This corresponds approximately to $12RT$ (~ 7 kcal) for the depth of the potential energy minima compared to the aqueous phases.

Relaxation time constant as a function of applied potential is given by combining Eqs. (166), (168), and (172). Thus

$$\tau(E) = \tau(0)\exp[0.005(EF/RT)^2]/\cosh(0.75EF/2RT), \qquad (174)$$

where $\omega = 0.005$ and $\delta = 0.75$. Accordingly a plot of $1/\tau$ against $\cosh(0.75EF/2RT)\exp[-0.005(EF/RT)^2]$ gave a straight line passing through the origin.

Initial membrane conductance as a function of potential is given by Eq. (163). As $G_m(0,0)$ is given by

$$(uF^2C) \bigg/ \int_r^{d-r} \exp\left(\frac{U_1(x)}{RT}\right) dx,$$

Eq. (163) together with Eqs. (164) and (168) can be expressed as

$$\frac{G_b(E,0)}{G_b(0,0)} = \frac{2RT}{\delta EF}\sinh\left(\frac{\delta EF}{2RT}\right) \bigg/ \exp\left[0.005\left(\frac{EF}{RT}\right)^2\right]. \qquad (175)$$

Experimental data for tetraphenyl borate ion and bacterial phosphatidyl ethanolamine membrane conformed to Eq. (175) and not to Eq. (133) when $\delta = 0.75$. Also equation

$$\frac{G(E,0)}{G(0,0)} = \frac{2RT}{0.65EF}\sinh\left(\frac{0.65EF}{2RT}\right), \qquad (176)$$

which corresponds to the Eyring rate theory ($\omega = 0$), and equation

$$\frac{G(E,0)}{G(0,0)} = 0.6\sinh\left(\frac{0.95EF}{2RT}\right) \bigg/ 0.95\sinh\left(\frac{0.6EF}{2RT}\right), \qquad (177)$$

which corresponds to a trapezoidal energy barrier with a horizontal plateau in the membrane from $x = 0.2d$ to $x = 0.8d$ and $\delta = 0.45$ (see Hladky, 1974b), predicted the experimental points well. But an accurate determination of δ ($= 0.75$) by Andersen and Fuchs ruled out the models underlying Eqs. (176) and (177) and showed that the potential energy barrier in the membrane to ion transport is related to the energy barrier due to the image force.

The other technique used to follow ion movement in lipid bilayer membranes is the charge-pulse method. In this method, the membrane is charged to an initial voltage by a current pulse of 10–100 nsec duration. At the end of the pulse, the external circuit is switched to infinite resistance and so redistribution of ions in the membrane and ion transport across the membrane lead to a decay of voltage E_m. From the time course of decay of this E_m, information about the kinetics of transport is obtained. The relevant equations describing these processes are given below:

$$E_m(t) = E_m(0)[a_1 \exp(-t/\tau_1) + a_2 \exp(-t/\tau_2)], \tag{178}$$

$$1/\tau_1 = \lambda_1 = P + \sqrt{P^2 - 4k_{ma}k_m b\beta C}, \tag{179}$$

$$1/\tau_2 = \lambda_2 = P - \sqrt{P^2 - 4k_{ma}k_m b\beta C}, \tag{180}$$

$$a_1 = [\lambda_1 - (2k_m + k_{ma})]/(\lambda_1 - \lambda_2), \tag{181}$$

$$a_2 = 1 - a_1, \tag{182}$$

$$P = \frac{k_{ma}}{2}[1 + 2(1 - \delta)^2 b\beta C] + k_m(1 + 2\delta^2 b\beta C), \tag{183}$$

$$b = F^2/4RTC_m. \tag{184}$$

$1/b$ becomes the amount of charge needed to charge the membrane capacitance to a voltage of $4RT/F^2$ ($1/b \approx 4 \times 10^{-13}$ mol/cm²).

When k_{ma} (desorption rate) is much smaller than k_m, Eqs. (179)–(181) and (183) give

$$\tau_1 \approx \frac{1}{2k_m(1 + 2\delta^2 b\beta C)}, \tag{185}$$

$$\tau_2 \approx \frac{1 + 2\delta^2 b\beta C}{2k_{ma} b\beta C}, \tag{186}$$

$$a_1 \approx \frac{2\delta^2 b\beta C}{1 + 2\delta^2 b\beta C}. \tag{187}$$

$E_m(t)$ recorded at long times ($t \gg \tau_1$) gives $E_m(t) \approx E_{(2)}\exp(-t/\tau_2)$, and so a plot of ln $E_m(t)$ against long times gives a straight line from which τ_2 and $E_{(2)}$ are obtained. If $\ln[E_m(t) - E_{(2)}\exp(-t/\tau_2)]$ is plotted as a function of time, again a straight line is obtained and from this values for $E_{(1)}$ and τ_1 may be derived. From voltage jump experiments, values for δ may be derived as described already. From experimental values of τ_1, a_1 (i.e., $E_{(1)}$) and δ values for k_m and β can be derived from Eqs. (185) and (187). Evaluation of k_{ma} cannot be made, because of the slow diffusion rate in water. Some of the values for the several parameters determined for several membrane systems are given in Table II. From the value of β, the free energy of adsorption may be estimated according to $\Delta G \approx -RT\ln(N_s/N_a)$ where $N_a = Cl_a$ is the concentration in moles per square centimeter of the ion in an energy minimum on the aqueous side of the membrane. This is roughly equal to about $12RT$.

The model of a lipophilic anion absorption and translocation described already is applicable at low concentrations. The number of absorbed ions and the conductance of the membrane increase linearly with aqueous concentration of the ions, and they move between the two energy wells with a single time constant. The voltage required to move a fraction of the ions from

TABLE II

Several Membrane Parameters Determined by the Charge-Pulse Relaxation Experiments[a]

C (nM)	τ_1 (ms)	τ_2 (s)	a_1	k_m (s^{-1})	N_s (pmol/cm^2)	β (cm)	$\beta/(d/2)$
\multicolumn{8}{c}{Dioleoyllecithin membranes ($d = 50$ Å) + Tetraphenyl borate $\delta \approx 0.9$}
30	23	4.0	0.67	7.0	0.5	2×10^{-2}	67×10^3
300	4.2	2.8	0.95	6.0	4.7	2×10^{-2}	63×10^3
3000	3.9	2.6	0.98	2.7	12.1	4×10^{-3}	16×10^3
\multicolumn{8}{c}{Dioleoyllecithin membranes + Dipicryl amine $\delta \approx 0.8$}
30	0.17	0.23	0.86	412	1.5	5×10^{-2}	200×10^3
300	0.094	0.23	0.95	268	4.4	1.5×10^{-2}	60×10^3

[a] From Benz et al. (1976b).

one energy well to the other is independent of concentration. This model breaks down at higher concentrations when both the absorbed ions and the membrane conductance no longer increase linearly with concentration. The time course of current deviates from a single relaxation, and greater potential is required to move a given fraction of charge between the two energy minima. These characteristics are now explained on the basis of a three-capacitor model. The equations relevant to this model as developed by Andersen et al. (1978) are outlined.

The electrostatic potential across the three capacitor model membrane is shown in Fig. 8. The adsorbed charges are smeared over the planes located

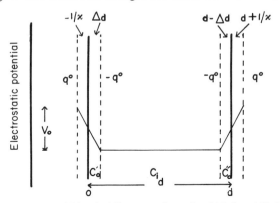

Fig. 8. Three-capacitor model for the bilayer membrane in which lipophilic ions are adsorbed at Δd and $d - \Delta d$. The corresponding counterions in the aqueous phase exist at $-1/\varkappa$ and $d + 1/\varkappa$. The three capacitors are C_0', C_0'', and C_i.

at $x = \Delta d$ and $x = d - \Delta d$. These roughly correspond to the Stern layers at the two surfaces of the membrane. The counterions forming the diffuse double layers are located at $x = -1/\varkappa$ and $x = d + 1/\varkappa$, where $1/\varkappa$ is the Debye length. In the absence of an applied potential, membrane is in equilibrium with symmetrical aqueous phase. If the bulk phase potential at left (') is set zero, the potential at the surface ('), V_0' is given by

$$V_0' = -q'/C_0 = -q_0/C_0. \tag{188}$$

Similarly at phase ("), the potential is given by

$$V_0'' = -q''/C_0 = -q_0/C_0. \tag{189}$$

q', q'' are charge densities in the two wells, and q_0 is the charge density at equilibrium. The total charge is given by

$$q' + q'' = 2q_0. \tag{190}$$

Application of a potential V_m increases the charge at the plane $x = d + 1/\varkappa$ by q_c while the same quantity of charge is removed from the plane at $x = -1/\varkappa$. Now the potentials are given by

$$V_0' = (q_c - q^0)/C_0, \tag{191}$$

$$V_i = (q_c - q_0 + q')/C_i, \tag{192}$$

$$V_0'' = (q_c + q^0)/C_0. \tag{193}$$

The total potential is given by

$$V_m = V_0' + V_i + V_0'' = q_c\left(\frac{2}{C_0} + \frac{1}{C_i}\right) + \frac{q' - q^0}{C_i}. \tag{194}$$

C_m, the specific membrane capacitance, is given by

$$C_m = \left(\frac{q_c}{V_m}\right)_{q'=q''=q^0 \text{ or } q^0 = 0}. \tag{195}$$

Thus Eqs. (194) and (195) give

$$1/C_m = 2/C_0 + 1/C_i. \tag{196}$$

C_m therefore becomes series combination of three capacitors, two outer and one inner. A term a is defined as

$$a = C_m/C_i, \tag{197}$$

where $0 \leq a < 1$. Equations (191)–(194) can be rewritten by using Eqs. (196) and (197). Hence

$$V_0' = (1 - a)(q_c - q_0)/2C_m, \tag{198}$$

VI. Model for Lipid-Soluble Ions

$$V_i = a(q_c + q' - q_0)/C_m, \tag{199}$$

$$V_0'' = (1-a)(q_c + q_0)/2C_m, \tag{200}$$

$$V_m = [q_c + a(q' - q^0)]/C_m. \tag{201}$$

Eliminating q_c from Eqs. (199) and (201) gives

$$V_i = aV_m - (1-a)a\,\Delta q/C_m, \tag{202}$$

where $\Delta q = q^0 - q'$.

According to Eqs. (151) and (152), the charge densities in the two energy wells are given by

$$q''/q' = \exp(FV_i/RT). \tag{203}$$

Differentiation of Eq. (201) with respect to time gives

$$C_m \frac{dV_m}{dt} = \frac{dq_c}{dt} + a\frac{d(q'-q^0)}{dt}.$$

But in a voltage clamp experiment $dV_m/dt = 0$ as V_m is held constant. Thus

$$I_c = \frac{dq_c}{dt} = -a\frac{dq'}{dt}. \tag{204}$$

I_c is the current measured in the external circuit. Integration of Eq. (204) gives

$$\Delta q_c = a(q^0 - q') = a(q'' - q^0). \tag{205}$$

When Eq. (205) is used, Eqs. (201) and (202) become respectively

$$\Delta q_c = q_c - V_m C_m \tag{206}$$

and

$$V_i = aV_m - (1-a)\Delta q_c/C_m. \tag{207}$$

When the membrane potential is changed from an initial value of zero to a new value V_m, current flows in the external circuit. This current is composed of two fractions [see Eq. (206)], one transient in nature, which charges the membrane capacity, and the rest, which decays with time because of charge transfer within the membrane. As the outer regions of the membrane are not involved, the charge translocation will increase with increase in V_m and reaches an upper limit $\Delta q_{c,\text{max}}$. Thus $q'' = \Delta q_{c,\text{max}} + \Delta q_c$ and $q' = \Delta q_{c,\text{max}} - \Delta q_c$, and so Eq. (203) together with Eq. (207) becomes

$$\frac{\Delta q_{c,\text{max}} + \Delta q_c}{\Delta q_{c,\text{max}} + \Delta q_c} = \exp\left[\frac{F}{RT}\left(aV_m - \frac{(1-a)\Delta q_c}{C_m}\right)\right]. \tag{208}$$

TABLE III

Variation of a and δ with Concentration of Tetraphenyl Borate for the Bilayer Membrane of Bacterial Phosphatidyl Ethanolamine[a]

Concentration TPB$^-$ (M)	$\Delta q_{c,max}$ (μC/cm^2)	a	Capacitance C_0 (μF/cm^2)	δ
10^{-8}	3.5×10^{-2}	0.93	15	0.86
10^{-7}	3.5×10^{-1}	0.97	35	0.71
10^{-6}	1.49	0.97	37	0.41

[a] From Andersen, et al. (1978).

The relation between external charge moved Δq_c and charge transferred within the membrane Δq is given by Eq. (205) as $\Delta q_c = a\, \Delta q$. Thus $\Delta q_{c,max}$ is given by

$$\Delta q_{c,max} = aq^0. \tag{209}$$

Variations of Δq_c measured as a function of V_m [see Eq. (173) applicable at low concentrations of the hydrophobic ion; error less than 10% at moderate concentrations] at several concentrations of tetraphenyl borate ion (TPB$^-$) were fitted (least-squares fit) to Eq. (208) by choosing appropriate values for $\Delta q_{c,max}$ and a. The values for these parameters and for δ determined according to Eq. (155) in which the boundary potentials are ignored are given in Table III. Although the values of a depend on the concentration of the ion, they are less dependent on $\Delta q_{c,max}$ than are the values of δ. The boundary capacitance values for C_0 are obtained from the relation

$$a = 1 - 2C_m/C_0. \tag{210}$$

Equation (210) follows from Eqs. (196) and (197). The value of C_m for bacterial phosphatidyl ethanolamine membranes was $0.53\,\mu$F/cm^2.

When a membrane is in equilibrium with aqueous solutions of identical composition (TPB$^-$), combination of Henry's law and the Boltzmann relation give

$$q^0 = F\beta[\text{TPB}^-]\exp(-FV_0/RT), \tag{211}$$

where

$$V_0 = V_0'' = -V_0' = q^0/C_0. \tag{212}$$

Substituting Eq. (212) into Eq. (211) gives

$$q^0 = F\beta(\text{TPB}^-)\exp[-Fq^0/(RTC_0)]. \tag{213}$$

VI. Model for Lipid-Soluble Ions

Experimentally determinable parameters are C_m and $\Delta q_{c,max}$, and so substituting these from Eqs. (209) and (210) into Eq. (213) gives

$$\Delta q_{c,max} = aF\beta(\text{TPB}^-)\exp\left[-\frac{F\Delta q_{c,max}(1-a)}{aRT2C_m}\right]. \quad (214)$$

Again this equation enables evaluation of a and β by measuring $\Delta q_{c,max}$ as a function of (TPB$^-$). Values of $a = 0.985$ and $\beta = 3.7 \times 10^{-2}$ cm and $C_0 = 70$ μF/cm^2 have been obtained for bacterial phosphatidyl ethanolamine membranes. Similar analysis gave values of $a = 0.99$ and $\beta = 2.9 \times 10^{-3}$ cm and $C_0 = 90$ μF/cm^2 for glyceryl monooleate membranes.

Other evidence that absorption of TPB$^-$ changes the electrostatic potential within the bilayer membrane comes from incorporation of probe molecules into the membrane. The change in electrostatic potential in the middle of the membrane ΔV may be calculated from

$$G'' = G'\exp(\Delta VF/RT), \quad (215)$$

where G'' is the conductance of membrane containing the probe molecule measured after addition of TPB$^-$ and G' is that measured before the addition of TPB$^-$. TPB$^-$ in general depresses the conductance of a negative probe such as 5,6-dichloro-2-trifluoromethylbenamidazole (DTFB). In the case of DTFB, values of ΔV calculated from Eq. (215) as a function of TPB$^-$ concentration agreed with the predictions of the equation of the three-capacitor model. This equation is obtained by combining Eqs. (211) and (212). Thus

$$V_0 = (F\beta/C_0)[\text{TPB}^-]\exp(-FV_0/RT). \quad (216)$$

The values of β and C_0 used in the calculations were 3.7×10^{-2} cm and 70 μF/cm^2, which were derived by curve fitting by using Eq. (214).

The other evidence validating the three-capacitor model comes from charge-pulse measurements, which give values for a and $\Delta q_{c,max}$ agreeing with those derived by the voltage clamp measurements (see Table III). In the charge-pulse experiments, a known quantity of charge q_c is injected as quickly as possible (near $t = 0$) into the aqueous phase. The membrane potential is then observed as a function of time. If the initial membrane potential $V_m(0)$ is measured just after charge injection when $q' = q'' = q^0$, i.e., before charge translocation in the membrane, then Eq. (195) can be written

$$V_m(0) = q_c/C_m. \quad (217)$$

Completion of charge translocation in the membrane reduces V_m to $V_m(\infty)$. So Eq. (201) can be written

$$V_m(\infty) = [q_c + a(q' - q^0)]/C_m. \quad (218)$$

Combining Eqs. (217) and (218) gives

$$\Delta V_m = V_m(0) - V_m(\infty) = a(q^0 - q)/C_m. \tag{219}$$

This should be equal to $a(q'' - q^0)/C_m$ since charge loss in one well is equal to charge gain in the other well.

$\Delta V_{m,max}$ is obtained when $q' \to 0$, i.e., all the charge is translocated. Thus from Eq. (219), one gets

$$\Delta V_{m,max} = aq^0/C_m. \tag{220}$$

Combining Eqs. (199) and (201) to eliminate $a(q' - q^0)$ gives

$$V_i = V_m - (1 - a)q_c/C_m. \tag{221}$$

Thus one can write

$$V_i(\infty) = V_m(\infty) - (1 - a)q_c/C_m. \tag{222}$$

Substituting Eq. (217) into Eq. (222) gives

$$V_i(\infty) = V_m(\infty) - (1 - a)V_m(0) = aV_m(0) - \Delta V_m. \tag{223}$$

So, Eq. (203) may be written as

$$q''/q' = \exp\{F[aV_m(0) - \Delta V_m]/RT\}. \tag{224}$$

Substitution of Eq. (220) into Eq. (219) gives

$$\frac{q''}{q'} = \frac{\Delta V_{m,max} + \Delta V_m}{\Delta V_{m,max} - \Delta V_m}$$

and so Eq. (224) becomes

$$\frac{\Delta V_{m,max} + \Delta V_m}{\Delta V_{m,max} - \Delta V_m} = \exp\left(\frac{F[aV_m(0) - \Delta V_m]}{RT}\right). \tag{225}$$

This can be written

$$\Delta V_m = \Delta V_{m,max} \tanh[(F/2RT)(aV_m(0) - \Delta V_m)]. \tag{226}$$

Determination of $\Delta V_{m,max}$ is difficult since membranes will not withstand more than 400 mV. In order to determine $\Delta V_{m,max}$, multiple charge injections are made so that in the final injection no fast relaxation is seen, i.e., transfer of charge from one well to the other is complete. Consequently

$$\Delta V_{m,max} = \sum_{n=1}^{\infty} (\Delta V_m)_n. \tag{227}$$

According to this three-capacitor model, apparent saturation in the absorption of lipid-soluble ions into the membrane is due to the generation of

VI. Model for Lipid-Soluble Ions

electrostatic boundary potentials. As opposed to this interpretation, Wang and Bruner (1978), who obtained results (dipicrylamine absorption by dioleoylphosphatidylcholine membrane) similar to those of Andersen (1978) and co-workers (1975, 1978) (TPB$^-$ absorption by bacterial phosphatidylethanolamine membranes), have attributed the absorption of dipicrylamine (DPA$^-$) entirely to a diffuse double layer capacitance. Equation (102) of Chapter 3 for a symmetrical electrolyte solution can be written

$$\sigma_s/2\sigma_0 = \sinh(zF\psi_s/2RT) \tag{228}$$

and

$$\sigma_0 = \sqrt{\varepsilon RTC^0/2\pi}, \tag{229}$$

where C^0 is the concentration of indifferent electrolyte in solution.

The concentration of DPA$^-$ indicated by (DPA$^-$) is related to the surface concentration (or surface charge density $|\sigma_s|$) by

$$|\sigma_s| = \beta(\text{DPA}^-)\exp(-|\psi_s|F/RT). \tag{230}$$

At low surface potentials $|\psi_s| \ll RT/F$, Eq. (230) simplifies to

$$|\sigma_s| \approx \beta(\text{DPA}^-). \tag{231}$$

When $|\psi_s| \gg RT/F$, Eq. (228) can be written

$$|\sigma_s|/\sigma_0 \approx \exp(z|\psi_s|F/2RT). \tag{232}$$

Eliminating $|\psi_s|F/RT$ between Eqs. (230) and (232) gives

$$|\sigma_s| \approx \sigma_0[\beta(\text{DPA}^-)/\sigma_0]^{z/(z+2)} \tag{233}$$

when $F|\psi_s|/RT \gg 1$.

The surface charge density σ_s of adsorbed DPA$^-$ is equated to the measured charge transport per unit area of the membrane [see Eq. (173)]. Thus a plot of $\log|\sigma_s|$ against $\log(\text{DPA}^-)$ should give, according to Eq. (231), a straight line of slope unity at low concentrations of DPA$^-$, and according to Eq. (233), a straight line of lower slope $[z/(z+2)]$ at higher concentrations of DPA$^-$. This transition occurs when $|\psi_s| \approx RT/F$ or $|\sigma_s| = \sigma_0$. When $\beta(\text{DPA}^-)$ is eliminated from Eqs. (231) and (233), a value for the point of intersection of the two straight lines, i.e., $|\sigma_s| = \sigma_0$, is obtained. Thus from a log–log plot of experimental data, σ_0 can be evaluated by projecting the point of intersection of the two lines onto the $|\sigma_s|$ axis. From Eq. (229), σ_0 can be calculated since C^0, the concentration of the indifferent electrolyte, is known. As the DPA$^-$ concentration is very small compared to C^0, it does not enter into the determination of σ_0 by Eq. (229).

Voltage clamp relaxation experiments carried out with dioleoyllecithin bilayer membranes and several indifferent electrolytes (e.g., NaCl, $MgCl_2$, $BaCl_2$; concentration range 10^{-4}–1.0 M) containing DPA^- at several concentrations (10^{-9}–10^{-5} M) showed that the data for all salt concentrations converged to a common line of unit slope at low concentrations of DPA^-. At higher DPA^- concentrations, the experimental points for each salt concentrations were well fitted by lines of slope $\frac{1}{3}$ (for Na^+) and $\frac{1}{2}$ (for Mg^{2+} and Ba^{2+}), these lines being displaced upward with increasing salt concentration.

Although the slopes were in agreement with theory, the values of σ_0 derived by the projection of the point of intersection of the two lines onto the $|\sigma_s|$ axis were lower than those calculated from Eq. (229) at high concentration of the indifferent electrolyte; however, there was agreement at low concentrations of indifferent electrolyte. This discrepancy observed at high concentration of several electrolytes was attributed to dielectric saturation of water. This interpretation has been questioned by Andersen et al. (1978) on two grounds: (1) In studies where both the double-layer potential and charge density have been measured directly, no evidence of dielectric saturation was found. (2) At high salt concentrations (10^{-1} M), the measured zeta potential produced by 10^{-6} M DPA^- was an order of magnitude too low. On these grounds, it is concluded that the DPA^- ions are located deep in the membrane and not confined to the outer surface of the membrane.

VII. Model for Carriers of Small Ions

A number of macrocyclic antibiotics such as valinomycin, enniatin A, B, monactin, etc., are able to increase the cation permeability of lipid bilayer membranes by several orders of magnitude. An alkali metal ion is insoluble in the unmodified bilayer membrane, whereas the macrocyclic compound is soluble in the membrane. Hence the alkali metal ion crosses the membrane in the form of a complex, which probably forms at the membrane–solution interface.

Following the treatment of Läuger and his colleagues (1981), carrier-mediated transport occurs by four main reaction steps: (1) binding of the ion by the carrier at one membrane interface, (2) translocation of the carrier–ion complex to the opposite interface, (3) release of the ion, and (4) transition of the carrier back to the original position (see Fig. 9).

A bilayer membrane is in contact on both sides with an aqueous solution of an univalent cation M^+ of concentration C_M. The carrier concentration is C_S. The carrier S combines with M^+ to form a complex MS^+ of concentra-

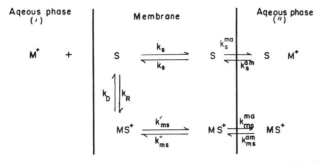

Fig. 9. Transport of cation M⁺ by a neutral carrier S. (After Läuger and Stark, 1970.)

tion C_{MS} in the aqueous phase. Thus the association constant K is given by

$$K = C_{MS}/C_S C_M \tag{234}$$

M⁺ is completely excluded from the membrane phase whereas S and MS⁺ may exchange between the aqueous phase (a) and the membrane phase (m). Thus

$$S(a) \underset{k_S^{ma}}{\overset{k_S^{am}}{\rightleftarrows}} S(m) \tag{235}$$

$$MS^+(a) \underset{k_{ms}^{ma}}{\overset{k_{ms}^{am}}{\rightleftarrows}} MS^+(m) \tag{236}$$

In addition, a chemical reaction occurs at the interface between an ion M⁺ in solution and a carrier in the membrane phase. This reaction is described by

$$M^+(a) + S(m) \underset{k_D}{\overset{k_R}{\rightleftarrows}} MS^+(m) \tag{237}$$

At equilibrium (i.e., no current)

$$K_h = k_R/k_D = N_{MS}/C_M N_S. \tag{238}$$

K_h is an heterogeneous equilibrium constant, N_S and N_{MS} are the interfacial concentrations (moles per square centimeter) of S and MS⁺ at equilibrium. But the dimensionless partition coefficients for S and MS⁺ give

$$\beta_S = \frac{N_S}{C_S} \frac{2}{d}, \tag{239}$$

$$\beta_{MS} = \frac{N_{MS}}{C_{MS}} \frac{2}{d}. \tag{240}$$

Substituting these and Eq. (234) into Eq. (238) gives

$$K_h = \frac{\beta_{MS}C_{MS}d/2}{C_M\beta_S C_S d/2} = \frac{\beta_{MS}}{\beta_S}K. \tag{241}$$

In general, the rate of reaction represented by (237) is high compared with reactions (235) and (236). The fluxes of S and MS$^+$ across the membrane are given by

$$J_S = k_S(N'_S - N''_S), \tag{242}$$

$$J_{MS} = k'_{MS}N'_{MS} - k''_{MS}N''_{MS}. \tag{243}$$

The charge carrier in the membrane is the charged species MS$^+$, and so the current density is given by

$$I = FJ_{MS}. \tag{244}$$

The rate constant k_S is the same for S in either direction, but for MS$^+$, the rate constants k'_{MS} and k''_{MS} depend on the voltage. Thus

$$k'_{MS} = k_{MS}\exp(-EF/2RT), \tag{245}$$

$$k''_{MS} = k_{MS}\exp(EF/2RT), \tag{246}$$

where $k_{MS} = v\exp(-\Delta G/RT)$. ΔG is the energy for zero voltage at the top of the barrier.

In the steady state, the sum of net chemical production and fluxes toward the interface vanish. Hence

$$\frac{dN'_S}{dt} = -k_R C_M N'_S + k_D N'_{MS} - J_S + k_S^{am}C_S - k_S^{ma}N'_S = 0, \tag{247}$$

$$\frac{dN''_S}{dt} = -k_R C_M N''_S + k_D N''_{MS} + J_S + k_S^{am}C_S - k_S^{ma}N''_S = 0, \tag{248}$$

$$\frac{dN'_{MS}}{dt} = k_R C_M N'_S - k_D N'_{MS} - J_{MS} + k_{MS}^{am}C_{MS} - k_{MS}^{ma}N'_{MS} = 0, \tag{249}$$

$$\frac{dN''_{MS}}{dt} = k_R C_M N''_S - k_D N''_{MS} + J_{MS} + k_{MS}^{am}C_{MS} - k_{MS}^{ma}N''_{MS} = 0, \tag{250}$$

when $E = 0$, $N'_S = N''_S = N_S$, and $N'_{MS} = N''_{MS} = N_{MS}$.

These six equations (242), (243), and (247)–(250) may be solved for the six unknowns (J_S, J_{MS}, N'_S, N'_{MS}, N''_S, and N''_{MS}). Lauger and co-workers (1981) have shown that substitution of the value of J_{MS} into Eq. (244) gives

$$I = \frac{Fdk_{MS}KC_M C^0 \beta_{MS}}{1 + KC_M} \frac{\sinh(EF/2RT)}{1 + A\cosh(EF/2RT)}, \tag{251}$$

VII. Model for Carriers of Small Ions

where

$$A = \frac{2k_{MS}(k_R C_M + 2k_S + k_S^{ma})}{(k_D + k_{MS}^{ma})(k_R C_M + 2k_S + k_S^{ma}) - k_R k_D C_M} \tag{252}$$

and

$$C^0 = C_S + C_{MS}. \tag{253}$$

When low voltages are applied, membrane conductance $G(0)$ is given by $(I/E)_{E\approx 0}$. Thus

$$G(0) = \frac{F^2}{2RT} dk_{MS} \frac{KC_M C^0 \beta_{MS}}{(1 + KC_M)(1 + A)} = \frac{F^2}{2RT} dk_{MS} \frac{C_{MS}^m}{1 + A}, \tag{254}$$

where $C_{MS}^m = [KC_M C^0/(1 + KC_M)]\beta_{MS}$ from Eqs. (234) and (253). The term A indicates the role of the interface in controlling charge transport. If the exchange of S and MS^+ across the interface is slow (i.e., $k_S^{ma} \approx k_{MS}^{ma} = 0$), then

$$A = \frac{2k_{MS}}{k_D} + \frac{k_{MS}}{k_S} \frac{k_R C_M}{k_D}. \tag{255}$$

If G is the conductance given by Eq. (251) as I/E, then combining Eqs. (251) and (254) gives

$$\frac{G}{G(0)} = \frac{2RT}{EF}(1 + A)\frac{\sinh(EF/2RT)}{1 + A\cosh(EF/2RT)}. \tag{256}$$

In the case of carriers, monactin and valinomycin, $K < 0.1 \, M^{-1}$ and so $KC_m \ll 1$. Equation (254) can be written, with the help of Eqs. (238) and (241),

$$G(0) = \frac{F^2}{2RT} dC^0 \beta_S k_{MS} \frac{k_R C_M}{k_D(1 + A)}. \tag{257}$$

The value for the parameter A can be deduced from the best fit of $G/G(0)$ [Eq. (256)] to the experimental values at different applied voltages. Plotting A as a function of C_M, values for $2k_{MS}/k_D$ and $k_{MS}k_R/k_S k_D$ may be derived. Further, a plot of $G(0)$ as a function of C^0, the total carrier concentration, enables evaluation of $\beta_S k_{MS} k_R/k_D$. Thus three combinations of five parameters are evaluated. A complete evaluation is possible if an additional measurement of current relaxation following a voltage step is carried out.

The relaxation current $I(t)$ due to redistribution of surface concentrations of complexed N_{MS} and uncomplexed N_S carrier molecules is given, according to Stark et al. (1971), by

$$I(t) = I_\infty[1 + \alpha_1 \exp(-t/\tau_1) + \alpha_2 \exp(-t/\tau_2)], \tag{258}$$

where

$$\tau_1 = 1/(a - b), \tag{259}$$

$$\tau_2 = 1/(a + b), \tag{260}$$

$$a = \tfrac{1}{2}[k_R C_M + k_D + 2k_S + 2k_{MS}\cosh(EF/2RT)], \tag{261}$$

$$b = \tfrac{1}{2}[(k_R C_M - k_D + 2k_S - 2k_{MS}\cosh(EF/2RT))^2 + 4k_R k_D C_M]^{1/2}, \tag{262}$$

$$\alpha_1 = (A/2)\cosh(EF/2RT) + B, \tag{263}$$

$$\alpha_2 = (A/2)\cosh(EF/2RT) - B, \tag{264}$$

$$B = \frac{\cosh(EF/2RT)}{4b}\left[A(k_R C_M + k_D + 2k_S - 2k_{MS}\cosh\frac{EF}{2RT}] - 4k_{MS}, \tag{265}$$

and

$$I_\infty = 2FN_S\frac{k_R C_M}{k_D}k_{MS}\frac{\sinh(EF/2RT)}{1 + A\cosh(EF/2RT)}. \tag{266}$$

In practice only one larger relaxation time constant τ_1 and amplitude α_1 are detected. These measurements are sufficient to calculate all the four rate constants and the partition coefficient β_S.

In the evaluation of τ_1 and α_1, the relaxation current

$$\ln\frac{I(t) - I(\infty)}{I(\infty)}$$

may be plotted against time $t > \tau_1$ giving a straight line whose slope gives τ_1, and α_1 is given by $[I(0) - I(\infty)]/I(\infty)$. Equations (259)–(265) show that both τ_1 and α_1 depend on applied potential. Experimental determination of this dependence conformed to the equations mentioned. In the case of bilayer membranes of lipids extracted from soybean and the carrier valinomycin, appropriate experiments performed with KCl maintaining an ionic strength of 1 M (with LiCl) gave the following values (see Gamble et al., 1973).

Fit of conductance data to Eq. (256) gave values for A ($A = 1.2$, $C_M = 1\ M$; $A = 0.4$, $C_M = 10^{-2}\ M$). From Eq. (255) the following values for the combination of rate constants were derived:

$$k_{MS}/k_D = 0.2, \qquad k_{MS}k_R/k_S k_D = 0.8\ M^{-1}.$$

A plot of $G(0)$ against C^0 gave the value $1.3 \times 10^4\ M^{-1}\text{s}^{-1}$ [see Eq. (257)] for $\beta_S k_{MS} k_R/k_D$. Measurement of relaxation current gave $\alpha_1 = 1.6$ and $\tau_1 = 36\ \mu\text{s}$ ($C_M = 1\ M$ and $E = 48$ mV). Values derived for two membrane systems by the voltage clamp method are given in Table IV.

VII. Model for Carriers of Small Ions

TABLE IV

Values for Several Kinetic Parameters of Carrier Transport in Bilayer Membranes

	Valinomycin/K$^+$	
Parameters	Phosphatidyl inositol[a]	Soybean lipids[b]
k_R ($M^{-1} s^{-1}$)	5×10^4	7×10^4
k_D (s^{-1})	5×10^4	4×10^4
k_S (s^{-1})	2×10^4	2×10^4
k_{MS} (s^{-1})	2×10^4	10^4
β_S	6×10^4	10^4

[a] Stark et al. (1971).
[b] Gamble et al. (1973).

The other method used to follow the kinetics of charge transport is the charge-pulse technique. The relevant equations derived by Benz and Lauger (1976) are the following.

The rate of fall of voltage E_m after a charge pulse is given by

$$\frac{dE_m}{dt} = \frac{I}{C_m} = \frac{J_{MS}F}{C_m}. \tag{267}$$

Equation (267) expressed in terms of conductance $G(E,t)$ becomes

$$\frac{d\ln E_m}{dt} = \frac{G(E_m,t)}{C_m} = \frac{J_{MS}F}{C_m E_m}. \tag{268}$$

Unlike the relaxation current, which is described by two exponentials [see Eq. (258)], $E_m(t)$ is described by three exponentials in the form

$$E_m(t) = E_m^0 [a_1 \exp(-\lambda_1 t) + a_2 \exp(-\lambda_2 t) + a_3 \exp(-\lambda_3 t)], \tag{269}$$

$$a_1 + a_2 + a_3 = 1. \tag{270}$$

The relaxation times $\tau_i = 1/\lambda_i$ and the relaxation amplitudes a_i ($i = 1, 2, 3$) are defined by the following relations:

$$P_1 = \lambda_1 + \lambda_2 + \lambda_3, \tag{271}$$

$$P_2 = \lambda_1 \lambda_2 + \lambda_1 \lambda_3 + \lambda_2 \lambda_3, \tag{272}$$

$$P_3 = \lambda_1 \lambda_2 \lambda_3, \tag{273}$$

$$P_4 = a_1 \lambda_1 + a_2 \lambda_2 + a_3 \lambda_3, \tag{274}$$

$$P_5 = a_1 \lambda_1^2 + a_2 \lambda_2^2 + a_3 \lambda_3^2. \tag{275}$$

The several rate constants and the total concentration $N^0 (= N'_S + N''_S + N'_{MS} + N''_{MS})$ of the carrier per unit area of the membrane are shown to be given by

$$k_{MS} = \frac{1}{2}\left(\frac{P_5}{P_4} - P_4\right), \tag{276}$$

$$k_D = \frac{1}{2k_{MS}}\left[\frac{P_1 P_5}{P_4} - P_2 + \frac{P_3}{P_4} - \left(\frac{P_5}{P_4}\right)^2\right], \tag{277}$$

$$k_S = \frac{1}{2k_D}\frac{P_3}{P_4}, \tag{278}$$

$$k_R = \frac{1}{C_M}(P_1 - P_4 - 2k_S - 2k_{MS} - k_D), \tag{279}$$

$$N^0 = \frac{2RTC_m}{F^2}\frac{P_4}{k_{MS}}\left(1 + \frac{k_D}{C_M k_R}\right). \tag{280}$$

Thus from five experimental parameters, values for the four rate constants and for the total carrier concentration N^0 may be derived.

Rearrangement of Eq. (238) gives

$$N_S/N_{MS} = k_D/k_R C_M. \tag{238'}$$

The total concentration of carrier molecules N^0 in the membrane is given by

$$N^0 = 2N_S + 2N_{MS}. \tag{281}$$

Solving Eqs. (238) and (281) for N_S and N_{MS} gives

$$N_S = \left(\frac{N^0}{2}\right)\bigg/\left(1 + \frac{k_R}{k_D}C_M\right) = \left(\frac{N^0}{2}\right)\bigg/(1 + K_h C_M), \tag{282}$$

$$N_{MS} = \frac{N_0}{2}\left(\frac{K_h C_M}{1 + K_h C_M}\right). \tag{283}$$

Relaxation times and voltage amplitudes can be evaluated from the experimental record of $E_m(t)$ in the following way. For $t \gg \tau_3$ (long times), the relation $E_m(t) \approx E_3 \exp(-t/\tau_3)$ holds. Thus a plot of $\ln E_m(t)$ against long times t should give a straight line from which τ_3 and E_3 are derived. Next $\ln[E_m(t) - E_3 \exp(-t/\tau_3)]$ is plotted against t. Again a straight line is obtained from which τ_2 and E_2 are derived. Repetition of this process with $\ln[E_m(t) - E_3 \exp(-t/\tau_3) - E_2 \exp(-t/\tau_2)]$ ultimately gives τ_1 and E_1. These values determined for a monoolein-n-decane membrane and valinomycin are given in Table V. The values for the several rate constants and N^0 and β_S evaluated from Eqs. (276)–(280) are given in Table VI.

TABLE V

Time Constants and Relaxation Amplitudes for Monoolein/n-decane and 10^{-7} M Valinomycin System[a]

	C_M (M)	τ_1 (μs)	τ_2 (μs)	τ_3 (μs)	a_1	a_2	a_3
KCl at 25°C	1	0.865	2.59	51.9	0.295	0.300	0.405
RbCl at 25°C	1	0.696	2.04	63.1	0.357	0.236	0.407
CsCl at 25°C	1	0.821	3.66	33.1	0.102	0.407	0.491
RbCl at 10°C	1	3.35	7.52	414	0.374	0.195	0.431

[a] From Benz and Lauger (1976).

From the values of the rate constants determined at the two temperatures, 25 and 10°C, for the Rb$^+$, the activation energies for the rate constants were estimated to be 13.5 (k_R), 20.7 (k_D), 16.2 (k_{MS}), and 14.5 (k_S) kcal/mol.

In the above procedures the kinetic parameters were deduced from the analysis of three relaxations. Feldberg and Kissel (1975) and Feldberg and Nakadomari (1977) have shown that it is possible by the charge-pulse method (data derived from high-voltage and low-voltage application to membrane) to deduce similar values for the same kinetic parameters even when the first and second relaxations escape analysis because they occur too fast. Benz and Lauger (1976) have shown that the third relaxation is associated with the time constant for the discharge of the effective capacitance through the conductive pathway.

Equations pertinent to steady-state analysis of the current–voltage behavior of a bilayer membrane when it is subject to a high voltage for a relatively short time are outlined.

TABLE VI

Values for the Rate Constants Derived from the Data of Table V

	k_R ($M^{-1}s^{-1}$)	k_D (s^{-1})	k_{ms} (s^{-1})	k_s (s^{-1})	N^0 (mol/cm^2)	β_s^a
KCl at 25°C	29×10^4	27×10^4	21×10^4	3.8×10^4	78×10^{-14}	7.5×10^3
RbCl at 25°C	37×10^4	24×10^4	27×10^4	3.5×10^4	68×10^{-14}	5.3×10^3
CsCl at 25°C	22×10^4	56×10^4	25×10^4	4.1×10^4	72×10^{-14}	10.3×10^3
RbCl at 10°C	11×10^4	3.7×10^4	6.3×10^4	0.98×10^4	65×10^{-14}	8.2×10^3

[a] Calculated from Eqs. (282) and (239). A value of $C_m = 0.39$ μF/cm^2 is used in the calculation of N^0.

A known quantity of charge is injected quickly into the membrane, which is made up of two elements, one capacitative in nature (C_m) existing in parallel with the other, which is conductive [$G(E,t)$]. Injection of charge leads to instantaneous development of a voltage across the capacitance and then decays through the conductive element. Study of the nature of this voltage decay gives information about the several processes related to membrane conductance. In the steady state when the several processes occur rapidly relative to the time constant of voltage decay, the relation (268) is valid provided C_m is independent of voltage.

Hladky (1972, 1974a,b, 1975, 1979) has shown that the flux J_{MS} in Eq. (268) is given by

$$J_{MS} = N_S \left[\frac{(k'_R/k'_D)k'_{MS}C'_M - (k''_R/k''_D)k''_{MS}C''_M}{1 + k'_{MS}/k'_D + k''_{MS}/k''_D + k'_R k'_{MS}C'_M/k'_D 2k_S + (k''_R k''_{MS}/k''_D 2k_S)C''_M} \right], \quad (284)$$

when $C'_M = C''_M = C_M$, $k'_R = k''_R = k_R$, and $k'_D = k''_D = k_D$, Eq. (284) simplifies to

$$J_{MS} = \frac{N_S C_M (k_R/k_D)(k'_{MS} - k''_{MS})}{1 + [(k'_{MS} + k''_{MS})/k_D][1 + (k_R/2k_S)C_M]}. \quad (285)$$

It is assumed that k_R, k_D and k_S are voltage independent. Feldberg and Kissel (1975) expressed the voltage dependence of k'_{MS} and k''_{MS}, assuming a double Eyring energy barrier (see Fig. 10), as

$$k'_{MS} = \frac{1}{2} k^*_{MS} \frac{\exp(EF/2RT)}{\cosh(\frac{1}{2} - n)EF/RT}, \quad (286)$$

$$k''_{MS} = \frac{1}{2} k^*_{MS} \frac{\exp(-EF/2RT)}{\cosh(\frac{1}{2} - n)EF/RT}, \quad (287)$$

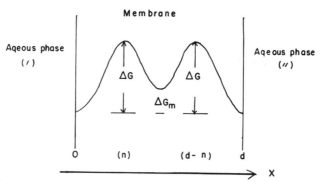

Fig. 10. Symmetrical Eyring energy barrier. 0, n, $d-n$, and d are the values x/d (d is the thickness of the membrane) at the left membrane–aqueous interface, the first barrier peak, the second barrier peak, and the right membrane–solution interface.

where k_{MS}^* is the standard rate constant including the exponential $\exp(-\Delta G/RT)$. The height of the barrier at the midpoint in the membrane ΔG_m cancels out. $\frac{1}{2}$ appears because the rate constant of the ion complex covers half the membrane. Combining Eqs. (285)–(287) gives

$$J_{MS} = \frac{[N_S C_M(k_R/k_D)k_{MS}^* \sinh(EF/2RT)]/[\cosh(\tfrac{1}{2}-n)EF/RT]}{1+[k_{MS}^*\cosh(EF/2RT)/k_D\cosh(\tfrac{1}{2}-n)EF/RT][1+(k_R/2k_S)C_M]}. \quad (288)$$

Combining Eq. (288) with Eq. (268) and then numerically integrating the resultant equation gives the voltage–time transient. The parameters n, k_{MS}^*/k_D, and $[1+(k_R/2k_S)C_M]$ determine the "shape" of the charge–time curve, whereas the parameter $N_S C_M(k_R/k_D)k_{MS}^*Ft/C_m$ determines its "scale."

In the case of a black lipid membrane formed from a solution of lipid in a hydrocarbon solvent, the black membrane will be in equilibrium with the unblackened region of the membrane. The carrier therefore will distribute itself between these two regions. The concentration of the carrier in the membrane may become buffered, and so the free-carrier concentration in the membrane (N_S) becomes constant. The other case is that the total concentration of the carrier in the membrane may remain constant. In the case of the latter, N_S is given by Eq. (282). The behavior of valinomycin in mediating K$^+$ or Cs$^+$ transport is an example of the first type. Behavior of actins (non-, mon-, din-, and trin-) in mediating NH$_4^+$ transport is an example of the latter (see below).

Application of a pulse to the membrane in equilibrium causes a shift in the surface concentrations of the carrier and the carrier complex as the whole system moves to a steady state. This results in a voltage that decays rapidly. Extrapolation of this decay to zero time gives a voltage intercept E_S that will be lower than that (E_i) predicted on the basis of charge q injected. That is,

$$E_i = q/C_m. \quad (289)$$

The difference $\Delta E = E_i - E_S$, called the intercept discrepancy by Feldberg and Kissel (1975), is related to N^0 and N_{MS} by Eqs. (283) and (290):

$$N_{MS} = \Delta E C_m/F. \quad (290)$$

When a high voltage is applied to the membrane system, the rate constant k'_{MS} is very high, whereas k''_{MS} is practically zero. The intercept discrepancy can be attributed to the transfer of N_{MS} from one side of the membrane to the other. This transfer will include part of N_{MS} that is due to conversion of N_S to N_{MS} by the reaction (237). The integral of the flux, $\int_0^t J_{MS}(E=\infty,t)\,dt$, gives the total charge, which can be normalized by dividing it by $N^0/2$ (total carrier in one membrane interface). Feldberg and Kissel (1975) calculated

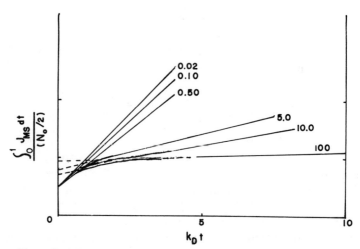

Fig. 11. Normalized flux integral as a function of time. The lines are for the condition $k'_{MS}/k_D = \infty$, $k''_{MS}/k_D = 0$, and $K_h C_M = 1.0$ and several values of $k_R C_M/k_S$. (After Feldberg and Kissel, 1975.)

this normalized flux integral as a function of normalized time ($k_D t$) for the conditions, $k'_{MS}/k_D = \infty$, $k''_{MS}/k_D = 0$ and $K_h C_M = 1.0$. The data obtained for several values of the parameter $k_R C_M/k_S$ are shown in Fig. 11.

When $K_h C_m = 1.0$, according to Eq. (283), $N_{MS}/(N^0/2) = 0.5$. This means that at $t = 0$, the intercept of all solid lines is 0.5 and corresponds to N_{MS} moved across the membrane. The dashed lines are the steady-state extrapolations to zero time. In theory with perfect electronics and large values of $k_R C_M/k_S$, three intercepts should be detected. The first voltage intercept given by Eq. (289), corresponding to quick transport of N_{MS} across the membrane, may be difficult to detect. But it can be calculated since q is known and C_m can be determined. The second intercept E_2 pertains to the completion of N_{MS} transport across the membrane (zero-time intercept of the steady-state straight line) and the onset of conversion of N_S to N_{MS}. The third intercept E_3 is the steady-state voltage decay extrapolated to zero time.

In practice E_3 is the measurable parameter and E_1 can be calculated as already indicated. From Fig. 11 it is seen that two limiting cases are well defined. (1) When $k_R C_M/k_S$ is large, the intercept approximates $N^0/2$. (2) When $k_R C_M/k_S$ is small, the intercept equals only N_{MS} ($= 0.5$).

Practical evaluations of the several kinetic parameters are carried out by the following curve-fitting procedures.

The capacitance of the membrane C_m is measured by following the rate of potential decay across the membrane through an external resistor. Integration of Eq. (268) gives

$$\ln(E/E_0) = (1/RC_m)t.$$

A plot of $\ln E$ against t gives a straight line whose slope is $(1/RC_m)$ and whose intercept is $\ln E_0$. As R and q are known, C_m can be evaluated both from the evaluation of the slope and the intercept. In this way a value of 0.45 $\mu F/cm^2$ was obtained by Feldberg and Kissel (1975) for the capacitance of a glyceromonooleate (GMO) membrane (solvent: n-decane).

These bilayer membranes doped with actins or valinomycin were used to collect charge-pulse data for the transport of NH_4^+ and K^+, respectively. The data pertained to decay of voltage following charge injection into the membrane employing a constant ionic strength (3 M) of the aqueous phase. The experimental points of the decay curve were fitted to the integrated form of Eq. (285) combined with Eq. (268). The following values for the several parameters of the equation fitted the experimental decay curves of the different membrane systems.

Membrane system	C_M (M/cm^3)	n	k_{MS}^*/k_D $(k_R C_M/2k_S) \approx 0$	$\dfrac{N_S C_M K_h k_{MS}^*}{C_m} F \cdot t$	t (s)
GMO + Nonactin + NH_4^+	3.0×10^{-6}	0.33	0.25	4.69	2×10^{-2}
GMO + Monactin + NH_4^+	3.75×10^{-6}	0.33	0.6	7.14	2×10^{-2}
GMO + Dinactin + NH_4^+	4.5×10^{-6}	0.33	0.8	11.1	2×10^{-2}
GMO + Trinactin + NH_4^+	5.9×10^{-6}	0.33	1.4	20.4	2×10^{-2}
GMO + Valinomycin + K^+	7.5×10^{-6}	0.26	$\dfrac{k_{MS}^*}{k_D}\left(1 + \dfrac{k_R C_M}{2k_S}\right)$ 1.6	12.5	5×10^{-2}

The values of n indicate that the positions of the peaks of the energy barriers in the membranes are different for the actin and valinomycin systems. This type of study extended to voltage-decay curves at several concentrations (C_M) revealed that in the case of actins the shape factor

$$\frac{k_{MS}^*}{k_D}\left(1 + \frac{k_R C_M}{2k_S}\right)$$

remained constant and the scaling factor $K_h k_{MS}^* N_S$ diminished with increasing C_M. On the other hand, in the case of valinomycin, the scaling factor remained constant and the shape factor increased with increase in C_M. These findings enabled Feldberg and Kissel (1975) to estimate values for the several kinetic parameters.

Inverting Eq. (282) and multiplying by $1/K_h k_{MS}^*$ gives

$$\frac{1}{K_h k_{MS}^* N_S} = \frac{1}{(N^0/2) K_h k_{MS}^*} + \frac{C_M}{(N^0/2) k_{MS}^*}. \qquad (291)$$

TABLE VII

Values for the Steady-State Parameters of the Actin-Mediated NH_4^+ Transport in Glyceromonooleate Membrane[a]

Parameter	Nonactin	Monactin	Dinactin	Trinactin
C_S (M/cm^3)	7.1×10^{-8}	7.1×10^{-8}	6.1×10^{-8}	6.2×10^{-8}
$K_h k_{MS}^*$ ($N^0/2$)	9×10^{-6}	1.3×10^{-5}	1.6×10^{-5}	1.2×10^{-5}
k_{MS}^*/k_D	0.28	0.6	0.8	1.0
K_h (cm^3/M)	4.8×10^3	7.5×10^3	12×10^3	10×10^3
k_{MS}^* ($N^0/2$)	1.9×10^{-9}	1.7×10^{-9}	1.3×10^{-9}	2.1×10^{-9}
k_R ($N^0/2$)	3.2×10^{-5}	2.2×10^{-5}	2.1×10^{-5}	1.2×10^{-5}

[a] From Feldberg and Kissel (1975).

According to this equation, the plot of $1/K_h k_{MS}^* N_S$ versus C_M should give a straight line, and hence the slope and the intercept should yield values for K_h and $N^0 k_{MS}^*$. In this way Feldberg and Kissel (1975) derived values for the several parameters which are given in Table VII.

In the case of valinomycin the values of

$$\frac{k_{MS}^*}{k_D}\left(1 + \frac{k_R C_M}{2k_S}\right)$$

that gave the best fit of experimental data points at several concentrations of C_M were plotted against C_M. From the straight line, the following values were estimated for the several parameters.

Ion	$K_h k_{MS} N_S$	k_{MS}^*/k_D	k_R/k_S	$k_S N_S$
Cs^+	1.8×10^{-6}	0.4	1.6×10^4	2.4×10^{-10}
K^+	4.0×10^{-6}	1.6	1.06×10^4	2.8×10^{-10}

$C_S = 8.85 \times 10^{-8}$ M/cm^3

The steady-state data given in Table VII for the actins and the above for valinomycin together with measurements of intercept discrepancy enabled estimation of values for the several kinetic parameters.

In the case of actins, because $k_R C_M/k_S$ was small, N_{MS} was evaluated from Eq. (290) and $N^0/2$ from Eq. (283) since K_h was known. In the case of valinomycin, because the value of $k_R C_M/k_S$ was large, two intercepts pertaining to E_2 and E_3 were detected. Furthermore, the voltage diminished below the value required to maintain $k'_{MS} = \infty$. In view of these difficulties the intercept discrepancy method gave rough estimates for N_S/C_S ($\sim 2 \times 10^{-8}$) and K_h ($\sim 3 \times 10^3$). The several values derived are given in Table VIII.

TABLE VIII

Values for the Several Kinetic Parameters of the GMO Membrane[a]

Parameter	Nonactin + NH_4^+	Monactin + NH_4^+	Dinactin + NH_4^+	Trinactin + NH_4^+	Valinomycin + K^+
C_S (M/cm^3)	7.1×10^{-8}	7.1×10^{-8}	6.1×10^{-8}	6.2×10^{-8}	8.85×10^{-8}
N_{MS}	1.2×10^{-13}	9.9×10^{-14}	8.6×10^{-14}	7.8×10^{-14}	
$(N^0/2)$	1.7×10^{-13}	1.3×10^{-13}	9.9×10^{-14}	9.5×10^{-14}	
$(N^0/2)/C_S$	2.5×10^{-6}	1.8×10^{-6}	1.6×10^{-6}	1.5×10^{-6}	
K_h					$\sim 3 \times 10^3$
N_S/C_S					$\sim 2 \times 10^{-8}$
k_R ($M^{-1} s^{-1}$)	1.8×10^8	1.7×10^8	2.1×10^8	1.2×10^8	1.7×10^9
k_D (s^{-1})	4.3×10^4	2.2×10^4	1.6×10^4	1.3×10^4	4.7×10^5
k_{MS}^* (s^{-1})	1.1×10^4	1.3×10^4	1.3×10^4	1.3×10^4	7.5×10^5
k_S (s^{-1})	$>3 \times 10^5$	$>2 \times 10^5$	$>3 \times 10^5$	$>1.5 \times 10^5$	1.6×10^5

[a] From Feldberg and Kissel (1975).

In the case of actins, only a lower limit to k_S could be estimated assuming that a 20% variation in the parameter $1 + k_R C_M/2k_S$ would affect the shape of the voltage decay curve (experimentally no such decay was noted). This means $k_S > (1/0.4)k_R C_{M,max}$, where $C_{M,max}$ is the maximum value of C_M used in the experiment.

Thus a simple experiment, with appropriate electronic system and suitable membrane system, has been shown to yield a wealth of information about the kinetics of carrier transport. Feldberg and Nakadomari (1977) examined both theoretically and experimentally the valinomycin-mediated transport of K^+ by the charge-pulse technique when the doped bilayer membrane was subject to both high and low voltages. In the high-voltage regime, the conditions

$$\frac{k'_{MS} - k''_{MS}}{k'_{MS} + k''_{MS}} \cong 1, \tag{292}$$

$$\frac{k'_{MS} + k''_{MS}}{k_D} \gg 1, \tag{293}$$

and that all of the charged complex N_{MS} ($= K_h C_M N_S$) moved across the membrane instantaneously prevailed. The time-dependent voltage $E^*(t, \infty)$ is given by

$$E^*(t, \infty) = \frac{q}{C_m} - \frac{F}{C_m} N_S \left[\left(\frac{B}{1+B} \right)^2 \right.$$
$$\left. \times \left\{ 2k_S \frac{(1+B)t}{B} - (\exp[-(2k_S - k_R C_M)t] - 1) \right\} + K_h C_M \right], \tag{294}$$

where
$$B = k_R C_M / 2k_S. \tag{295}$$

In the steady state when the exponential term tends to zero, Eq. (294) becomes

$$E(t, \infty) = \frac{q}{C_m} - \frac{F}{C_m} N_S \left[\left(\frac{B}{1+B} \right)^2 \left\{ 2k_S \frac{(1+B)t}{B} + 1 \right\} + K_h C_M \right]. \tag{296}$$

According to Eq. (296), $E(t, \infty)$ decays linearly with time, and its extrapolation to zero time gives the intercept voltage $E(0, \infty)$ as

$$E(0, \infty) = \frac{q}{C_m} - \frac{F N_S}{C_m} \left[\left(\frac{B}{1+B} \right)^2 + K_h C_M \right]. \tag{297}$$

Thus the intercept discrepancy as defined is given by

$$\Delta E(0, \infty) = \frac{q}{C_m} - E(0, \infty). \tag{298}$$

Substituting Eq. (297) into Eq. (298) gives

$$\Delta E(0, \infty) = \frac{F N_S}{C_m} \left[\left(\frac{B}{1+B} \right)^2 + K_h C_M \right]. \tag{299}$$

This can be written

$$\Delta E(0, \infty) = \Delta E_1(0, \infty) + \Delta E_2(0, \infty), \tag{300}$$

where

$$\Delta E_1(0, \infty) = \frac{F}{C_m} N_S K_h C_M \tag{301}$$

and

$$\Delta E_2(0, \infty) = \frac{F}{C_m} N_S \left(\frac{B}{1+B} \right)^2. \tag{302}$$

Equation (301) follows directly from Eq. (294) when $t = 0$.

Differentiation of Eq. (296) with respect to time gives

$$-\frac{dE(t, \infty)}{dt} = \frac{F}{C_m} N_S \frac{B}{1+B} 2k_S = \frac{F}{C_m} \frac{k_R C_M N_S}{1 + k_R C_M / 2k_S}. \tag{303}$$

But

$$J_{MS} = -\frac{C_m}{F} \frac{dE(t, \infty)}{dt} = \frac{k_R C_M N_S}{1 + k_R C_M / 2k_S}. \tag{304}$$

Application of the conditions of Eqs. (292) and (293) to Eq. (285) gives Eq. (304). Thus the intercept-discrepancy method is in agreement with the Lauger–Stark (1970) and Hladky (1972, 1979) steady-state model in the high-voltage limit.

As opposed to the high-voltage regime where charge distribution on the membrane surface does not change when the steady-state condition is reached, in the low-voltage regime continuous readjustment of surface concentrations take place with change in voltage. The conditions that prevail are

$$k'_{MS} + k''_{MS} = 2k_{MS}, \tag{305}$$

$$k'_{MS} - k''_{MS} = k_{MS}EF/RT. \tag{306}$$

Starting from the kinetic equations of Stark et al. (1971), Feldberg and Nakadomari (1977) have derived an equation to describe the low-voltage $[E(t, L)]$ behavior in the steady state. The equation is given by

$$\frac{d \ln E(t, L)}{dt} = -\left(\frac{F^2}{RTC_m} \frac{k_R C_M N_S W}{1 + A}\right) \Big/ (1 + Q), \tag{307}$$

where

$$W = k_{MS}/k_D, \tag{308}$$

$$A = 2W(1 + B) = \frac{2k_{MS}}{k_D}\left(1 + \frac{k_R C_M}{2k_S}\right), \tag{255'}$$

and

$$Q = \frac{1}{2}\frac{F^2 N_S}{RTC_m}\left(\frac{A}{1+A}\right)^2\left[\left(\frac{B}{1+B}\right)^2 + K_h C_M\right]. \tag{309}$$

According to Eq. (307), a plot of $\ln E(t, L)$ against time will be a straight line, and that equation is valid even when the transient originated in the high-voltage region. Consequently, from the same transient, a value can be derived for the ratio r, which is expressed as

$$r = \frac{d \ln E(t, L)/dt}{dE(t, \infty)/dt}. \tag{310}$$

Substitution of Eqs. (307) and (303) into Eq. (310) gives

$$r = \frac{\frac{1}{2}(F/RT)A}{(1 + A)(1 + Q)}. \tag{311}$$

So far the theory of the charge-pulse method has not considered the voltage dependence of the rate constants. Formally treating the membrane as an image force potential energy barrier for the diffusing ions, the voltage-dependent rate constants for the transfer of charged complex may be written [see Eq. (168)]

$$k'_{MS} = k_{MS}\exp(-\delta EF/2RT)\exp[-\omega(EF/RT)^2],$$
$$k''_{MS} = k_{MS}\exp(\delta EF/2RT)\exp[-\omega(EF/RT)^2]. \tag{312}$$

If the reaction (237) is confined to a plane in the membrane, then $(1-\delta)/2$ is the fraction of voltage that acts between the reaction plane in the membrane and the aqueous solution and so the reaction rate constants k'_R, k'_D, and k''_R, k''_D become voltage dependent. Thus

$$k'_R = k_R\exp[-(1-\delta)EF/4RT], \qquad k''_R = k_R\exp[(1-\delta)EF/4RT],$$
$$k'_D = k_D\exp[(1-\delta)EF/4RT], \qquad k''_D = k_D\exp[-(1-\delta)EF/4RT]. \tag{313}$$

Feldberg and Nakadomari (1977) redefined the voltage dependence of the rate constants as

$$k'_{MS} = k_{MS}\exp(\delta EF/2RT), \tag{314}$$
$$k''_{MS} = k_{MS}\exp(-\delta EF/2RT), \tag{315}$$
$$k'_R = k_R\exp[b(1-\delta)/2RT], \tag{316}$$
$$k'_D = k_D\exp[-(1-b)(1-\delta)/2RT], \tag{317}$$
$$k''_R = k_R\exp[-b(1-\delta)/2RT], \tag{318}$$
$$k''_D = k_D\exp[1-b)(1-\delta)/2RT], \tag{319}$$

where b locates the "on–off" reaction plane. It is the fraction of distance between the plane of approach of aqueous permeable ion and the plane where the ion complex resides ($0 \leq b \leq 1$, $b \approx 0.5$).

The steady-state flux of the complex is given by

$$J_{MS} = k'_R C_M N_S - k'_D N'_{MS} \tag{320}$$

for the (') interface,

$$J_{MS} = k'_{MS} N'_{MS} - k''_{MS} N''_{MS} \tag{321}$$

for the membrane interior, and

$$J_{MS} = k''_D N''_{MS} - k''_R C_M N_S \tag{322}$$

for the (″) interface.

Eliminating N'_{MS} and N''_{MS} with the help of Eqs. (320) and (322) from Eq. (321) gives on rearrangement

$$J_{MS} = C_M N_S \frac{k'_{MS} k'_R/k'_D - k''_{MS} k''_R/k''_D}{1 + k'_{MS}/k'_D + k''_{MS}/k''_D}. \tag{323}$$

Substituting Eqs. (314)–(319) into Eq. (323) gives on simplification

$$J_{MS} = \frac{C_M N_S K_h k_{MS}[\exp(EF/2RT) - \exp(-EF/2RT)]}{1 + (k_{MS}/k_D)\{\exp[(1-b+b\delta)EF/2RT] + \exp[-(1-b+b\delta)EF/2RT]\}} \quad (324)$$

or

$$J_{MS} = \frac{2C_M N_S K_h k_{MS} \sinh(EF/2RT)}{1 + (2k_{MS}/k_D)\cosh[(1-b+b\delta)EF/2RT]}. \quad (325)$$

At low voltage, Eq. (325) becomes

$$J_{MS} = \frac{K_h C_M N_S k_{MS} E(t,L)(F/RT)}{1 + 2k_{MS}/k_D} = \frac{k_R C_M N_S W E(t,L)(F/RT)}{1 + 2W}. \quad (326)$$

Thus at low voltages, the flux of the ion complex becomes independent of both b and δ. At high voltages, Eq. (324) becomes

$$J_{MS} = k_R C_M N_S \exp[b(1-\delta)E(t,\infty)(F/2RT)]. \quad (327)$$

Thus

$$\frac{dE(t,\infty)}{dt} = -\frac{FJ_{MS}}{C_m} = -\frac{F}{C_m} k_R C_M N_S \exp\left[b(1-\delta)E(t,\infty)\left(\frac{F}{2RT}\right)\right]$$

or

$$\exp\left[-b(1-\delta)\left(\frac{F}{2RT}\right)E(t,\infty)\right]dE(t,\infty) = -\frac{F}{C_m} k_R C_M N_S \, dt. \quad (328)$$

Integration of Eq. (328) gives

$$\left[\frac{\exp[-b(1-\delta)(F/2RT)E(t,\infty)]}{b(1-\delta)(F/2RT)}\right]_{E(0,\infty)}^{E(t,\infty)} = \left[\frac{F}{C_m} k_R C_M N_S t\right]_{t=0}^{t},$$

$$\exp\left[-b(1-\delta)\left(\frac{F}{2RT}\right)E(t,\infty)\right] - \exp\left[-b(1-\delta)\left(\frac{F}{2RT}\right)E(0,\infty)\right] \quad (329)$$

$$= \frac{F^2}{2RT} \frac{b(1-\delta)}{C_m} k_R C_M N_S t.$$

The experimental data for KCl derived by the charge-pulse experiment using glycerol monooleate membrane and valinomycin have been analyzed using the equations given above. A high-voltage charge pulse was applied to the doped membrane using as low a concentration of KCl as possible ($\sim 10^{-2}$ M) and keeping the decay time constant to less than a millisecond. The experimental data points in the time less than 600 μs conformed to Eq. (329) when the appropriate value for $b\delta$ was guessed. With a limiting value of 0.045 for $b(1 - \delta)/2$, the first term on the left-hand side of Eq. (329) plotted against time gave a straight line passing through the experimental points. Since b is assumed to have a value of 0.5, $\delta \approx 0.82$. This means 82% of the voltage applied is effective in translocating the ion complex in the membrane. From the slope of the straight line, a limiting value for dE/dt can be estimated. This value would correspond to a complexation rate that is not assisted by the applied field. At high concentrations of KCl (~ 1 M), similar fitting procedures gave again values for dE/dt. Although the theory for high concentrations is not well founded, the linear plot [Eq. (329)] was good provided the time did not exceed 50 μs. A plot of $[1/(dE/dt)]$ against $[1/C_M]$ according to Eq. (304) was linear giving values for the slope and the intercept:

$$\text{slope} = C_m/Fk_R N_S \quad (\text{V}^{-1}\,\text{s}\,\text{mol}\,\text{cm}^{-3}),$$

$$\text{intercept} = C_m/2Fk_S N_S \quad (\text{V}^{-1}\,\text{s}).$$

Thus values for the several parameters obtained were

$$k_R N_S = 1.45 \times 10^{-4} \quad (\text{cm}\,\text{s}^{-1}),$$

$$k_S N_S = 1.36 \times 10^{-8} \quad (\text{mol}\,\text{s}^{-1}\,\text{cm}^{-2}),$$

and

$$k_R/2k_S = 5.3 \times 10^3 \quad (\text{cm}^3\,\text{mol}^{-1}).$$

These values served as a guide to fit the data of intercept discrepancy measured as a function of C_M to Eq. (299). In this curve fitting the values used were

$$C_m = 64 \ \mu\text{F/cm}^2, \quad K_h = 10^3 \ \text{cm}^3\,\text{mol}^{-1},$$

$$k_R/k_S = 1.08 \times 10^4 \ \text{cm}^3\,\text{mol}^{-1}, \quad N_S = 5.5 \times 10^{-13} \ \text{mol}\,\text{cm}^{-2}.$$

From these data, values for the kinetic parameters were deduced as

$$k_R \ (\text{cm}^3\,\text{mol}^{-1}\,\text{s}^{-1}) = 2.6 \times 10^8,$$

$$k_D \ (\text{s}^{-1}) = 2.6 \times 10^5,$$

$$k_S \ (\text{s}^{-1}) = 2.6 \times 10^4.$$

An estimate of k_{MS} requires a value for $W(= k_{MS}/k_D)$. Although unambiguous determination is difficult, it could be deduced by measuring r [see Eqs. (310) and (311)] as a function of C_M. These data points could be fitted by choosing an appropriate value for W, which was about 4, and so $k_{MS} \approx 4k_D$.

A carrier-mediated ion flux J_M at low concentrations of C_M increases linearly with C_M. At high concentrations, the carrier becomes saturated, and J_M finally approaches a maximal value $J_{M,\text{max}}$. The maximum number of ions that may be transported per second by a single carrier molecule (turnover number f) is given by $f = J_{M,max}/N$, where N is the number of carrier molecules. According to Lauger (1972), this is given by

$$f = \frac{1}{1/k_{MS} + 1/k_S + 2/k_D}.$$

In view of the data of Table VI, valinomycin is able to translocate 10^4–10^5 K$^+$ per second through the bilayer.

VIII. Models for Channel-Forming Ionophores

Some examples of ionophores already considered are valinomycin and actins, which carry ions across the membrane by forming complexes. Others enhance ion permeability in lipid bilayer membranes by forming channels that span the whole width of the membrane. Examples of such compounds that have received considerable study are gramicidin A, amphotericin B, nystatin, EIM (excitability-inducing material), alamethicin, monazomycin, hemocyanin, and suzukacillin. All these at low concentrations produce discrete current fluctuations when the doped membrane is subject to a potential. The quantitative aspects of the different characteristics of the channels produced by these several ionophores are outlined.

A. Channels Produced by Gramicidin A

Gramicidin A is a linear pentadecapolypeptide. It is surface active and so if a very small amount of it (a few pM) is added to the aqueous phase, it is adsorbed to the bilayer membrane, which when subject to a fixed applied potential produces fluctuations in membrane current in a well defined steplike manner (see Fig. 12). Each upward step is followed by an equal downward transition indicating the opening and closing of a single conducting

Fig. 12. Representation of current fluctuations through phosphatidyl ethanolamine bilayer membrane doped with gramicidin A. Aqueous phases were 1.0 M KCl. Membrane potential 400 mV. Current changes occur in random. (After Andersen, 1977.)

channel. The height of each transition is fairly uniform, but the duration (time interval between opening and closing of a channel) varies. Hladky and Haydon (1972) found the magnitudes of both conductance step and duration were dependent on temperature. The activation energy for channel conductance was about 5 kcal/mol ($Q_{10} \approx 1.9$) corresponding to the ion conductance in an aqueous medium. On the other hand, the rate of termination of the channel (reciprocal of the duration of the channel) had an activation energy of about 19 kcal/mol ($Q_{10} \approx 10$).

Increase in the concentration of gramicidin increased the frequency of occurrence of the steps leading to overlapping, and when this occurred fluctuations could not be resolved. Currents were found to be multiples of a basic unit of a few picosiemens ($\sim 10^{-12}\,\Omega^{-1}$).

From statistical studies of doped bilayer membranes containing small quantities of gramicidin, it was found that the probability of finding 0, 1, 2, etc., channels open at any one time was given by the Poisson distribution. Although it was difficult to examine the frequency of opening of channels, it was, however, found that the frequency of opening of channels was greater the thinner the membrane and the higher the applied potential.

Bamberg and Läuger (1973) investigated the kinetics of gramicidin channel formation by relaxation measurements. The model of the gramicidin channel used was that it contained a dimer, which is formed by head-to-head association of two gramicidin monomers. The hole along the helix axis is lined by the peptide C=O moieties. The hydrophobic residues were on the exterior surface of the helix. On the basis of this model, the opening and closing of the channels in the membrane are due to an association–dissociation reaction

$$\text{Gr} + \text{Gr} \underset{k_\text{D}}{\overset{k_\text{R}}{\rightleftharpoons}} \text{Gr}_2 \tag{330}$$

where Gr_2 is the dimer channel. Assuming that the total gramicidin concentration N_t in the membrane remained constant, the concentrations of monomers (N_m: mol/cm^2) and dimers (N_d: mol/cm^2) are given by

$$N_\text{m} + 2N_\text{d} = N_\text{t} = \text{const.} \tag{331}$$

At time $t < 0$, the system is in equilibrium with the concentrations N_m^0 and N_d^0. Therefore

$$N_\text{d}^0/(N_\text{m}^0)^2 = k_\text{R}^0/k_\text{D}^0 = K_0. \tag{332}$$

At time $t = 0$ when a voltage is applied, the rate constants are changed to k_R and k_D, and the system reaches a new equilibrium with concentrations N_m^∞ and N_d^∞. Thus

$$N_\text{d}^\infty/(N_\text{m}^\infty)^2 = k_\text{R}/k_\text{D} = K_\infty. \tag{333}$$

The rate of change of N_D is given by

$$\frac{dN_D}{dt} = k_R N_m^2 - k_D N_d. \tag{334}$$

Taking the fraction of the dimer as

$$Y = N_d/N_t, \tag{335}$$

and using Eq. (331), Eq. (334) becomes

$$\frac{1}{k_D}\frac{dY}{dt} = 4N_t K_\infty Y^2 - (1 + 4N_t K_\infty)Y + N_t K_\infty. \tag{336}$$

The solution of Eq. (336) is given by

$$Y(t) = Y_\infty + (Y_0 - Y_\infty)\frac{L\exp(-t/\tau)}{1 + L - \exp(-t/\tau)}, \tag{337}$$

where

$$L = \frac{\sqrt{1 + 8N_t K_\infty}}{4N_t K_\infty (Y_\infty - Y_0)}, \tag{338}$$

$$\tau = 1/(k_D\sqrt{1 + 8N_t K_\infty}). \tag{339}$$

Substituting from Eqs. (331) and (333) in Eq. (339) gives

$$\tau = \frac{1}{k_D + 4k_R N_m^\infty}. \tag{340}$$

The values of Y_∞ and Y_0 follow from Eqs. (331)–(333) and (335) as

$$Y_0 = \frac{1}{8N_t K_0}(1 + 4N_t K_0 - \sqrt{1 + 8N_t K_0}), \tag{341}$$

$$Y_\infty = \frac{1}{8N_t K_\infty}(1 + 4N_t K_\infty - \sqrt{1 + 8N_t K_\infty}). \tag{342}$$

At any given applied potential E_m, the current density $I(t)$ is proportional to the number of dimer channels N_d and so

$$I(t) = N_d \gamma E_m, \tag{343}$$

where γ is the single channel conductance. In view of Eqs. (343) and (335), Eq. (337) can be expressed as

$$\frac{I(t) - I(0)}{I(\infty) - I(0)} = 1 - L\frac{\exp(-t/\tau)}{1 + L - \exp(-t/\tau)}. \tag{344}$$

When $L \gg 1$, i.e., from Eq. (338) $|Y_\infty - Y_0| \to 0$ or $N_t K_\infty \approx 0$, Eq. (344) reduces to

$$\frac{I(t) - I(0)}{I(\infty) - I(0)} \approx 1 - \exp\left(-\frac{t}{\tau}\right). \tag{345}$$

The stationary conductance G_∞ is related to the single-channel conductance by the relation

$$G_\infty = N N_d^\infty \gamma, \tag{346}$$

where N is Avogadro's number. From Eq. (333), Eq. (346) becomes

$$G_\infty = (k_R/k_D)(N_m^\infty)^2 N\gamma. \tag{347}$$

Thus substituting for N_m^∞ in Eq. (340) gives

$$\tau = \frac{1}{k_D + 4\sqrt{k_D k_R G_\infty/N\gamma}}. \tag{348}$$

Thus a plot of $1/\tau$ against $\sqrt{G_\infty}$ (i.e., measurement of τ and G_∞ at several concentrations of gramicidin) should give a straight line whose intercept on the $1/\tau$ axis is equal to k_D and whose slope is equal to $4\sqrt{k_D k_R/N\gamma}$. Thus from independent measurements of single-channel conductance, values for k_D and k_R can be determined.

According to the model considered, disappearance of a channel may be equated to the dissociation of a dimer. In this case the reciprocal of the dissociation constant k_D corresponds to the mean life time τ_c of a channel. Some of the values of k_D, k_R, τ_c, and K_∞ determined as functions of temperature together with the values for the activation energies are given in Table IX.

The values of τ_c and $1/k_D$ agree within about 40% in the temperature range 10–40°C, in which $1/k_D$ varies 20-fold. The energy for the formation of the channel (i.e., E_R) is fairly high. This may mean that either a high-energy-consuming rearrangement of the lipid structure around the dimer is involved or that several hydrogen bonds between gramicidin and water must be broken before dimerization can occur.

The relaxation amplitude a is given by

$$a = (I_\infty - I_0)/I_0 = (Y_\infty - Y_0)/Y_0. \tag{349}$$

Substituting for Y_∞ and Y_0 from Eqs. (341) and (342) gives in the limit of small concentrations of gramicidin, i.e., $N_t K_\infty, N_t K_0 \ll 1$,

$$a \approx (K_\infty - K_0)/K_0 \qquad (G_\infty \to 0). \tag{350}$$

If a is known in the limit of low conductance (around 5 according to Bamberg and Lauger, 1973), a value for K_0 can be calculated from Eq. (350)

TABLE IX

Values for the Kinetic Parameters of the Gramicidin Channel in Dioleoyllecithin in n-Decane[a]

Parameter	Temperature (°C)		
	10	25	40
k_R (cm^2 mol^{-1} s^{-1})	2.3×10^{13}	20×10^{13}	68×10^{13}
k_D (s^{-1})	0.25	1.6	4.5
$K_\infty = k_R/k_D$ (cm^2 mol^{-1})	8.8×10^{13}	12×10^{13}	15×10^{13}
γ (pS)	6.5	12	23
τ_c (s)	3.8	1.1	0.15
$1/k_D$ (s)	4.0	0.63	0.22

Activation energies
$E_{k_R} = 20$ kcal/mol
$E_{k_D} = 17$ kcal/mol
$E_\gamma = 7.3$ kcal/mol

[a] From Bamberg and Läuger (1974).

using a value of $K_\infty = 10^{15}$ cm^2 mol^{-1} determined at a voltage of 205 mV. Thus the equilibrium constant K_0 at zero voltage is 1.5×10^{14} cm^2 mol^{-1}. Furthermore, one can calculate N_t, which appears in Eqs. (341) and (342), from the relation that follows from Eqs. (331), (333) and (346). Thus

$$N_t = 2G_\infty/N\gamma + \sqrt{G_\infty/K_\infty N\gamma}. \tag{351}$$

The other aspect of the gramicidin channel that has interested several investigators is the mechanism of ion transport in this peptide-lined pore. The single-channel conductance indicates a flux of 10^7 ions per second. The channel is selective to monovalent cations rather than anions. The conductance sequence for alkali cations is the same as for aqueous diffusion, with H$^+$ having the highest value.

The gramicidin A channel is a narrow pore of width about 4 Å, consisting of an array of coordination sites associated with the peptide carbonyls. An ion or a water molecule goes through the pore by jumping from site to site. The interactions with the sites lead to a series of potential energy minima along the route of the ion. On top of it is the dielectric interaction of the ion with the water phase and the membrane lipid giving rise to a broad energy barrier with peak in the middle of the membrane. So the binding site for the ion should exist close to either end of the channel since that position is energetically favored.

Study of single-channel conductance as a function of concentration has thrown light on the probable number of ions present in the channel. If the channel were empty, the conductance would increase linearly at all concentrations. This is not observed. If the channel had only one ion at a time, the conductance would reach a limiting value corresponding to the channel being always occupied. If a second ion could enter the channel, interactions would occur, thereby leading to drastic changes in conductance. These interactions are indicated by observations such as concentration-dependent permeability ratios, blocking effect by ions such as thallium, divalent ions, peculiar shape of conductance dependence on concentration, and most important, deviations from the Ussing flux ratio relation. These and other observations have been analyzed on the basis of two models. Model 1 is simpler and is based on the presence in the membrane of a central barrier with two binding sites, one in each of the two energy minima at the ends of the pore. Model 2 is very complex and is based on the presence of one barrier in the center of the membrane and four binding sites, two of which exist at one end of the pore in the two energy minima and the other two exist at the other end of the pore in the corresponding two energy minima. To develop tractable equations, several simplifying assumptions have been made. The one-barrier–four-sites model developed by Eisenman *et al.* (1978) is too complex, and their papers must be consulted for details. Here an outline of the two-site model considered by several investigators (see review by Finkelstein and Andersen, 1981) is presented.

1. Channel Occupancy by a Single Ion

Single-ion occupancy of a channel is indicated if (a) single-channel conductance should increase with an increase in the activity of the ion and attain saturation, and (b) the Behn–Ussing flux ratio equation [see Eq. (112) of Chapter 4] should be followed. An example is the Na^+ in the gramicidin channel. The single-channel conductance versus Na activity curve followed Eq. (138) of Chapter 4, i.e.,

$$\gamma_{Na} \propto \frac{[Na^+]}{K_{Na} + [Na^+]}$$

and

$$K_{Na} = \frac{[\text{channel}][Na^+]}{[\text{channel} - Na^+]}.$$

Finkelstein and Andersen (1981) found values of $K_{Na} = 0.31\ M$ and $\gamma_{Na,max} = 14.6$ pS to fit the experimental curve in the case of NaCl and gramicidin A channel in the bacterial phosphatidylethanolamine/n-decane membrane.

When there is flux interaction in the two directions, the Ussing flux ratio equation becomes

$$(\vec{J}_j/\overleftarrow{J}_j) = \exp(nz_j \Delta EF/RT),$$

where n gives the number of ions simultaneously present in the channel. Applying this equation to the case of Na$^+$, single channel conductance γ_{Na} measured when $\Delta E \to 0$ is given by [see Eq. (115) of Chapter 4]

$$\gamma_{Na} = (nF^2/RT)J_{Na}. \tag{352}$$

Andersen and Porcopio (1980) found this equation to be valid when $n = 1$ in the case of Na$^+$ and gramicidin A–doped bacterial phosphatidyl ethanolamine/n-decane membrane system. When the value of 14.6 pS is substituted for γ_{Na}, $J_{Na} \approx 2 \times 10^6$ ions per second.

2. Channel Occupancy by Two Ions

The lines of evidence on which two-ion occupancy of the gramicidin A channel is based are (a) the single-channel conductance versus activity curve reaches a maximum and then decreases with increase in concentration; (b) a minimum in the single-channel conductance versus activity curve occurs at low ion concentration, indicating at least two binding sites in the channel; (c) the exponent n in the Ussing flux ratio equation is larger than unity; (d) the permeability ratios become functions of ion concentration; (e) replacement of a small amount of alkali metal cation (Na$^+$, K$^+$, or Cs$^+$) by Ag$^+$ or Tl$^+$ produces a marked decrease in single-channel conductance.

The kinetic model of two ions per channel is indicated in the state diagram of Fig. 13. Single-cation (e.g., Cs$^+$) of concentrations C_1 and C_2 are present in the left- and right-hand solutions, respectively. The channel can exist in four states with probabilities P_0 (empty), an ion in the left well (P_L), an ion in the right well (P_R), and both wells occupied (P_2). Of the several rate constants k_j, only k is considered to be voltage dependent. Hence the ion flux

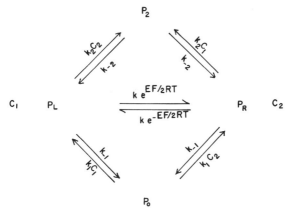

Fig. 13. State diagram of two ions per channel (see text for description).

through the channel is given by

$$J = k[P_L \exp(EF/2RT) - P_R \exp(-EF/2RT)]. \quad (353)$$

The solution to the probability of ion occupying different states can be realized by following the King–Altman procedure (see Section II).

The probability P_L is proportional to the sum of the several rate constants as determined by the King–Altman procedure indicated by (see Fig. 13 to follow the arrows)

P_L probability:

Thus P_L is proportional to

$$C_1 k_1 k_{-2}[2k_{-1} + k_2(C_1 + C_2)] + 2k \exp(-EF/2RT) k_1 k_{-2}(C_1 + C_2). \quad (354)$$

Similarly P_R, P_2, and P_0 are proportional, respectively, to

$$C_2 k_1 k_{-2}[2k_{-1} + k_2(C_1 + C_2)] + 2k \exp(EF/2RT) k_1 k_{-2}(C_1 + C_2), \quad (355)$$

$$C_1 C_2 k_1 k_2[2k_{-1} + k_2(C_1 + C_2)] + kk_1 k_2\{C_1^2 \exp(EF/2RT) \\ + C_2^2 \exp(-EF/2RT) + C_1 C_2[\exp(EF/2RT) + \exp(-EF/2RT)]\}, \quad (356)$$

and

$$k_{-1} k_{-2} k_2 (C_1 + C_2) + 2k_{-1}^2 k_{-2} \\ + 2kk_{-1} k_{-2} [\exp(EF/2RT) + \exp(-EF/2RT)]. \quad (357)$$

If $\Sigma = P_0 + P_L + P_R + P_2$ $(=1)$, then

$$P_L = \frac{\text{Eq. (354)}}{\Sigma}, \quad P_R = \frac{\text{Eq. (355)}}{\Sigma},$$

$$P_2 = \frac{\text{Eq. (356)}}{\Sigma}, \quad P_0 = \frac{\text{Eq. (357)}}{\Sigma}.$$

For the simple case when $C_1 = C_2 = C$, Σ becomes

$$\Sigma = 2\{k_{-1} + k[\exp(EF/2RT) + \exp(-EF/2RT)] + Ck_2\} \\ \times (k_{-1} k_{-2} + 2Ck_1 k_{-2} + C^2 k_1 k_2). \quad (358)$$

Substituting the values of P_L and P_R into Eq. (353) gives on simplification

$$J = 2Ckk_1 k_{-2}(k_{-1} + Ck_2)[\exp(EF/2RT) - \exp(-EF/2RT)]/\Sigma$$

$$= \frac{kk_1 k_{-2} C(k_{-1} + Ck_2)[\exp(EF/2RT) - \exp(-EF/2RT)]}{\{k_{-1} + k[\exp(EF/2RT) + \exp(-EF/2RT)] + Ck_2\}} \\ \times (k_{-1} k_{-2} + 2Ck_1 k_{-2} + C^2 k_1 k_2)$$

or

$$J = \frac{2Ck(k_{-1} + Ck_2)\sinh(EF/2RT)}{[k_{-1} + 2k\cosh(EF/2RT) + Ck_2](K_1 + 2C + C^2/K_2)}, \quad (359)$$

where $K_1 = k_{-1}/k_1$ and $K_2 = k_{-2}/k_2$, the dissociation constants for the first and second ion.

The conductance γ at small values of E is given by

$$\gamma = \frac{I}{E} = \frac{JF}{E} = \frac{F^2}{RT}\frac{2C}{K_1 + 2C + C^2/K_2}\frac{k(k_{-1} + k_2C)}{2(2k + k_{-1} + k_2C)}. \quad (360)$$

The single-channel conductance versus Cs^+ concentration curve was found in the case of diphytanoyl phosphatidylcholine/n-decane system (25-mV applied potential) by Finkelstein and Andersen (1981) to conform to Eq. (360) with the following values for the several kinetic parameters:

$$k_1 = 1.34 \times 10^8 \quad \text{mol}^{-1}\text{s}^{-1}; \quad k_{-1} = 7.8 \times 10^6 \quad \text{s}^{-1},$$

$$k = 8.9 \times 10^7 \quad \text{s}^{-1}; \quad k_2 = 1.4 \times 10^8 \quad \text{mol}^{-1}\text{s}^{-1},$$

$$k_{-2} = 5.2 \times 10^7 \quad \text{s}^{-1}; \quad K_1 = 0.06 \quad M; \quad K_2 = 0.37 \quad M.$$

According to the model of Hladky *et al.* (1979), the values derived for the gramicidin A-doped glycerol monooleate membrane system were

$$k_1 = 1.8 \times 10^8 \quad \text{mol}^{-1}\text{s}^{-1}, \quad k_{-1} = 2.9 \times 10^5 \quad \text{s}^{-1},$$

$$k = 8.2 \times 10^7 \quad \text{s}^{-1}, \quad k_2 = 1.6 \times 10^8 \quad \text{mol}^{-1}\text{s}^{-1},$$

$$k_{-2} = 1.6 \times 10^8 \quad \text{s}^{-1}.$$

According to Läuger (1973) and Finkelstein and Andersen (1981), the maximum conductance of a channel occupied by a single ion is given by

$$\lambda_{\text{max},1} = \frac{F^2}{RT}\frac{k_{-1}k}{2(2k + k_{-1})}, \quad (361)$$

whereas the maximum conductance of the channel occupied by two ions is given by

$$\lambda_{\text{max},2} = \frac{F^2}{RT}\frac{k(k_{-1} + k_{-2})}{2(2k + k_{-1} + k_{-2})}. \quad (362)$$

Before $\lambda_{\text{max},2}$ is reached the conductance decreases at high concentrations, this decrease according to Finkelstein and Andersen (1981) is described by

$$\lambda = \frac{F^2}{RT}\frac{K_2 k}{C}. \quad (363)$$

When $k_{-1}, k_{-2} \gg k$, i.e., ion getting out of the channel does not affect ion transport across the membrane, Eqs. (361) and (362) give

$$\lambda_{max,1} = \lambda_{max,2} = \frac{F^2}{RT} \frac{k}{2}. \tag{364}$$

This equation combined with Eq. (352) gives the maximum flux of the ion. Thus

$$J_{max\ 1\ or\ 2} = \frac{1}{n} \frac{k}{2}. \tag{365}$$

From the value of k already given for the Cs^+ transport across diphytanoyl phosphatidylcholine membrane, its flux is given by

$$J_{max,Cs} = 8.9 \times 10^7/(2 \times 2) = 2.2 \times 10^7 \quad \text{ions s}^{-1}.$$

Gramicidin A pore is so narrow that an ion cannot pass another ion as they pass through the membrane channel. In this single-file transport, ion–ion interaction is very significant and gives rise to flux coupling particularly in channels with many binding sites. The simplest single-file model is a pore with two binding sites. Some simple properties of single-file transport can be assessed by measuring flux ratios by using isotopically labeled permeants. According to Ussing equation a value of one for n indicates that there is no flux coupling (independence principle followed). If $n < 1$, negative flux coupling occurs and this phenomenon is noted in the case of a carrier with a single binding site. Positive coupling, i.e., $n > 1$, may occur when the carrier has more than one binding site for the mobile species; but the phenomenon is very frequent in the case of channels wherein single filing and multiple occupancy occur.

Ordinary (equilibrium) permeability P_e is related to conductance at low potentials by the relation

$$P_e = \left(\frac{RT}{F^2}\gamma\right)\frac{1}{C}.$$

Substituting for γ from Eq. (360) gives

$$P_e = \left\{\frac{2Ck(k_{-1} + k_2C)}{[K_1 + 2C + C^2/K_2]2[2k + k_{-1} + k_2C]}\right\}\frac{1}{C}. \tag{366}$$

In the case of permeability P^* determined from the flux of isotopically labeled ions, i.e., $P^* = J^*/\Delta C^*$, it has been shown by Andersen (quoted by Finkelstein and Andersen, 1981) that P^* is given by

$$P^* = \left\{\frac{2Ck(2k_{-1} + k_2C)}{[K_1 + 2C + C^2/K_2]2[2(2k + k_{-1}) + k_2C]}\right\}\frac{1}{C}. \tag{367}$$

In the case of single-ion occupancy of the channel $P_e = P^*$; but in the case of multi-ion occupancy $P_e \geq P^*$ and n in the Ussing flux equation ≥ 1.0. Combining Eqs. (366) and (367) gives

$$\frac{P^*}{P_e} = f = \frac{1}{n} = \frac{(2k_{-1} + k_2 C)(2k + k_{-1} + k_2 C)}{[2(2k + k_{-1}) + k_2 C](k_{-1} + k_2 C)}. \tag{368}$$

This equation expresses the magnitude of interaction among ions moving through the channel. The flux ratio exponent determined for Cs^+ in the gramicidin A channels as a function of cesium ion concentration conformed to Eq. (368).

B. Channels Produced by Alamethicin

Alamethicin is a cyclopeptide of molecular weight 1691. It forms channels exhibiting interesting patterns of current fluctuations. The conductance of the alamethicin-doped bilayer membrane is zero at zero potential and increases with the applied potential. The conductance is very steep, e-fold change for a change in potential between 4 and 8 mV. Discrete fluctuations of channels induced by alamethicin have been noted. A sequence of conductance jumps is shown in Fig. 14, which represents formation of single-pore structures, their transitions to several conductance states and final disappearance. The pores facilitate passage of both cations and anions exhibiting weak ionic selectivity. The values of conductance seen are not integral multiples of each other. In each step conductance increases by a large amount. At a given potential, the conductance is very sensitive to both aqueous-phase alamethicin concentration and electrolyte concentration. Increase in either concentration shifted the log concentration–voltage curves toward smaller voltages.

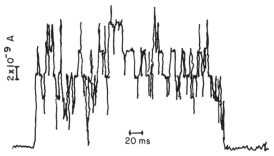

Fig. 14. Pictorial representation of multiple step fluctuations of an alamethicin channel in glycerol monooleate–cholesterol–decane. (After Ehrenstein and Lecar, 1977.)

Three models have been proposed to explain the alamethicin data. Baumann and Mueller (1974) and Boheim (1974) proposed an oligomer model. According to this, transitions to the next higher or lower channel state would occur by uptake or release of one monomer. The voltage-dependent step is the insertion of the monomers into the membrane after their reaction with the cations, which are released at the other side of the membrane. The free monomers can form dimers, trimers, etc. (i.e., oligomerize) before returning to the initial membrane side. Gordon and Haydon (1972, 1975, 1976) suggested a model in which alamethicin formed structures (oligomers) with parallel conductance units which open and close in a statistical manner. As opposed to the Baumann–Mueller–Boheim model where the pore expands by the addition of monomers, in the Gordon–Haydon model separate channels of different sizes are considered to exist. This was proposed to account for the impermeability of large cations such as tetraethylammonium ion, which would, according to the Baumann–Mueller–Boheim model, go through the alamethicin channel. A third model has been proposed by Hall (1975). In this model stable two-dimensional micelles are inserted into the bilayer membrane by the electric field and fluctuated among various configurations corresponding to discrete conductance changes. This is based on the microlevel conductance variations observed at different ionic strengths. At low ionic strength, according to the aggregation model, because of the electrostatic repulsion between an aggregate and a monomer, the higher conductance states should become less probable. The experimental fact being otherwise (highly probable), existence of micelles, which because of repulsion allow molecules into more open configurations, is favored.

Kinetic descriptions of the first two models are outlined in the following.

The statistics of current fluctuations as given by Boheim (1974) is described with reference to fluctuations shown in the scheme of Fig. 15.

The single-channel conductance γ_v in state v is given by

$$\gamma_v = I_v/E, \tag{369}$$

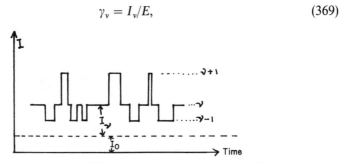

Fig. 15. Schematic representation of fluctuation of membrane current as a function of time at constant potential of a single conducting channel. Three conducting states of the channel are given by $v - 1$, v, and $v + 1$. I_0 is the current through the undoped bilayer membrane.

where I_v is the total current minus that due to undoped membrane. If t_v is the total time period, the channel exists in state v, and t_{obs} is the observation time, which is large, then the probability of existence of the state v is given by

$$P_v = t_v/t_{obs}. \tag{370}$$

During the observation time t_{obs} if state v occurs z_v times, then the mean life time τ_v of state v is given by

$$\tau_v = t_v/z_v. \tag{371}$$

If $\sigma_{v,v+1}$ is the mean number of transitions occurring per second (i.e., frequency) from state v to state $v+1$ and $\sigma_{v+1,v}$ is the corresponding number of transitions occurring from state $v+1$ to v, then under steady-state conditions the relation

$$\sigma_{v,v+1} = \sigma_{v+1,v} \tag{372}$$

is valid. It is assumed that the transitions to $v \pm 2$ from v are forbidden. So state v comes into existence either by a transition from $v+1$ to v or by a transition from state $v-1$ to v. Disappearance of state v occurs by transitions either from v to $v+1$ or v to $v-1$. Thus the total number of transitions occurring in time t_{obs} is given by

$$z_v/t_{obs} = \tfrac{1}{2}(\sigma_{v,v-1} + \sigma_{v,v+1} + \sigma_{v-1,v} + \sigma_{v+1,v}). \tag{373}$$

In view of Eq. (372), Eq. (373) becomes

$$z_v = (\sigma_{v,v-1} + \sigma_{v,v+1})t_{obs}. \tag{374}$$

Substituting from Eqs. (371) and (370) into Eq. (374) gives on rearrangement

$$P_v = \tau_v(\sigma_{v,v-1} + \sigma_{v,v+1}). \tag{375}$$

This is always valid since it is not based on any model and so can be used to check the internal consistency in the analysis of results.

If N is the monomer concentration in the membrane (mole per square centimeter) and N_v is the concentration of the aggregate in the vth conductance state, then the association and dissociation reactions of the aggregates may be represented by

$$N_{v-1} + N \underset{k_{v,v-1}}{\overset{k_{v-1,v}}{\rightleftarrows}} N_v \tag{376}$$

$$N_v + N \underset{k_{v+1,v}}{\overset{k_{v,v+1}}{\rightleftarrows}} N_{v+1} \tag{377}$$

Thus

$$\frac{dN_v}{dt} = k_{v-1,v} N_{v-1}N - k_{v,v-1}N_v - k_{v,v+1}N_vN + k_{v+1,v} N_{v+1}. \tag{378}$$

In the steady state when $dN_v/dt = 0$, each reaction represented by Eqs. (376) and (377) is in the steady state, Thus

$$k_{v-1,v}N_{v-1}N = k_{v,v-1}N_v, \tag{379}$$

$$k_{v,v+1}N_vN = k_{v+1,v}N_{v+1}, \tag{380}$$

and

$$\frac{N_v}{N_{v-1}N} = \frac{k_{v-1,v}}{k_{v,v-1}} = K_v, \tag{381}$$

$$\frac{N_{v+1}}{N_vN} = \frac{k_{v,v+1}}{k_{v+1,v}} = K_{v+1}, \tag{382}$$

where K_v, K_{v+1} are the equilibrium constants which may be expressed by probabilities of Eq. (370) provided time average is equal to the aggregate average. That is

$$P_v = t_v/t_{\text{obs}} = N_v/N_P, \tag{383}$$

where $N_P = \sum_{v=0}^{\infty} N_v$ is the total concentration of both conducting and nonconducting channels per square centimeter. Thus Eqs. (381) and (382) may be written

$$K_vN = P_v/P_{v-1}, \tag{384}$$

$$K_{v+1}N = P_{v+1}/P_v, \tag{385}$$

and so

$$K_v/K_{v+1} = (P_v)^2/(P_{v-1})(P_{v+1}). \tag{386}$$

Values for K_vN, $K_{v+1}N$, etc., and for K_v/K_{v+1} may be derived since P_v, P_{v-1}, P_{v+1}, etc., can be evaluated from experimental current records. Furthermore, $k_{v,v-1}N_v$ is the number of transitions per square centimeter of membrane per second and so $k_{v,v-1}N_v/N_P$ is the number of transitions per channel per second. Thus

$$\sigma_{v,v-1} = k_{v,v-1}N_v/N_P. \tag{387}$$

Substituting Eq. (383) in Eq. (387) gives

$$k_{v,v-1} = \sigma_{v,v-1}/P_v. \tag{388}$$

With the help of Eq. (388), the rate constants for transitions to lower state may be calculated from experimental data.

Time resolution of the measuring system is limited. Consequently, transitions of short duration may be suppressed in the current records. It is therefore difficult to determine mean life times τ_v. This is circumvented by assuming that at $t = 0$, all channels of the aggregates exist in state v and transitions occur only from state v to states $v - 1$ and $v + 1$. Then Eq. (378) becomes

$$\frac{dN_v}{dt} = -k_{v,v-1}N_v - k_{v,v+1}N_vN, \qquad (389)$$

when N is very large (i.e., constant), the solution of Eq. (389) is

$$N_v(t) = N_v(0)\exp(-t/\tau_v), \qquad (390)$$

where

$$\tau_v = \frac{1}{k_{v,v-1} + k_{v,v+1}N}. \qquad (391)$$

Statistically $N_v(t)$ is the number of events $n_v(t)$ with a lifetime longer than t in the fluctuation pattern of a single channel. As $n_v(t)$ may be counted from the current records τ_v may be evaluated from a plot of $\ln n_v(t)$ versus time.

The values for the several parameters derived from the analysis of single-channel data for a phosphtidylcholine membrane doped with alamethicin are given in Table X. The special features to be noted are (a) although the single-channel conductances γ_v increase with increase in the state v, the

TABLE X

Values of Several Parameters Derived from Single-Channel Data for a Phosphatidylcholine Membrane Doped with Alamethicin at 100 mV[a]

State v	γ_v ($\times 10^{-10}$ S) [Eq. (369)]	P_v (%) [Eq. (370)]	$\sigma_{v,v+1}$ (s^{-1})	τ_v (ms) [Eq. (390)]	P_v (%) [Eq. (375)]	$k_{v,v-1}$ (s^{-1}) [Eq. (388)]
1	—	0.06	$\sigma_{1,2} = 0.24$	—	—	—
2	2.4	3.13	$\sigma_{2,3} = 4.89$	5.2	2.67	$k_{2,1} = 7.7$
3	6.5	29.1	$\sigma_{3,4} = 21.6$	9.7	25.7	$k_{3,2} = 16.8$
4	11.3	49.0	$\sigma_{4,5} = 24.6$	9.5	43.9	$k_{4,3} = 44.1$
5	16.5	17.9	$\sigma_{5,6} = 4.72$	5.2	15.0	$k_{5,4} = 137$
6	21.5	0.89	—	2.2	0.92	$k_{6,5} = 530$

Total events $z = \sum_v z_v = 1384$
Eq. (384)

γ_2/γ_1 —	NK_1 —	
$\gamma_3/\gamma_2 = 2.71$	$NK_2 = 52.1$	
$\gamma_4/\gamma_3 = 1.74$	$NK_3 = 9.3$	$K_2/K_3 = 5.6$
$\gamma_5/\gamma_4 = 1.46$	$NK_4 = 1.68$	$K_3/K_4 = 5.5$
$\gamma_6/\gamma_5 = 1.33$	$NK_5 = 0.366$	$K_4/K_5 = 4.6$
	$NK_6 = 0.050$	$K_5/K_6 = 7.3$

[a] From Boheim (1974).

ratios γ_{v+1}/γ_v are approximately equal; (b) the ratio K_v/K_{v+1} is independent of both voltage (values at other voltages were similar, see Boheim, 1974) and the state v; (c) the probabilities calculated from Eq. (370) and derived from Eq. (375) are about equal and so attest to the consistency of analysis.

Furthermore, assuming that the molecular structure of the alamethicin in the pore to correspond to an elongated loop where carbonyl oxygen form one side of the molecule and the hydrophobic amino acid residues form the opposite side, Boheim estimated the inner channel radii r_v by using the formula

$$r_v = \sqrt{\left(\frac{v+2}{2\pi}\right)^2 a^2 - \frac{v+2}{\pi} A_c} \qquad (v = 0, 1, 2, \ldots), \qquad (392)$$

where $a = 9.6$ Å (diameter of hole normal to the loop estimated from CPK molecule model) and $A_c = 14$ Å^2—cross-sectional area of the loop assumed constant. The values of $(r_{v+1}/r_v)^2$, i.e., the ratios of channel volumes agreed approximately with the ratio of the consecutive pore conductances (i.e., γ_{v+1}/γ_v).

Single-channel conductances and mean lifetimes τ_v measured as a function of temperature for states $v = 3, 4, 5$ gave values of 3, about 4 and 12 kcal/mol, respectively.

The different steps involved in pore formation are described by the several equilibrium constants $K_m, K_0, K_1, K_v, \ldots$. Interfacial equilibrium is given by

$$N_S \underset{k_{ms}}{\overset{k_{sm}}{\rightleftarrows}} N$$

and so

$$K_m = \frac{k_{sm}}{k_{ms}} = \exp\left[\frac{\alpha_m EF - \Delta G_m}{RT}\right], \qquad (393)$$

where N_S is the alamethicin monomer concentration at the membrane interface (monomers per square centimeter) and N that in the membrane, $\alpha_m EF$ is the energy per mole due to transfer of one elementary charge across the fraction α_m of the membrane thickness (corresponds to orientation of dipoles), and ΔG_m is the change in free energy per mole associated with the transfer of monomers from the interface into the membrane.

Dimerization reaction in the membrane is given by

$$N + N \underset{k_{om}}{\overset{k_{mo}}{\rightleftarrows}} N_0$$

and so

$$K_0 = \frac{k_{mo}}{k_{om}} = \frac{1}{N^*} \exp\left[\frac{\alpha_p EF - \Delta G_0}{RT}\right], \qquad (394)$$

where N_0 is the concentration of nonconducting dimers and N^* is the standard concentration in moles per square centimeter. $\alpha_p EF$ is the voltage-dependent part of energy (per mole) associated with uptake and release of monomer by the pore and ΔG_0 is the change in energy per mole associated with dimerization in the membrane.

Oligomerization reaction in the membrane is given by Eqs. (376) and (377) and K_v of Eq. (381) becomes

$$K_v = \frac{1}{N^*} \exp\left[\frac{\alpha_p EF - \Delta G_v}{RT}\right], \tag{395}$$

where ΔG_v is the change in free energy per mole associated with uptake and release of a monomer by the oligomer. Equation (393) can be written

$$N = N_s \exp\left[\frac{\alpha_m EF - \Delta G_m}{RT}\right], \tag{396}$$

Combining Eqs. (395) and (396) gives

$$K_v N = \frac{N_s}{N^*} \exp\left[\frac{\alpha EF - (\Delta G_v + \Delta G_m)}{RT}\right], \tag{397}$$

where $\alpha = \alpha_m + \alpha_p$. If ΔG_R is the energy per mole required to enlarge the channel in the membrane against the forces of compression, then

$$\Delta G_v = \Delta G_0 + v \Delta G_R. \tag{398}$$

It is assumed that ΔG_v varies linearly with v. Making the substitution that

$$\Gamma = \frac{N_s}{N^*} \exp\left(-\frac{\Delta G_0 + \Delta G_m}{RT}\right), \tag{399}$$

and combining it with Eqs. (397) and (398) gives

$$K_v N = \Gamma \exp\left(\frac{\alpha EF - v \Delta G_R}{RT}\right). \tag{400}$$

Hence taking logarithms gives

$$\ln(K_v N) = \ln \Gamma + \frac{\alpha EF}{RT} - \frac{v \Delta G_R}{RT}. \tag{401}$$

A plot of $\ln K_v N$ against E should give a straight line with slope equal to $\alpha F/RT$. Boheim's (1974) data for alamethicin-doped phosphatidylcholine membrane gave an average value of 0.93 for α.

Equation (401) for the pore state $v + 1$ may be written

$$\ln(K_{v+1} N) = \ln \Gamma + \frac{\alpha EF}{RT} - \frac{(v+1) \Delta G_R}{RT}. \tag{402}$$

Subtracting Eq. (402) from Eq. (401) gives

$$\ln(K_v/K_{v+1}) = \Delta G_R/RT.$$

The value for $\ln(K_v/K_{v+1})$ evaluated from the plots of $\ln K_v N$ against E (displacements of lines for the states v and $v + 1$ on the $\ln K_v N$ axis) was 1.61. Thus $K_v/K_{v+1} = 5.0$ and $\Delta G_R = 0.91$ kcal/mol for the doped phosphatidylcholine membrane. Use of the appropriate values given above in Eq. (400) gives a value of 30 for Γ. Increase or decrease in the value of Γ evaluated in this manner for other doped membranes such as phosphatidylserine or sphingomyelin will give information about the differences in the nature of the membrane surface.

Equation (400) with the help of Eq. (381) may be written

$$N_v = N_{v-1}\Gamma \exp\frac{\alpha EF - v\Delta G_R}{RT}. \tag{403}$$

When $v = 0, 1, 2, 3, \ldots$, Eq. (403) gives

$$N_0 = N\Gamma \exp\left(\frac{\alpha EF}{RT}\right),$$

$$N_1 = N_0\Gamma \exp\left(\frac{\alpha EF - \Delta G_R}{RT}\right) = N\Gamma^2 \exp\left(\frac{2\alpha EF - 2\Delta G_R/2}{RT}\right),$$

$$N_2 = N_1\Gamma \exp\left(\frac{\alpha EF - 2\Delta G_R}{RT}\right) = N\Gamma^3 \exp\left(\frac{3\alpha EF - 3(2\Delta G_R/2)}{RT}\right),$$

$$\vdots$$

$$N_v = N_{v-1}\Gamma \exp\left(\frac{\alpha EF - v\Delta G_R}{RT}\right)$$

$$= N\Gamma^{v+1} \exp\left[(v+1)\left(\frac{\alpha EF - v\Delta G_R/2}{RT}\right)\right]. \tag{404}$$

Equation (404) is the general expression for the oligomeric concentration in the membrane. The time dependence of N_v is given by Eq. (378), and those of N and N_0 are given by

$$\frac{dN}{dt} = k_{sm}N_s - k_{ms}N - k_{mo}N^2 + k_{om}N_0$$

$$+ \sum_{v=1}^{\infty} (-k_{v-1,v}NN_{v-1} + k_{v,v-1}N_v),$$

$$\frac{dN_0}{dt} = k_{mo}N^2 - k_{om}N_0 - k_{01}NN_0 + k_{10}N_1.$$

In addition to single-channel experiments (microlevel), multichannel or multipore experiments (macro) can be carried out. The success of the oligomeric model depends on how well the equations used for the analysis of single-channel data can explain the data obtained by multipore experiments. Multipore experiments are generally done by measuring membrane currents at constant voltages. Voltage jumps lead to a pattern of currents which either increase exponentially in the case of a jump from zero voltage to a voltage E or decay exponentially in the case of switching off from E to 0 voltage. In these experiments, the $I-E$ curves assume patterns shown in Fig. 16.

Generally two current relaxations, one fast and the other slow, are seen after a voltage jump. Both are exponential in nature and the time constants (τs) are of the order of milliseconds (τ_f) and seconds (τ_s) for the fast and slow relaxations, respectively.

τ_v given by Eq. (391) of the most probable pore state is nearly independent of voltage. Similarly τ_f observed in multipore experiments is weakly voltage dependent in contrast to τ_s, which is very voltage dependent. Both τ_v and τ_f are nearly of the same order of magnitude (milliseconds) and have the same activation energy (12 kcal/mol).

The slow relaxation time τ_s observed in multipore experiments relates to the mean lifetime of a fluctuating burst in view of the fact that the mean lifetime of bursts is one or two orders of magnitude larger than the mean lifetime of the pore states.

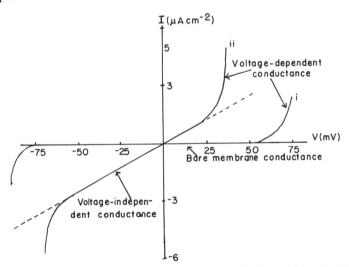

Fig. 16. General characteristics of current–voltage curves. Addition of suzukacillin to one compartment at short time (about 5 min) gives curve (i) and at long times (about 60 min) gives curve (ii). (After Boheim et al., 1976.)

The relation between the measured current density I (A/cm²) and the pore density N_P in the membrane of a many pore system is given by

$$I = E\bar{\gamma}N_P, \tag{405}$$

where $\bar{\gamma}$ is the mean conductance of a single pore and is given by

$$\bar{\gamma} = \sum_{v=0}^{\infty} P_v \gamma_v \tag{406}$$

and

$$N_P = \sum_{v=0}^{\infty} N_v, \tag{407}$$

the total concentration of pores per square centimeter of membrane.
Thus

$$I = N_P E \sum_{v=0}^{\infty} P_v \gamma_v. \tag{408}$$

According to this equation, the slow relaxation is due to the variation of N_P with time and the fast relaxation is due to P_v.

The time course of the dimerization reaction, i.e., the time dependence of $N_P(t)$ during a relaxation experiment, is given by

$$\frac{dN_P}{dt} = k_{m0}N^2 - k_{0m}N_0,$$

but $N_0 = P_0 N_P$, and so

$$\frac{dN_P}{dt} = k_{m0}N^2 - k_{0m}P_0 N_P \tag{409a}$$

or

$$\frac{dN_P}{dt} = \mu - k_b N_P, \tag{409b}$$

where

$$\mu = k_{m0}N^2 \tag{410}$$

and

$$k_b = k_{0m}P_0. \tag{411}$$

μ is the pore formation rate (mol cm^{-2} s^{-1}) and k_b is pore decay rate (s^{-1}). Integration of Eq. (409) for the condition when $t = 0$ and $N_P = 0$ gives

$$N_P = \frac{\mu}{k_b}[1 - \exp(-k_b t)]. \tag{412}$$

Thus at long times Eq. (412) becomes

$$N_{P(\infty)} = \mu/k_b. \tag{413}$$

Both μ and k_b are functions of voltage. At any applied voltage, μ is given by the initial rate of rise of current and k_b is the rate of closing. Their behavior is described by

$$\mu = \mu_0 \exp(E/E_\mu), \tag{414}$$

$$k_b = k_{b0} \exp(-E/E_{k_b}). \tag{415}$$

For a bacterial phosphatidyl ethanolamine membrane, Eisenberg et al. (1973) found $E_\mu = 6.7$ mV and $E_{k_b} = 9.6$ mV. At a concentration of 6.5×10^{-7} g/ml alamethicin, μ_0 was 10^3 pores cm^{-2} s^{-1} and k_{b0} was 20 s^{-1}. With an increase in voltage, forward rate increased and the reverse rate decreased.

Increase of N_P with voltage is given by

$$N_P = \frac{\mu}{k_b} = \frac{\mu_0}{k_{b0}} \exp\left[E\left(\frac{1}{E_\mu} + \frac{1}{E_{k_b}}\right)\right]. \tag{416}$$

Many-pore conductance increase with voltage is described by

$$G = G_0 \exp(E/E_0), \tag{417}$$

and for a alamethicin-doped phosphatidyl ethanolamine membrane E_0 was found to be 3.94 mV and independent of alamethicin concentration.

Comparison of Eqs. (416) and (417) show that

$$\frac{1}{E_0} = \frac{1}{E_\mu} + \frac{1}{E_{k_b}}.$$

In the many-channel experiments (current high), it is difficult to demonstrate fluctuations of a pore since the steps are not resolved. However, fluctuations in conductance expected at high levels can be calculated. Fluctuations can arise either from fluctuations in conductance of individual pores or from fluctuations in the number of pores. If G is the membrane conductance and G_n is the conductance of a membrane containing exactly n open pores, then

$$G_n = \sum_{i=0}^{n} \gamma_i, \tag{418}$$

where γ_i is the conductance of the ith pore.

To calculate variance (i.e., standard deviation squared, σ_G^2), average conductance and average of the square of the conductance must be calculated. At constant n, the average of the conductance is given by

$$\bar{G}_n = \sum_{i=0}^{n} \gamma_i = n\bar{\gamma}, \tag{419}$$

where $\bar{\gamma}$ is the mean conductance of a pore and is given by Eq. (406). If n is allowed to vary, then

$$\bar{G} = \bar{G}_n = \bar{n}\bar{\gamma}. \tag{420}$$

At fixed n, the average of the square of the conductance is given by

$$\overline{G_n^2} = \sum_{i=1}^{n} \gamma_i \sum_{j=1}^{n} \gamma_j = n\overline{\gamma^2} + (n^2 - n)(\bar{\gamma})^2. \tag{421}$$

When n is allowed to vary, Eq. (421) may be written

$$\overline{G^2} = \bar{n}\overline{\gamma^2} + \overline{(n^2 - n)}(\bar{\gamma})^2. \tag{422}$$

For a Poisson distribution that gives the probability of n pores open at any given time, $\overline{n^2} - \bar{n}^2 = \bar{n}$, and so Eq. (422) becomes

$$\overline{G^2} = \bar{n}\overline{\gamma^2} + (\bar{n}\bar{\gamma})^2. \tag{423}$$

As $\sigma_G^2 = \overline{G^2} - (\bar{G})^2$, combining Eqs. (423) and (420) gives

$$\sigma_G^2 = \bar{n}\overline{\gamma^2}. \tag{424}$$

Dividing Eq. (424) by Eq. (420) gives

$$\sigma_G^2/\bar{G} = \overline{\gamma^2}/\bar{\gamma}, \tag{425}$$

and as $\sigma_\gamma^2 = \overline{\gamma^2} - \bar{\gamma}^2$, Eq. (425) becomes

$$\sigma_G^2/\bar{G} = \bar{\gamma} + \sigma_\gamma^2/\bar{\gamma}. \tag{426}$$

This equation is independent of n, and relates multichannel distributions to single-channel distributions. Eisenberg et al. (1973) checked and found this equation to be valid for alamethicin-doped bacterial phosphatidyl ethanolamine membrane.

For the condition $\alpha = \alpha_m + \alpha_P \approx \alpha_m$, combination of Eqs. (396) and (404) gives

$$N_P = \sum_{v=0}^{v=\infty} N_v = \sum_{v=0}^{\infty} N_s \Gamma^{v+1} \exp\left[(v+2)\frac{\alpha EF}{RT} - \frac{\Delta G_m + (v+1)v\Delta G_R/2}{RT}\right]. \tag{427}$$

Introducing a mean value $\bar{v}$ (not necessarily an integral number), Eq. (427) becomes

$$N_P \approx N_s \Gamma^{\bar{v}+1} \exp\left[(\bar{v}+2)\frac{\alpha EF}{RT} - \frac{\Delta G_m + (\bar{v}+1)\bar{v}\Delta G_R/2}{RT}\right] = N_{\bar{v}}. \tag{428}$$

At $E = 0$, $N_{P(0)}$ is the initial pore density corresponding to a current I_0 and $N_{P(\infty)}$ is the pore density corresponding to the final steady-state current

I_∞ after a voltage jump from 0 to E. The voltage dependence of the pore density can be written as

$$N_{P(\infty)} - N_{P(0)} \approx N_{\bar{v}(\infty)} - N_{\bar{v}(0)} \approx \exp[(\bar{v} + 2)\alpha EF/RT]. \quad (429)$$

The fast relaxation time τ_f is short (2–10 ms) compared to the slow relaxation time τ_s (21–1500 ms). If the initial pore density $N_{P(0)}$ stays constant in the time τ_f, then I_0/E gives the voltage dependence of the mean conductance of the pore $[\bar{\gamma}(E)]$. Here I_0 is the value of the current obtained by extrapolation to time zero of the plot of $\ln[I_\infty - I(t)]$ against time.

The voltage dependence of the steady-state relaxation current can be described by

$$\frac{I_\infty - I_0}{E} \approx \exp\left[\alpha_\infty \frac{EF}{RT}\right] \quad (430)$$

and

$$\bar{\gamma}(E) \approx \frac{I_0}{E} \approx \exp\left[\alpha_0 \frac{EF}{RT}\right]. \quad (431)$$

Equation (405) gives the relation

$$\frac{I_\infty - I_0}{E} = \bar{\gamma}(E)\left[N_{P(\infty)} - N_{P(0)}\right]. \quad (432)$$

Substituting from Eq. (428) into Eq. (432) gives

$$N_{\bar{v}(\infty)} - N_{\bar{v}(0)} \approx \frac{1}{\bar{\gamma}(E)} \frac{I_\infty - I_0}{E}. \quad (433)$$

In view of Eqs. (430) and (431), Eq. (433) becomes

$$N_{\bar{v}(\infty)} - N_{\bar{v}(0)} \approx \frac{I_\infty - I_0}{I_0} \approx \exp\left[(\alpha_\infty - \alpha_0)\frac{EF}{RT}\right]. \quad (434)$$

Hence, equating Eqs. (434) and (429) gives the relation

$$\alpha_\infty - \alpha_0 = (\bar{v} + 2)\alpha. \quad (435)$$

Assuming that k_{mo} in Eq. (410) is roughly independent of voltage, one can write, on substituting from Eq. (396), the approximate relation

$$\mu = k_{mo}N^2 \approx N_S^2 \exp[2\alpha EF/RT]. \quad (436)$$

Similarly, substituting Eq. (405) into Eq. (413) gives the approximate relation

$$\mu = k_b N_{P(\infty)} = k_b \frac{I_\infty}{E\bar{\gamma}(E)} \approx \exp\left[\frac{(\alpha_\mu - \alpha_0)EF}{RT}\right]. \quad (437)$$

TABLE XI

Conductances and Rate Constants for Suzukacillin-Doped Phosphatidylcholine Membranes[a,b]

E (mV)	(a) at 25°C			(b) at 11°C		
	I_∞/E (μS cm^{-2})	$k_b = 1/\tau_b$ (s^{-1})	I_0/E (μS cm^{-2})	I_∞/E (μS cm^{-2})	k_b (s^{-1})	I_0/E (μS cm^{-2})
20				5.5	1.7	1.6
35	13.0	14.1	7.9	78.0	0.44	2.4
40	24.4	11.3	9.0			
45	47.8	8.6	10.3			
50	107.0	6.0	11.6	1860	0.092	

[a] From Boheim et al. (1976).
[b] 1.0 M KCl and (a) 5×10^{-7} g/ml suzukacillin on both sides; (b) 3×10^{-7} g/ml suzukacillin in one compartment only.

Equating Eqs. (436) and (437) gives

$$\alpha_\mu - \alpha_0 = 2\alpha. \quad (438)$$

Boheim and colleagues (1976) using suzukacillin-doped (a polypeptide related to alamethicin) phosphatidylcholine membrane made the appropriate measurements to validate the relations given above. Their data are given in Table XI.

Plots of $\ln[(I_\infty - I_0)/E]$ versus E according to Eq. (430) gave values of (a) $\alpha_\infty = 4.7$, (b) $\alpha_\infty = 5.0$. Similarly plots of $\ln(I_0/E)$ versus E [see Eq. (431)] gave for both the cases (a) and (b) a value of 0.7 for α_0. Plots of $\ln[k_b I_\infty/E]$ versus E [see Eq. (437)] gave for α_μ values of 2.6 for case (a) and 2.4 for case (b). These values introduced into Eq. (438) gave for α values of 0.95 for case (a) and 0.85 for case (b). These compare very well with the value of 0.93 derived by single-channel experiments carried out with alamethicin.

Again the several values of the α_i substituted into Eq. (435) gave $\bar{v} = 2.2$ for case (a) and $\bar{v} = 3.0$ for case (b). Thus these values show that the most probable states existing in the voltage range of the relaxation experiments are the second (i.e., tetramer) and the third (pentamer). Under similar conditions for alamethicin, the most probable states were third (pentamer) and fourth (hexamer). In view of these results, it is the dimerization step of the monomers incorporated into the membrane that is rate limiting. Incorporation of monomers must be very fast compared to the dimerization process. All this refers to the pore formation state (μ process). In the k_b process (pore closure) the pore state probabilities (P_v, $v = 0, 1, 2, \ldots$) attain their final values quickly. The growth process (oligomerization) occurs more rapidly than the nucleation reaction.

In addition to the phenomena described above, suzukacillin-doped membrane exhibits inactivation. That is, application of a voltage leads to a current that reaches a maximum and then decreases with time. This means that the oligomers build up very fast initially from the inserted monomers decay back into nonconducting dimers or trimers. Thus in the inactivated state only lower aggregates exist.

Interesting discussions related to several other aspects of kinetics of aggregation and nucleation exist in the papers of Boheim (1974), Mueller (1975a,b, 1979), and Gordon and Haydon (1972, 1975, 1976), and so those papers must be consulted for further details.

The steady-state conductance of the doped membrane depends on voltage, alamethicin concentration, and the concentration of the electrolyte used. In these studies, a characteristic voltage E_c as the voltage at which a fixed conductance G_c is reached is defined. Dependence on alamethicin concentration can be described by

$$E_c = -E_{al} \ln(C_{al}/C_0), \tag{439}$$

where C_0 is some reference concentration. E_{al} is the slope, which is found to be 36 mV by Eisenberg, *et al.* (1973) for the doped bacterial phoaphatidyl ethanolamine membrane. E_c decreased with increase in C_{al}. Similarly, E_c decreased with the electrolyte concentration also. This dependence is given by

$$E_c = -E_{El} \ln(C_{El}/C_0'), \tag{440}$$

where E_{El} was found to be 16 mV. Combining Eqs. (439) and (440) gives

$$E_c = E_i - E_{al} \ln C_{al} - E_{El} \ln C_{El}, \tag{441}$$

where E_i is the characteristic voltage at unit concentration of alamethicin and of salt. E_i thus depends on salt and membrane area.

The doped membrane conductance is given by Eq. (417) and $E_0 = 3.94$ mV. The number of pores may be written

$$n = n_c \exp[(E - E_c)/E_0], \tag{442}$$

where n_c is the number of pores at $E = E_c$.

Substituting Eq. (441) into Eq. (442) gives

$$n = n_c \exp\left[\frac{E - E_i + E_{El} \ln C_{El} + E_{al} \ln C_{al}}{E_0}\right]. \tag{443}$$

Substituting the numerical values for E_0, E_{El}, and E_{al} given already, Eq. (443) becomes

$$n = n_c \exp\left[\frac{E - E_i}{3.94}\right] \exp\left[\frac{16}{3.94} \ln C_{El}\right] \exp\left[\frac{36}{3.94} \ln C_{al}\right]$$

or

$$n = n_c C_{El}^4 C_{al}^9 \exp[(E - E_i)/3.94]. \tag{444}$$

Yantorno et al. (1982), following Gordon and Haydon (1975), expressed Eq. (417) by the relation

$$G = G_0 \exp(\mu_d E/\kappa T), \tag{445}$$

where μ_d is the dipole moment and E is the electric field intensity in statvolts (esu, cgs). A plot of $\kappa T \ln G$ versus E gave a straight line whose slope was equal to the dipole moment. A value of ~ 680 DU was obtained for the dipole moment. In addition, the dipole moment of alamethicin in lipophilic solvents was determined to be ~ 70 DU (Debye unit). Comparison of these two values show that 9 to 10 (i.e., 680/70) molecules of alamethicin are involved in the conductance of the alamethicin-doped bilayer membrane.

In the Gordon–Haydon mechanism (called the a mechanism) alluded to in the very beginning, the alamethicin-conducting complex is supposed to enter and leave different energy states pertaining to different levels of conductance. It can be represented by

$$nA \underset{}{\overset{K_a}{\rightleftarrows}} A_n \underset{}{\overset{K_a(E)}{\rightleftarrows}} A_n^*$$

where the asterisk denotes the conducting species. Only the equilibrium constant $K_a(E)$ would be field dependent. The Bauman–Mueller–Boheim mechanism (called the b mechanism) can be represented by

$$nA(S) \underset{}{\overset{K_b(E)}{\rightleftarrows}} nA \underset{}{\overset{K_b}{\rightleftarrows}} A_n^*$$

where A(S) are alamethicin molecules on the membrane surface and are molecules driven into the membrane by the electric field. These molecules would not conduct appreciably. A_n^* are the conducting species and n can have integral values up to 9. The papers of Gordon and Haydon (1975, 1976) may be consulted for arguments favoring the a mechanism.

C. Channels Produced by Monazomycin

Discrete conductance fluctuations of a monazomycin channel have been reported by Bamberg and Janko (1976). The histogram of conductance amplitudes showed the channel to have two distinct conductance states. The voltage dependence of conductance fluctuations has not been determined. So it is impossible to relate single-channel properties to many-channel voltage-dependent properties. However, interesting work that exists is described in the following.

VIII. Models for Channel-Forming Ionophores

As with alamethicin, the monazomycin-doped bilayer membrane conductance depends on both monomer and salt concentrations. The conductance G_m changes e-fold per 6 mV. The conductance is proportional to the number of open (conducting) channels. Let this number be On, and let Cd be the number of closed channels. The total number of channels is $N = $ On + Cd. The fraction f of channels open is given by

$$f = \frac{\text{On}}{\text{On} + \text{Cd}} = \frac{\text{On}}{N}.$$

Let γ be the conductance of a single open channel. If the conductance of the closed channel and that of the unmodified membrane are zero, then

$$G_m = f\gamma N. \tag{446}$$

Assuming that the chemical energy difference between an open channel and a closed channel is ΔG and that n monazomycin molecules (each positively charged) are required to form a pore, then the Boltzmann distribution gives

$$\text{On} = \text{Cd} \exp\left[\frac{\Delta G + nEF}{RT}\right],$$

and so

$$f = \frac{\text{On}}{\text{On} + \text{Cd}} = \frac{\exp[(\Delta G + nEF)/RT]}{1 + \exp[(\Delta G + nEF)/RT]}. \tag{447}$$

Substituting for f in Eq. (446) gives

$$G_m = \gamma N \frac{\exp[(\Delta G + nEF)/RT]}{1 + \exp[(\Delta G + nEF)/RT]}. \tag{448}$$

For small values of E (i.e., $(\Delta G + nEF) \ll 0$ in the monazomycin system ΔG has a high negative value), Eq. (448) becomes

$$G_m = \gamma N \exp[(\Delta G + nEF)/RT]. \tag{449}$$

Combining γ, N, and ΔG, which are unknown constants, Eq. (449) becomes

$$G_m = \text{const} \exp[nEF/RT]. \tag{450}$$

Experimental data when plotted as $\log G_m$ versus E gave a straight line. The slope was steep, increasing e-fold per 5.7 mV. The average value of n for a doped phosphatidyl ethanolamine membrane was 4.4. At a given voltage, doubling the monazomycin concentration gave a 30-fold increase in membrane conductance. This means $G_m \propto [\text{Mona}^+]^S$, where S is about 5. Therefore the steady-state conductance of the doped membrane may be written

$$G_m = \text{const}(C_{\text{El}})(\text{Mona}^+)_{\text{int}}^S \exp(nEF/RT), \tag{451}$$

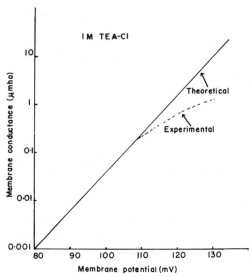

Fig. 17. Steady-state conductance-voltage curve for a monazomycin-doped (4μg/ml) bilayer membrane. (After Heyer *et al.*, 1976.)

where C_{El} is the electrolyte concentration, S and n are empirical constants, and n is nearly equal to 5. The voltage-dependent conductance at low conductances follow the relation

$$G_m \propto \exp(nEF/RT).$$

At higher conductances, the $\log G_m$ versus E curve bends toward the voltage axis (see Fig. 17). The kinetic response of current to voltage is S-shaped (see Fig. 18). The deviation of the curve (Fig. 17) from linearity indicates the onset of inactivation. This deviation becomes larger and so does the inactivation observed in the current-versus-time records. The inactivation is, according to the model of Heyer *et al.* (1976), due to depletion of monazomycin from the membrane-solution interface as the monazomycin moves across the membrane to the opposite (trans) side from the side of addition of the

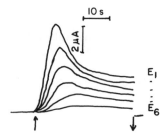

Fig. 18. Current responses of a monazomycin-doped bilayer membrane to increasing positive voltage steps $E_6, E_5, \ldots, E_1$. The voltage was on at the arrow ↑ and off at the arrow ↓. (Schematic after Heyer *et al.*, 1976.)

compound (cis). According to this model (depletion model)

$$G_{m(ideal)} = \text{const}(C_{El})(\text{Mona}^+)^S_{bulk} \exp(nEF/RT). \quad (452)$$

Here the interfacial concentration is equal to the bulk concentration of monazomycin. The aqueous unstirred layers are not rate limiting. Dividing Eq. (451) by Eq. (452) gives

$$[G_m/G_{m(ideal)}]^{1/S} = [(\text{Mona}^+)_{int}/(\text{Mona}^+)_{bulk}]. \quad (453)$$

According to this equation, deviation of the curve from linearity of Fig. 17 is due to a decrease in $[\text{Mona}^+]_{int}$.

In the steady state the flux of monazomycin across the cis-unstirred layer must be equal to its flux across the membrane. Thus

$$J^{mona} = J^{mona}_m = J^{mona}_{aq} = \frac{D_{mona}A}{\delta}\{[\text{Mona}^+]_{bulk} - [\text{Mona}^+]_{int}\}, \quad (454)$$

where D_{mona} is the diffusion coefficient of Mona^+ in water, A is the area of the membrane, and δ is the thickness of unstirred layer on the cis side. Combination of Eqs. (454) and (453) gives

$$J^{mona} = \frac{D_{mona}A}{\delta}[\text{Mona}^+]_{bulk}\left[1 - \left\{\frac{G_m}{G_{m(ideal)}}\right\}^{1/S}\right]. \quad (455)$$

Substituting for $G_{m(ideal)}$ from Eq. (452) into Eq. (455) gives on rearrangement

$$G_m = \text{const}(C_{El})\left[(\text{Mona}^+)_{bulk} - \frac{J^{mona}\delta}{D_{mona}A}\right]^S \exp\left(\frac{nEF}{RT}\right). \quad (456)$$

When $J^{mona} \to 0$, Eq. (456) becomes Eq. (452). Experimental data showed that monazomycin flux varied linearly with G_m. That is,

$$J^{mona} = BG_m. \quad (457)$$

Substituting Eq. (457) into Eq. (456) gives

$$G_m = \text{const}(C_{El})\left[(\text{Mona}^+)_{bulk} - \frac{BG_m\delta}{D_{mona}A}\right]^S \exp\left(\frac{nEF}{RT}\right). \quad (458)$$

According to Eq. (458) a flux of monazomycin (proportional to G_m) generates the right amount of monazomycin depletion at the cis interface to cause the experimental curve to deviate from the ideal straight line. To test this point experimentally, Eq. (458) may be recast by equating it to Eq. (452). That is, to make G_m equal to $G_{m(ideal)}$, a higher voltage must be applied and so E in Eqs. (458) and (452) are changed to E_2 and E_1, respectively. Thus

$$(\text{Mona}^+)^S_{bulk} \exp\left(\frac{nE_1F}{RT}\right) = \left[(\text{Mona}^+)_{bulk} - \frac{BG_m\delta}{D_{mona}A}\right]^S \exp\left(\frac{nE_2F}{RT}\right)$$

or

$$\exp\left(\frac{-n\Delta EF}{RT}\right) = \left[1 - \frac{BG_m\delta}{D_{mona}A[\text{Mona}^+]_{bulk}}\right]^S, \quad (459)$$

where $\Delta E = E_2 - E_1$. ΔE therefore is the voltage displacement of the experimental ($\log G_m$)-versus-E curve from the ideal straight line.

If $n = S (= 5)$, Eq. (459) becomes

$$\exp(-\Delta EF/RT) = 1 - B'G_m/[\text{Mona}^+]_{bulk}, \quad (460)$$

where $B' = B\delta/(D_{mona}A)$.

Evaluation of B' can be made separately. The values are $D_{mona} = 3.5 \times 10^{-6}$ cm^2 s^{-1} based on its molecular weight of 1200, $A = 10^{-2}$ cm^2, $\delta = 2 \times 10^{-2}$ cm, and B was evaluated from the experimental data fitted to Eq. (457). With these values experimental curve of membrane conductance versus voltage could be fitted to Eq. (460). Furthermore, several predictions of Eq. (460) were found to be experimentally true. For example, ΔE at any given value of G_m diminished with increase in the concentration of monazomycin. Also inactivation (measured by the ratio of peak to steady-state conductance in the current-versus-time curves) diminished with increase in the concentration of monazomycin.

A detailed kinetic study of monazomycin-induced voltage-dependent conductance has been carried out by Muller and colleagues (1981), who showed that the numerical solutions of an empirical equation, viz.,

$$\frac{dG}{dt} = kG_m\left[1 - \left(\frac{G}{G_\infty}\right)^D\right], \quad (461)$$

fitted both the linear and logarithmic current-versus-time records. In the equation, k is a rate constant, G_∞ is the steady-state conductance, and D is an empirical constant ($D = 0.7$). k and G are functions of membrane potential and monazomycin concentration as defined by the relations

$$k = k^*\left\{\frac{[\text{Mona}^+]}{[\text{Mona}^{+*}]}\right\}^X \exp\left(\frac{XEF}{RT}\right), \quad (462)$$

$$G_\infty = G_\infty^*\left\{\frac{[\text{Mona}^+]}{[\text{Mona}^{+*}]}\right\}^S \exp\left(\frac{SEF}{RT}\right), \quad (463)$$

where k^* and G_∞^* are, respectively, k and G_∞ in the presence of a reference concentration of monazomycin $[\text{Mona}^{+*}]$ when $E = 0$. X and S are constants with values of 2.6 and 5.8, respectively. With the help of models, the equations given above have been derived. The original papers (Muller et al., 1981) must be consulted for details.

In the models considered, the aggregation reactions are in equilibrium during conductance changes following simple first-order differential equations. Baumann and Mueller also propose a scheme that is based on coupled insertion and aggregation reactions. But the rate constants for aggregation step by step are specified as discussed under alamethicin reactions.

D. Channels Formed by Other "Channel Formers"

Some of these other channel formers are excitability-inducing material (EIM), Hemocyanin, amphotericin B, nystatin, black-widow venom. *Paramecium* mitochondria, proteins from *E. coli*, and *Ps. aeruginosa*.

EIM-doped membranes were the first in which discrete conductance jumps caused by the formation of channels were observed. Although EIM is not well characterized chemically, its presence in the bilayer membrane gives reproducible electrical effects and produces channels of uniform conductance. With very small EIM present in the bilayer, steady-state transitions caused by the opening and the closing of single channels can be observed. Simple fluctuations occur in EIM-doped oxidized cholesterol membranes. At positive potentials, channels show two states of conductance, 80 and 400 pS in 0.1-M KCl. The random transitions between the two states for various potentials show that the jump frequency is maximal at about 60 mV. The channel exists at the upper level preferentially at low potentials, but stays for most of its time in the lower level at high potentials. The amplitude of fluctuations increases with potentials.

Ehrenstein and Lecar (1977) proposed a two-state model (open and closed) for the channel. The rate constants for the opening and closing of single conducting channels may be determined from records of membrane current obtained at constant voltage for membranes with no more than six channels. The means lifetime (or dwell time t_d) is given by Eq. (371). if fluctuations conform to a Poisson process, then the probability of a transition per unit time is constant. For such a process $4_d \propto \exp(-kt_d)$, where k is the rate constant for transition in conductance. The average dwell time is given by

$$\bar{t}_d = \frac{1}{n}\sum_{i=1}^{n} t_{d_i} = \left(\int_0^\infty t_d \exp(-kt_d)\,dt_d\right) \bigg/ \left(\int_0^\infty \exp(-kt_d)\,dt_d\right). \quad (464)$$

Integration of Eq. (464) gives

$$\bar{t}_d = 1/k,$$
$$k = 1/\bar{t}_d = n/(\sum t_{d_i}). \quad (465)$$

If $k_1(E)$ is the rate constant for the opening of a closed channel and $k_2(E)$ is the rate constant for the closing of an open channel (E is the applied potential), then these can be determined from the distribution of dwell times. For single channels, this is simple. But EIM-doped membranes having single channels are observed less frequently. Often a minute amount of EIM gives several channels. In this case k_1 and k_2 may be determined by averaging the dwell times. $\log k_1$ and $\log k_2$ determined for several EIM-doped bilayer membranes varied linearly with voltage. Thus

$$k_1(E) = k_0 \exp[-A(E - E_0)], \tag{466}$$

$$k_2(E) = k_0 \exp[B(E - E_0)], \tag{467}$$

where k_0 is the transition rate constant measured at voltage E_0 for which the rate of opening and closing are equal. The average experimental values of the several parameters derived for EIM-doped oxidized cholesterol membranes by Ehrenstein et al. (1974) were

$$A = 0.07 \quad \text{mV}^{-1}, \qquad B = 0.05 \quad \text{mV}^{-1},$$

$$E_0 = 58 \quad \text{mV}, \qquad k_0 = \tfrac{1}{8} \quad \text{s}^{-1}.$$

Based on this two-state model an expression for the membrane conductance was derived by Ehrenstein et al. (1974). Let n_o be the number of open channels, each with conductance γ_o, and let n_c be the number of closed channels with conductance γ_c. Then total channels both open and closed N is given by

$$N = n_o + n_c. \tag{468}$$

If $\bar{n}_o$ and $\bar{n}_c$ are equilibrium values, then

$$\bar{n}_o/\bar{n}_c = k_1(E)/k_2(E). \tag{469}$$

Substituting for the rate constants from Eqs. (466) and (467) gives

$$\bar{n}_o/\bar{n}_c = \exp[-(A + B)(E - E_0)]. \tag{470}$$

Combining Eq. (470) with Eq. (468) gives on simplification

$$\frac{\bar{n}_o}{N} = \frac{1}{1 + \exp[(A + B)(E - E_0)]}. \tag{471}$$

The steady-state conductance $\bar{\gamma}$ may be expressed as

$$\bar{\gamma} = \bar{n}_o \gamma_o + \bar{n}_c \gamma_c. \tag{472}$$

Combining Eqs. (468), (471), and (472) gives

$$\bar{\gamma}(E) = N\left[\gamma_c + \frac{\gamma_o - \gamma_c}{1 + \exp[(A + B)(E - E_0)]}\right]. \tag{473}$$

The maximum conductance is given by $N\gamma_o$. Thus dividing Eq. (473) by $N\gamma_o$ gives the normalized conductance γ_{rel}. Thus

$$\gamma_{rel}(E) = \frac{\gamma_c}{\gamma_o} + \frac{1 - \gamma_c/\gamma_o}{1 + \exp[(A + B)(E - E_0)]}. \quad (474)$$

For a first-order process, the relaxation time τ is given by $1/(k_1 + k_2)$. Thus $\tau = 1/[k_1(E) + k_2(E)]$. Substituting Eqs. (466) and (467) gives

$$\tau(E) = \frac{1}{k_0\{\exp[-A(E - E_0)] + \exp[B(E - E_0)]\}}. \quad (475)$$

Equations (474) and (475) are based on single channels. The prediction of Eq. (474) is that the relative conductance versus voltage should be a sigmoidal curve whose steepness is determined by $A + B$. The midpoint conductance (halfway between maximum and minimum) is at $E = E_0$. Equation (475) predicts a bell-shaped curve for the relaxation time. Many channel experimental data [g_{rel} and $\tau(E)$] were fitted to Eqs. (474) and (475) with the following values for the parameters:

$$A = 0.41 \text{ mV}^{-1}, \quad B = 0.025 \text{ mV}^{-1},$$
$$E_0 = 59 \text{ mV}, \quad k_0 = 1/6.6 \text{ s}^{-1}.$$

These many-channel data are in rough agreement with single-channel data given already.

The transition between the two states of the channel is dependent on voltage. Presumably the transition occurs by the movement of charged groups in the membrane. The difference in the energy between the two states may be written

$$\Delta W = \Delta W_i + qE, \quad (476)$$

where ΔW_i is the intrinsic energy difference between the two states and q is the charge moved. If a constant E_0 is defined as

$$E_0 = -\Delta W_i/q, \quad (477)$$

then substituting Eq. (477) into Eq. (476) gives

$$\Delta W(E) = q(E - E_0). \quad (478)$$

When the system is at equilibrium, $\bar{n}_o$ and $\bar{n}_c$ are related by a Boltzmann distribution

$$\bar{n}_o/\bar{n}_c = \exp(-\Delta W/\kappa T). \quad (479)$$

The fraction of channels open at a given potential, $f(E)$, is given by

$$f(E) = \frac{\bar{n}_o}{N} = \frac{1}{1 + \exp[q(E - E_0)/\kappa T]}. \quad (480)$$

This equation shows a sigmoid voltage dependence. The transition between the two states of the channel may be regarded as a jump over a voltage-independent energy barrier. Then the rate of opening the channel is given by

$$k_1 = k_0 \exp(-\Delta W/2kT). \tag{481}$$

The rate of closing is given by

$$k_2 = k_0 \exp(\Delta W/2kT). \tag{482}$$

Hence

$$\bar{n}_o/n_c = k_1/k_2 = \exp(-\Delta W/kT), \tag{483}$$

giving Eq. (479). Thus

$$k_1, k_2 = k_0 \exp[\mp q(E - E_0)/2kT]. \tag{484}$$

The relaxation time is given by

$$\tau = \frac{1}{k_1 + k_2}$$

$$= \frac{1}{k_0\{\exp[-q(E - E_0)/2\kappa T] + \exp[q(E - E_0)/2\kappa T]\}}$$

$$= \frac{1}{2k_0 \cosh[q(E - E_0)/2\kappa T]}, \tag{485}$$

$$\tau(E) = \tau_{\max} \operatorname{sech}[q(E - E_0)/2\kappa T],$$
$$\tau_{\max} = 1/2k_0. \tag{486}$$

The $\tau(E)$-versus-E relation is bell shaped, with the peak of the $\tau - E$ curve occurring at E_0, the midpoint of the sigmoid $G_{\text{rel}} - E$ curve (see Fig. 19).

Comparing Eqs. (475) and (485) gives $A = q/2\kappa T$. When values of $A = 0.068$ mV^{-1} ($= 68$ V^{-1}) and $T = 20°$C are substituted, $q = 3.4$ electronic charges. With E_0 value of 60 mV, ΔW_i attains a value of 4.7 kcal/mol.

Instead of oxidized cholesterol, use of other lipids to form bilayer membranes can show more than two conductance states in the conductance of EIM-doped membranes. So the two-state model proposed for oxidized cholesterol membranes is a simple one since a model of three states has been proposed to explain the existence of inactivation (see Ehrenstein et al. 1978). When the voltage is changed from a high positive value to a high negative value, transitions between the two closed states go through the open state, resulting in inactivation. At zero voltage a stable open configuration exists for the three-state channel. For a discussion, Ehrenstein et al. (1978) and Arndt and Roper (1975) should be consulted.

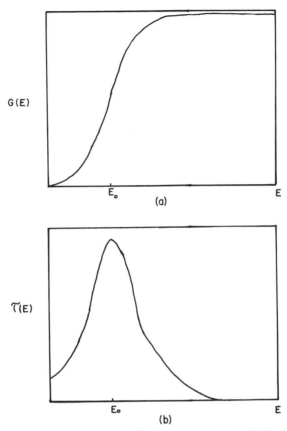

Fig. 19. (a) Conductance–voltage relation for the two-state channel. The curve is given by Eq. (480). (b) Relaxation-time–voltage curve for the simple two-state channel. The curve is given by Eq. (486). (After Ehrenstein and Lecar, 1977.)

The open- and closed-channel conductances measured as functions of temperature show that while the open-channel conductance increased with increasing temperature, the closed-channel conductance decreased with increasing temperature. From these observations it may be inferred that the open and closed states of EIM channels have different molecular configurations.

Hemocyanin is a large (molecular weight $> 10^6$) oxygen-transporting protein, and its incorporation into a lipid bilayer membrane produces a cation-selective voltage-dependent conductance appearing in discrete steps. With positive potentials, the channel exhibits lower conductances as opposed to higher conductances seen at negative potentials. Hemocyanin channels have two voltage-dependent processes. One process in oxidized cholesterol membrane has a relaxation time on the order of seconds and the other has a relaxation time of 10^{-4} s.

Voltage dependence of many-channel conductance agrees with the voltage dependence of single-channel conductance, indicating that the channels conduct independently. At a given membrane potential the conductance increases with increasing salt concentration and attains saturation. The saturation curves assume different shapes for different potentials. A model for saturation proposed by Latorre and colleagues (1974, 1975) assumes that no more than a single ion can occupy a given channel at a given time. If the time the ion spends in the channel is long and the ion entry rate is fast, the channel will be occupied for a large fraction of the time. When the channel is occupied by an ion, another ion is forbidden to enter the channel. Under these conditions, the channel conductance γ is given by

$$\gamma \propto (C_{EI})P_a = \text{const}(C_{EI})P_a, \tag{487}$$

where P_a is the probability that the channel is available for the ion. Therefore $1 - P_a$ is the probability that the channel is occupied by an ion and is given by $\bar{t}k$, where $\bar{t}$ is the mean time an ion spends in the channel and k is the rate of ion entry into the channel. k in units of γ in mhos and an applied potential of 0.1 V is given by

$$k = \frac{1 \text{ mho} \times 0.1\gamma}{1.6 \times 10^{-19} \text{ (C)}}$$

$$= 0.625 \times 10^{18}\gamma \text{ (s}^{-1}).$$

Thus $P_a = 1 - 0.625 \times 10^{18}\bar{t}\gamma$, and substituting this into Eq. (487) gives on rearrangement

$$\gamma = \frac{\text{const}[(C_{EI})}{1 + 0.625 \times 10^{18} \text{ const } \bar{t}(C_{EI})}.$$

This equation describes saturation. When C_{EI} is low, $\gamma \propto C_{EI}$, and when it is high, conductance becomes independent of C_{EI}. At low concentration, the constant becomes the conductance parameter, which was found to be 3.5 times bigger at -100 mV than at $+100$ mV, but $\bar{t}$ was about the same (~ 4 ns). A barrier model that can conform to this description is one in which the potential profile in the channel is constant, but the barriers to ion flow at the inner and outer edges of the channel are different and vary with potential. So structural changes causing fast voltage-dependent conductance occur at the edges of the hemocyanin channel.

The channel conductances observed in the case of other channel formers in lipid bilayer membranes are given in Table XII.

TABLE XII

Channel Conductances in Doped Bilayer Membranes

Compound	Lipids	Conductance level (pS)	Electrolyte KCl (M)	Reference
Amphotericin B	Brain lipids + cholesterol	2	0.5	Ermishkin et al., 1976
Nystatin	Brain lipids + cholesterol	2	3	Ermishkin et al., 1976
Amphotericin B	Brain lipids + cholesterol	6.5	2	Ermishkin et al., 1977
Black-widow spider venom	Lecithin + cholesterol	360	0.1	Finkelstein et al., 1976
Paramecium mitochondria	Asolecithin	500	0.1–1.0	Schein et al., 1976
Protein from *E. Coli*	Monoolein or cholesterol	2000	1.0	Benz et al., 1979
Protein from *Ps. aeruginosa*	Egg lecithin	5600	1.0	Benz and Hancock, 1981
Protein from *Ps. aeruginosa*	Oxidized cholesterol	5900	1.0	Benz and Hancock, 1981

References

Alvarez, O., Diaz, E., and Latorre, R. (1975). *Biochim. Biophys. Acta* **389**, 444.
Andersen, O. S. (1977). *In* "Renal Function" (G. H. Giebisch and E. F. Purcell, eds.), p. 71. Josiah Macy, Jr. Found., New York.
Andersen, O. S. (1978). *In* "Membrane Transport In Biology" (G. Giebisch, D. C. Tosteson, and H. H. Ussing, eds.), Vol. 1, p. 369. Springer-Verlag, Berlin and New York.
Andersen, O. S., and Fuchs, M. (1975). *Biophys. J.* **15**, 795.
Andersen, O. S., and Procopio, J. (1980). *Acta Physiol. Scand.*, Suppl. **481**, 27.
Andersen, O. S., Feldberg, S., Nakadomari, H., Levy, S., and McLaughlin, S. (1978). *Biophys. J.* **21**, 35.
Armstrong, C. M. (1975). *In* "Membranes" (G. Eisenman, ed.), Vol. 3, p. 325. Dekker, New York.
Arndt, R. A., and Roper, L. D. (1975). *J. Theor. Biol.* **54**, 249.
Bamberg, E., and Janko, K. (1976). *Biochim. Biophys. Acta* **426**, 447.
Bamberg, E., and Läuger, P. (1973). *J. Membr. Biol.* **11**, 177.
Bamberg, E., and Läuger, P. (1974). *Biochim. Biophys. Acta* **367**, 127.
Bamberg, E., and Läuger, P. (1977). *J. Membr. Biol.* **35**, 351.
Baumann, G., and Mueller, P. (1974). *J. Supramol. Struct.* **2**, 538.
Benz, R. (1978). *J. Membr. Biol.* **43**, 367.
Benz, R., and Hancock, R. E. W. (1981). *Biochim. Biophys. Acta* **646**, 298.
Benz, R., and Läuger, P. (1976). *J. Membr. Biol.* **27**, 171.
Benz, R., Gisin, B. F., Ting-Beall, H. P., Tosteson, D. C., and Läuger, P. (1976a). *Biochim. Biophys. Acta* **455**, 665.

Benz, R., Läuger, P., and Janko, K. (1976b). *Biochim. Biophys. Acta* **455**, 701.
Benz, R., Frohlich, O., and Läuger, P. (1977). *Biochim. Biophys. Acta* **464**, 465.
Benz, R., Janko, K., and Läuger, P. (1979). *Biochim. Biophys. Acta* **551**, 238.
Bezanilla, F., and Armstrong, C. M. (1972). *J. Gen. Physiol.* **60**, 588.
Boheim, G. (1974). *J. Membr. Biol.* **19**, 277.
Boheim, G., and Kolb, H. A. (1978). *J. Membr. Biol.* **38**, 99.
Boheim, G., Janko, K., Leibfritz, D., Ooka, T., Konig, W. A., and Jung, G. (1976). *Biochim. Biophys. Acta* **433**, 182.
Bruner, L. J. (1975). *J. Membr. Biol.* **22**, 125.
Ciani, S. M., Eisenman, G., Laprade, R., and Szabo, G. (1973). *In* "Membranes" (G. Eisenman, ed.), Vol. 2, p. 61. Dekker, New York.
Cornish-Bowden, A. (1979). "Fundamentals of Enzyme Kinetics." Butterworth, London.
Donavan, J. J., and Latorre, R. (1979). *J. Gen. Physiol.* **73**, 425.
Ehrenstein, G., and Lecar, H. (1977). *Q. Rev. Biophys.* **10**, 1.
Ehrenstein, G., Lecar, H., and Nossal, R. (1970). *J. Gen. Physiol.* **55**, 119.
Ehrenstein, G., Blumenthal, R., Latorre, R., and Lecar, H. (1974). *J. Gen. Physiol.* **63**, 707.
Ehrenstein, G., Lecar, H., and Latorre, R. (1978). *In* "Membrane Transport Processes" (D. C. Tosteson, Yu. V. Ovchinnikov, and R. Latorre, eds.), Vol. 2, p. 175. Raven Press, New York.
Eisenberg, M., Hall, J. E., and Mead, C. A. (1973). *J. Membr. Biol.* **14**, 143.
Eisenman, G., Sandblom, J., and Neher, E. (1978). *Biophys. J.* **22**, 307.
Ermishkin, L. M., Kasumov, Kh. M., and Potzeluyev, V. M. *Nature* (1976). **262**, 698.
Ermishkin, L. M., Kasumov, Kh. M., and Potzeluyev, V. M. (1977). *Biochim. Biophys. Acta* **470**, 357.
Eyring, H., Lumbry, R., and Woodbury, J. W. (1949). *Rec. Chem. Prog.* **10**, 100.
Feldberg, S. W., and Kissel, G. (1975). *J. Membr. Biol.* **20**, 269.
Feldberg, S. W., and Nakadomari, H. (1977). *J. Membr. Biol.* **31**, 81.
Finkelstein, A., and Andersen, O. S. (1981). *J. Membr. Biol.* **59**, 155.
Finkelstein, A., Rubin, L. L., and Tzeng, M. C. (1976). *Science* **193**, 1009.
Gamble, F., Gliozzi, A., and Robello, M. (1973). *Biochim. Biophys. Acta* **330**, 325.
Gavach, C., and Sandeaux, R. (1975). *Biochim. Biophys. Acta* **413**, 33.
Gordon, L. G. M. (1974). *In* "Drugs and Transport Processes" (B. A. Callingham, ed.), p. 251. Univ. Park Press, Baltimore, Maryland.
Gordon, L. G. M., and Haydon, D. A. (1972). *Biochim. Biophys. Acta* **255**, 1014.
Gordon, L. G. M., and Haydon, D. A. (1975). *Philos. Trans. R. Soc. London, Ser. B* **270**, 433.
Gordon, L. G. M., and Haydon, D. A. (1976). *Biochim. Biophys. Acta* **436**, 541.
Hall, J. E. (1975). *Biophys. J.* **15**, 934.
Haydon, D. A., and Hladky, S. B. (1972). *Q. Rev. Biophys.* **5**, 187.
Heyer, E. J., Muller, R. U., and Finkelstein, A. (1976). *J. Gen. Physiol.* **67**, 703, 731.
Hille, B. (1975a). *In* "Membranes" (G. Eisenman, ed.), Vol. 3, p. 255. Dekker, New York.
Hille, B. (1975b). *J. Gen. Physiol.* **66**, 535.
Hille, B. (1979). *In* "Membrane Transport Processes" (C. F. Stevens and R. W. Tsien, eds.), p. 5. Raven Press, New York.
Hille, B., and Schwarz, W. (1978). *J. Gen. Physiol.* **72**, 409.
Hladky, S. B. (1972). *J. Membr. Biol.* **10**, 67.
Hladky, S. B. (1974a). *In* "Drugs and Transport Processes" (B. A. Callingham, ed.), p. 193. Univ. Park Press, Baltimore, Maryland.
Hladky, S. B. (1974b). *Biochim. Biophys. Acta* **352**, 71.
Hladky, S. B. (1975). *Biochim. Biophys. Acta* **375**, 350.
Hladky, S. B. (1979). *In* "Current Topics in Membranes and Transport" (F. Bonner and A. Kleinzeller, eds.), Vol. 12. p. 53. Academic Press, New York.

Hladky, S. B., and Haydon, D. A. (1972). *Biochim. Biophys. Acta* **274**, 294.
Hladky, S. B., Urban, B. W., and Haydon, D. A. (1979). *In* "Membrane Transport Processes" (C. F. Stevens and R. W. Tsien, eds.), Vol. 3, p. 89. Raven Press, New York.
Hodgkin, A. L., and Huxley, A. F. (1952). *J. Physiol. (London)* **116**, 449.
Hoffman, R. A., Long, D. O., Arndt, R. A., and Roper, L. D. (1976). *Biochim. Biophys. Acta* **455**, 780.
Ketterer, B., Neumcke, B., and Läuger, P. (1971). *J. Membr. Biol.* **5**, 225.
King, E. L., and Altman, C. (1956). *J. Phys. Chem.* **60**, 1375.
Knoll, W., and Stark, G. (1975). *J. Membr. Biol.* **25**, 249.
Kotyk, A., and Janacek, K. (1977). *Biomembranes* **9**, pp. 208–220.
Laidler, K. J., and Bunting, P. S. (1973). "The Chemical Kinetics of Enzyme Action," 2nd ed. Oxford Univ. Press (Clarendon), London and New York.
Laprade, R., Ciani, S. M., Eisenman, G., and Szabo, G. (1975). *In* "Membranes" (G. Eisenman, ed.), Vol. 3, p. 127. Dekker, New York.
Latorre, R., Ehrenstein, G., and Lecar, H. (1972). *J. Gen. Physiol.* **60**, 72.
Latorre, R., Alvarez, O., and Verdugo, P. (1974). *Biochim. Biophys. Acta* **367**, 361.
Latorre, R., Alvarez, O., Ehrenstein, G., Espinoza, M., and Reyes, J. (1975). *J. Membr. Biol.* **25**, 163.
Läuger, P. (1972). *Science* **178**, 24.
Läuger, P. (1973). *Biochim. Biophys. Acta* **311**, 423.
Läuger P. (1979). *In* "Membrane Transport Processes" (C.F. Stevens and R. W. Tsien, eds.), Vol. 3, p. 17. Raven Press, New York.
Läuger, P. (1980). *J. Membr. Biol.* **57**, 163.
Läuger, P., and Neumcke, B. (1973). *In* "Membranes" (G. Eisenman, ed.), Vol. 2, p. 1. Dekker, New York.
Läuger, P., and Stark, G. (1970). *Biochim. Biophys. Acta* **211**, 458.
Läuger, P., Benz, R., Stark, G., Bamberg, E., Jordan, P. C., Fahr, A., and Brock, W. (1981). *Q. Rev. Biophys.* **14**, 513.
Lecar, H., Ehrenstein, G., and Latorre, R. (1975). *Ann. N. Y. Acad. Sci.* **264**, 304.
Levitt, D. G. (1978). *Biophys. J.* **22**, 221.
Mueller, P. (1975a). *Ann. N. Y. Acad. Sci.* **264**, 247.
Mueller, P. (1975b). *Int. Rev. Sci.: Biochem. Ser.* **3**, 75.
Mueller, P. (1979). *In* "The Neurosciences: Fourth Study Program" (F. O. Schmitt and F. G. Worden, eds.), p. 641. MIT Press, Cambridge, Massachusetts.
Muller, R. U., and Finkelstein, A. (1972). *J. Gen. Physiol.* **60**, 263.
Muller, R. U., and Peskin, C. S. (1981). *J. Gen. Physiol.* **78**, 201.
Muller, R. U., Orin, G., and Peskin, C. S. (1981). *J. Gen. Physiol.* **78**, 171.
Neher, E. (1975). *Biochim. Biophys. Acta* **401**, 540.
Neher, E., Sandblom, J., and Eisenman, G. (1978). *J. Membr. Biol.* **40**, 97.
Neumcke, B., and Läuger, P. (1969). *Biophys. J.* **9**, 1160.
Parlin, R. B., and Eyring, H. (1954). *In* "Ion Transport across Membranes" (H. T. Clarke, ed.), p. 106. Academic Press, New York.
Roy, G. (1975). *J. Membr. Biol.* **24**, 71.
Sandblom, J., Eisenman, G., and Neher, E. (1977). *J. Membr. Biol.* **31**, 383.
Schein, S. J., Colombini, M., and Finkelstein, A. (1976). *J. Membr. Biol.* **30**, 99.
Stark, G. (1978). *In* "Membrane Transport in Biology" (G. Giebisch, D. C. Tosteson, and H. H. Ussing, eds.), Vol. 1, p. 447. Springer-Verlag, Berlin and New York.
Stark, G., and Benz, R. (1971). *J. Membr. Biol.* **5**, 133.
Stark, G., Ketterer, B., Benz, R., and Läuger, P. (1971). *Biophys. J.* **11**, 981.
Szabo, G. (1976). *In* "Extreme Environments: Mechanisms of Microbial Adaptation" (M. R. Heinrich, ed.) p. 321. Academic Press, New York.

Szabo, G., Eisenman, G., Laprade, R., Ciani, S. M., and Krasne, S. (1973). *In* "Membranes" (G. Eisenman, ed.), Vol. 2, p. 179. Dekker, New York.
Urban, B. W., and Hladky, S. B. (1979). *Biochim. Biophys. Acta* **554**, 410.
Urban, B. W., Hladky, S. B., and Haydon, D. A. (1980). *Biochim. Biophys. Acta* **602**, 331.
Wang, C. C., and Bruner, L. J. (1978). *J. Membr. Biol.* **38**, 311.
Woodbury, J. W., White, S. H., Mackey, M. C., Hardy, W. L., and Chang, D. B. (1970). *In* "Physical Chemistry" (H. Eyring, D. Henderson, and W. Jost, eds.), Vol. 9, Part B. p. 903. Academic Press, New York.
Woodhull, A. M. (1973). *J. Gen. Physiol.* **61**, 687.
Yantorno, R., Takashima, S., and Mueller, P. (1982). *Biophys. J.* **38**, 105.
Zwolinski, B. J., Eyring, H., and Reese, C. E. (1949). *J. Phys. Colloid Chem.* **53**, 1426.

Chapter **6**

STEADY-STATE THERMODYNAMIC APPROACH TO MEMBRANE TRANSPORT

The kinetic approach to membrane transport described in Chapter 5 requires models based on detailed information about the behavior of the system. Such detailed information about several membrane systems is neither available nor obtainable. Consequently, to get an insight into such membrane phenomena, a "black-box" approach has proved very useful. Several of the equations derived from the application of the formalism of the thermodynamics of irreversible process to such phenomena are outlined in this chapter.

I. Basic Principles

The fundamental quantity characterizing irreversible processes is the inner entropy production $d_i S/dt$. The total change in entropy of an open system is composed of two parts:

$$dS = d_e S + d_i S, \qquad (1)$$

where $d_e S$ arises from interaction with the external surroundings of the system and $d_i S$ is produced within the system itself by irreversible processes. $d_e S$ may be positive or negative, but $d_i S$ is always a positive quantity,

$$d_i S \geq 0. \qquad (2)$$

Under steady-state conditions, $d_i S = -d_e S$. The fundamental theorem of nonequilibrium thermodynamics is that forces and fluxes are so chosen that they conform to the equation

$$\phi = T\theta = \sum_j J_j X_j, \qquad (3)$$

where ϕ is the dissipation function and θ the rate of entropy production (d_iS/dt) due to irreversible processes. The X_j ($j = 1, 2, \ldots, n$) are the forces causing the flows or fluxes J_j ($j = 1, 2, 3, \ldots, n$). The set of phenomenological equations may be written as follows:

$$J_1 = L_{11}X_1 + L_{12}X_2 + L_{13}X_3 + \cdots + L_{1n}X_n,$$
$$J_2 = L_{21}X_1 + L_{22}X_2 + L_{23}X_3 + \cdots + L_{2n}X_n,$$
$$J_3 = L_{31}X_1 + L_{32}X_2 + L_{33}X_3 + \cdots + L_{3n}X_n, \qquad (4)$$
$$\vdots$$
$$J_n = L_{n1}X_1 + L_{n2}X_2 + L_{n3}X_3 + \cdots + L_{nn}X_n,$$

or

$$J_j = \sum_{k=1}^{n} L_{jk}X_k. \qquad (5)$$

The L_{ik} are the phenomenological coefficients. The straight coefficients L_{jj} appear on the diagonal matrix of forces in Eq. (4). These equations show that J_1 may be caused by the forces $X_2, X_3, \ldots, X_n$ if the "coupling" or "cross" coefficients $L_{12}, L_{13}, \ldots, L_{1n}$ are not zero. The other implication of these equations is that the dependence of flows on nonconjugated forces is also linear. The linearity holds only when the system is close to the equilibrium state. When the set of Eq. (4) is solved for the forces, the following equations in which forces are represented as linear functions of the flows are obtained. Thus

$$X_1 = R_{11}J_1 + R_{12}J_2 + R_{13}J_3 + \cdots + R_{1n}J_n,$$
$$X_2 = R_{21}J_1 + R_{22}J_2 + R_{23}J_3 + \cdots + R_{2n}J_n,$$
$$X_3 = R_{31}J_1 + R_{32}J_2 + R_{33}J_3 + \cdots + R_{3n}J_n, \qquad (6)$$
$$\vdots$$
$$X_n = R_{n1}J_1 + R_{n2}J_2 + R_{n3}J_3 + \cdots + R_{nn}J_n,$$

or

$$X_j = \sum_{k=1}^{n} R_{jk}J_k. \qquad (7)$$

The L coefficients ($L_{jk} = J_j/X_k)_{1,2,\ldots}$ are flows per unit force and represent generalized conductances. The R coefficients ($R_{jk} = X_j/J_k)_{1,2,\ldots}$ are force per unit flow and represent generalized resistances or frictions. Practical considerations and convenience dictate which set of equations, either (4) or (6), to use. Furthermore, one system of coefficients can always be transformed to

I. Basic Principles

the other. Thus, for a simple case of two flows and two forces, one can write

$$J_1 = L_{11}X_1 + L_{12}X_2, \quad X_1 = R_{11}J_1 + R_{12}J_2,$$
$$J_2 = L_{21}X_1 + L_{22}X_2, \quad X_2 = R_{21}J_1 + R_{22}J_2. \quad (8)$$

The relations between L_{jk} and R_{jk} are given by

$$L_{11} = \frac{R_{22}}{|R|}, \quad L_{12} = \frac{-R_{12}}{|R|}, \quad L_{21} = \frac{-R_{21}}{|R|}, \quad L_{22} = \frac{R_{11}}{|R|}, \quad (9)$$

where the determinant $|R| = R_{11}R_{22} - R_{12}R_{21}$. In general,

$$L_{jk} = |R|_{jk}/|R|, \quad (10)$$

where $|R|$ is the determinant of the matrix of R_{jk} coefficients and $|R|_{jk}$ is the minor of the determinant of R_{jk}. Similarly, one has

$$R_{jk} = |L|_{jk}/|L|. \quad (11)$$

Phenomenological coefficients are difficult to apply to transport processes since in a system of n components, there are n^2 coefficients to be determined. However, Onsager (1931) has shown that the matrix of phenomenological coefficients is symmetric. That is,

$$L_{jk} = L_{kj} \quad (j \neq k). \quad (12)$$

Examination of the available experimental data has shown that the Onsager reciprocal relation [Eq. (12)] may be regarded as one of the laws of nature. In view of this, only $n(n + 1)/2$ coefficients are required to be determined to characterize the transport processes in a membrane system of n components.

The phenomenological coefficients of Eq. (4) are independent. The absolute values of the cross coefficients are limited by the absolute values of the straight coefficients because of θ (rate of entropy production) being always positive. For the simple case of two flows and two forces, θ is given by

$$\theta = J_1X_1 + J_2X_2. \quad (13)$$

Introducing the L coefficients of Eq. (8) into Eq. (13) gives on rearrangement

$$\theta = L_{11}X_1^2 + X_1X_2(L_{12} + L_{21}) + L_{22}X_2^2 > 0. \quad (14)$$

It follows that

$$L_{11}X_1^2 \geq 0, \quad L_{22}X_2^2 \geq 0 \quad (15)$$

since only one force X_1 or X_2 can be made to operate. Consequently, L_{11} and L_{22} must be positive. As Eq. (14) is a quadratic, a positive solution leads to the condition that the determinant

$$|L| = \begin{vmatrix} L_{11} & L_{12} \\ L_{21} & L_{22} \end{vmatrix} \geq 0. \tag{16}$$

Therefore the limitation imposed on the magnitudes of the cross coefficients is

$$L_{11}L_{22} - L_{21}L_{12} \geq 0. \tag{17}$$

In view of Eq. (12), Eq. (17) becomes $L_{11}L_{22} \geq L_{12}^2$.

For the general case of any number of forces and flows, the conditions given above may be expressed as

$$L_{jj} \geq 0, \quad |L| \geq 0, \quad \text{and} \quad L_{jj}L_{kk} \geq L_{jk}^2. \tag{18}$$

The principles given above may be applied to describe the several transport phenomena in membranes. The first important task is to evaluate θ, which will allow writing the appropriate forms of the forces X_j. This is based on the use of the laws of conservation of mass and energy and the Gibbs equation.

Consider a membrane acting as a barrier to the flow of chemical species between two subsystems I and II in contact with the two faces of the membrane. The two subsystems are kept so well stirred that uniform values of the variables such as concentration, temperature, pressure, and electrical potential prevail in each subsystem and any gradient ($\Delta\mu$, ΔT, ΔP, ΔE) of these variables maintained act only across the membrane. Conservation of mass of the system

Subsystem I	Membrane	Subsystem II

is expressed by

$$dn_k^I + dn_k^{II} = 0 \quad (k = 1, 2, \ldots, n) \tag{19}$$

where n_k is the number of molecules of the species k.

Conservation of energy is expressed by

$$dE^I = d_e E^I + d_i E^I, \tag{20}$$

$$dE^{II} = d_e E^{II} + d_i E^{II}, \tag{21}$$

and

$$d_i E^I + d_i E^{II} = 0. \tag{22}$$

$d_e E^I$ is the energy exchanged by I with its surroundings and $d_i E^I$ the energy exchanged with II. Thus the total energy absorbed by the system as a whole

I. Basic Principles

is given by

$$dE = d_e E^{I} + d_e E^{II}, \tag{23}$$

where

$$d_e E^{I} = d_e Q^{I} - P^{I} dV^{I}, \tag{24}$$

$$d_e E^{II} = d_e Q^{II} - P^{II} dV^{II}. \tag{25}$$

$d_e Q^{I}$ is the heat absorbed by I from its surroundings.

The fundamental principle of the theory of thermodynamics of irreversible process is that the Gibbs equation

$$T\,dS = dE + P\,dV - \sum_k \mu_k\,dn_k \tag{26}$$

is valid even for systems not in thermodynamic equilibrium. Applying Eq. (26) to the two subsystems gives

$$\begin{aligned} T^{I}\,dS^{I} &= dE^{I} + P^{I}\,dV^{I} - \sum_k \mu_k^{I}\,dn_k^{I}, \\ T^{II}\,dS^{II} &= dE^{II} + P^{II}\,dV^{II} - \sum_k \mu_k^{II}\,dn_k^{II}. \end{aligned} \tag{27}$$

The change of entropy of the total system is given by

$$dS = dS^{I} + dS^{II}. \tag{28}$$

Substituting Eqs. (27), (20), and (21) into Eq. (28) gives

$$\begin{aligned} dS =\ & \frac{d_e E^{I} + P^{I}\,dV^{I}}{T^{I}} + \frac{d_e E^{II} + P^{II}\,dV^{II}}{T^{II}} \\ & + \frac{d_i E^{I}}{T^{I}} + \frac{d_i E^{II}}{T^{II}} - \left[\left(\sum_k \mu_k^{I}\,dn_k^{I}\right)\Big/T^{I}\right] - \left[\left(\sum_k \mu_k^{II}\,dn_k^{II}\right)\Big/T^{II}\right]. \end{aligned} \tag{29}$$

The first two terms of Eq. (29) represent $d_e S^{I} + d_e S^{II}$, the gain in entropy due to interaction with the surroundings, since they are equal to $d_e Q^{I}/T^{I} + d_e Q^{II}/T^{II}$ in view of Eqs. (24) and (25). The remaining four terms of Eq. (29) represent $d_i S$, the entropy production due to irreversible processes within the system. Thus

$$d_i S = \frac{d_i E^{I}}{T^{I}} + \frac{d_i E^{II}}{T^{II}} - \sum_k \frac{\mu_k^{I}\,dn_k^{I}}{T^{I}} - \sum_k \frac{\mu_k^{II}\,dn_k^{II}}{T^{II}}. \tag{30}$$

Substitution of Eqs. (22) and (19) into Eq. (30) gives

$$d_i S = -d_i E^{I}\,\Delta\!\left(\frac{1}{T}\right) + \sum_{k=1}^{n} dn_k^{I}\,\Delta\!\left(\frac{\mu_k}{T}\right), \tag{31}$$

where Δ represents the difference of quantities in subsystems II and I. Comparison of Eqs. (3) and (31) leads to

$$T\frac{d_i S}{dt} = J_E X_E + \sum_{k=1}^{n} J_k X_k, \tag{32}$$

where J_E is the energy flux and J_k is the flux of species k. But these are given by

$$J_E = -\frac{d_i E^I}{dt}, \tag{33}$$

$$J_k = -\frac{dn_k^I}{dt}. \tag{34}$$

Substituting Eqs. (33) and (34) into Eq. (32) gives

$$\frac{d_i S}{dt} = -\frac{d_i E^I}{dt}\frac{1}{T} X_E - \sum_{k=1}^{n} \frac{dn_k^I}{dt}\frac{1}{T} X_k. \tag{35}$$

Equating Eqs. (35) and (31) gives

$$X_E = T\Delta\left(\frac{1}{T}\right) \quad \text{and} \quad X_k = -T\Delta\left(\frac{\mu_k}{T}\right). \tag{36}$$

Considering linearity between fluxes and forces, one can write

$$J_j = \sum_{k=1}^{n} L_{jk} X_k + L_{jE} X_E, \tag{37}$$

$$J_E = \sum_{k=1}^{n} L_{Ek} X_k + L_{EE} X_E. \tag{38}$$

The Onsager relations

$$L_{jk} = L_{kj} \quad \text{and} \quad L_{kE} = L_{Ek} \tag{39}$$

are valid between the phenomenological coefficients.

The coefficient L_{jE} in Eq. (37) can be eliminated by defining the energy of transfer E_k^*, which is the energy carried by unit flow of k (i.e., $J_k = 1$) at uniform temperature (i.e., $T = 0$ and $X_E = 0$). Thus

$$L_{jE} = \sum_{k=1}^{n} L_{jk} E_k^* \quad (j = 1, 2, \ldots, n). \tag{40}$$

J_j is therefore given by

$$J_j = \sum_{k=1}^{n} L_{jk}(X_k + E_k^* X_E). \tag{41}$$

Substitution of Eq. (36) into Eq. (41) gives

$$J_j = \sum_{k=1}^{n} L_{jk}\left[-T\Delta\left(\frac{\mu_k}{T}\right) + E_k^* T\Delta\left(\frac{1}{T}\right)\right] \quad (42)$$

or

$$J_j = \sum_{k=1}^{n} L_{jk}\bar{X}_k. \quad (43)$$

Equation (43) defines $\bar{X}_k$ when the forces are infinitesimal

$$\bar{X}_k = -Td\left(\frac{\mu_k}{T}\right) + E_k^* T d\left(\frac{1}{T}\right)$$

or

$$\bar{X}_k = -d\mu_k - \frac{dT}{T}(E_k^* - \mu_k). \quad (44)$$

But $d\mu_k = \bar{V}_k dP - \bar{S}_k dT + RT d\ln a_k + z_k F d\psi$, where z_k is the valence including the sign of the species k and ψ is the electric potential. Therefore

$$\bar{X}_k = -\bar{V}_k dP - RT d\ln a_k - z_k F d\psi - \frac{dT}{T}(E_k^* - H_k^*), \quad (45)$$

where $H_k^* = \mu_k + T\bar{S}_k$. The difference between E_k^* and H_k^* is called the heat of transfer Q_k^*. This represents the heat absorbed in I and liberated into II when 1 mol of k is transported from I to II at constant temperature and pressure, apart from the change in heat content associated with the removal of k. Q_k^* may be positive, negative, or zero.

The general expression for material fluxes is given by

$$J_j = \sum_{k=1}^{n} L_{jk}\left(-\bar{V}_k dP - RT d\ln a_k - z_k F d\psi - Q_k^* \frac{dT}{T}\right). \quad (46)$$

II. Electrical Parameters

Certain measurable parameters can be expressed in terms of phenomenological coefficients.

(1) Current density $I = F\sum_j z_j J_j$. Thus substituting in Eq. (46) gives

$$I = \sum_j \sum_k z_j F L_{jk}(-\bar{V}_k dP - RT d\ln a_k - Q_k^* d\ln T) - \sum_j \sum_k z_j z_k L_{jk} F^2 d\psi. \quad (47)$$

(2) Electrical conductance L_e

$$k_{sp} = -\left(\frac{I}{d\psi}\right)_{dP=0, dT=0, d\ln a_k=0}$$
$$= \sum_j \sum_k z_j z_k F^2 L_{jk} = L_e F^2 \qquad (48)$$

(unit cube), where $L_E = \sum_j \sum_k z_j z_k L_{jk}$. For a membrane of thickness d cm and area 1 cm², $k_{sp} = L_e F^2 d$.

(3) Transport number $\bar{t}_k$

$$\bar{t}_k = \left(\frac{I_k}{I}\right)_{dP=0, dT=0, d\ln a_k=0}$$
$$= \frac{F z_k J_k}{I} = \left(F \sum_j z_j z_k F L_{jk} \, d\psi\right) \bigg/ \left(\sum_j \sum_k z_j z_k F^2 L_{jk} \, d\psi\right)$$
$$= \left(F^2 \sum_j z_j z_k L_{jk}\right) \bigg/ k_{sp} = \left(\sum_j z_j z_k L_{jk}\right) \bigg/ L_e. \qquad (49)$$

Staverman introduced a parameter called the reduced electrical transport number $\bar{t}_k^r$. This is also called mass transport number which is given by

$$\bar{t}_k^r = \frac{\bar{t}_k}{z_k} = F^2 \sum_j z_j L_{jk} \bigg/ k_{sp} = \sum_j z_j L_{jk} \bigg/ L_e. \qquad (50)$$

This parameter is useful in describing the flow of uncharged molecules (solvent) in electrical transport measurements. These satisfy the relation $\sum_k z_k \bar{t}_k^r = 1$.

From Eq. (47) expressions for electrical potentials arising from different driving forces may be derived. The general expression for the potential gradient at any point in the membrane is obtained by substituting Eqs. (48) and (50) into Eq. (47). Thus

$$d\psi = -\frac{1}{k_{sp}} - \frac{1}{F} \sum_k \bar{t}_k^r (\bar{V}_k \, dP + RT \, d\ln a_k + Q^* \, d\ln T). \qquad (51)$$

The potential gradient is composed of two parts, one a purely electrical part (I/k_{sp}) and the second the chemical part given by the second term. Three special cases are of interest.

(a) At uniform temperature and pressure, when $I = 0$,

$$-F \frac{d\psi}{dx} = \sum_k \bar{t}_k^r RT \, d\ln a_k \bigg/ dx$$

or

$$\psi^{II} - \psi^I = -\frac{RT}{F} \int_I^{II} \sum_k \bar{t}_k^r \, d\ln a_k. \qquad (52)$$

The right-hand side of Eq. (52) is the classic isothermal diffusion potential derived many years ago by Nernst (1888, 1889) and more recently by Staverman (1952) and Kirkwood (1954). The important aspect of Eq. (52) is that uncharged components are included in the summation.

(2) At uniform temperature and chemical potential,

$$-F\frac{d\psi}{dx} = \sum_k \bar{t}_k^r \bar{V}_k \frac{dP}{dx}$$

or

$$\psi^{II} - \psi^{I} = -\frac{1}{F}\int_I^{II} \sum_k \bar{t}_k^r \bar{V}_k \, dP. \tag{53}$$

This is the streaming potential in the membrane arising from the maintenance of a pressure gradient across the membrane. In the pressure range normally used, the $\bar{t}_k^r$ and $\bar{V}_k$ are independent of pressure. So Eq. (53) may be written

$$\frac{\Delta\psi}{\Delta P} = -\frac{1}{F}\sum_k \bar{t}_k^r \bar{V}_k. \tag{54}$$

(3) At uniform pressure and activity,

$$-F\frac{d\psi}{dx} = \sum_k \bar{t}_k^r Q_k^* \, d\ln T \Big/ dx$$

or

$$\psi^{II} - \psi^{I} = -\frac{1}{F}\int_I^{II} \sum_k \bar{t}_k^r Q_k^* \, d\ln T. \tag{55}$$

This is the thermal membrane potential in the membrane arising from the maintenance of a temperature gradient across the membrane. If $\bar{t}_k^r$ and Q_k^* are assumed independent of temperature over a small temperature range ΔT, then Eq. (55) may be written

$$F\frac{\Delta\psi}{\Delta T} = -\sum_k \bar{t}_k^r Q_k^* \Big/ T. \tag{56}$$

III. Electrokinetic Phenomena

Again, the application of the principles of irreversible thermodynamics to electrokinetic effects in membranes is illustrated.

Consider two compartments I and II separated by a porous plug or an ion-exchange membrane. Let each compartment contain n ($k = 1, 2, \ldots, n$)

components carrying $z_k F$ charge per unit mass. Temperature and concentration are kept uniform throughout the system. Conservation of mass and charge are given by

$$dn_k^I + dn_k^{II} = 0 \quad (k = 1, 2, \ldots, n), \tag{57}$$

$$\sum_{k=1}^{n} z_k F \, dn_k^I + \sum_{k=1}^{n} z_k F \, dn_k^{II} = 0. \tag{58}$$

Flow of charge (electric current) from I to II is given by

$$I = -\sum_{k=1}^{n} z_k F \frac{dn_k^I}{dt} = \sum_{k=1}^{n} z_k F \frac{dn_k^{II}}{dt}. \tag{59}$$

The conservation of energy in the system gives

$$dE = dE^I + dE^{II} = dQ - P^I \, dV^I - P^{II} \, dV^{II} + (\psi^{II} - \psi^I) I \, dt. \tag{60}$$

Equation (27) gives the total entropy change in the system. Substituting Eqs. (59) and (60) into Eq. (27) gives

$$T \, dS = dQ - \sum_k \mu_k^I \, dn_k^I - \sum_k \mu_k^{II} \, dn_k^{II} - \psi^I \sum_k z_k F \, dn_k^I - \psi^{II} \sum_k z_k F \, dn_k^{II}. \tag{61}$$

Substituting from Eq. (57) into Eq. (61) gives

$$T \, dS = dQ - \sum_{k=1}^{n} (\mu_k^I + z_k F \psi^I - \mu_k^{II} - z_k F \psi^{II}) \, dn_k^I. \tag{62}$$

But $dS = d_e S + d_i S$ and $d_e S = dQ/T$, and so the second term on the right-hand side of Eq. (62) is equal to $T \, d_i S$. Therefore

$$d_i S = -\sum_{k=1}^{n} \Delta(\mu_k + z_k F \psi) dn_k^I \bigg/ T. \tag{63}$$

As $\Delta \mu_k = \bar{V}_k \Delta P$, the dissipation function is given by

$$T \Phi = T \frac{d_i S}{dt} = -\sum_k \bar{V}_k \, dP \frac{dn_k^I}{dt} - \sum_k z_k F \Delta \psi \frac{dn_k^I}{dt}. \tag{64}$$

The volume flow J_v [see Eq. (110)] is given by

$$J_v = -\sum_k \bar{V}_k \frac{dn_k^I}{dt}, \tag{65}$$

and the current I is given by Eq. (59). Hence, substituting Eqs. (59) and (65) into Eq. (64) gives an expression for the dissipation function. Thus

$$\phi = J_v \Delta P + I \Delta \psi. \tag{66}$$

III. Electrokinetic Phenomena

Linear phenomenological equations for the dependence of flows on forces when the same solution exists in the two compartments are

$$J_v = L_{11} \Delta P + L_{12} \Delta \psi, \tag{67}$$

$$I = L_{21} \Delta P + L_{22} \Delta \psi. \tag{68}$$

The dependence of forces on fluxes are written

$$\Delta P = R_{11} J_v + R_{12} I, \tag{69}$$

$$\Delta \psi = R_{21} J_v + R_{22} I. \tag{70}$$

The Onsager relations are $L_{12} = L_{21}$ and $R_{12} = R_{21}$. The four quantities (i.e., two forces and two flows) interacting with one another under a set of given conditions (i.e., two taken at a time keeping the third constant at zero) give 24 separate permutations. Half of these are reciprocals, and so 12 separate electrokinetic relations can be expected. These are given in Table I.

The straight coefficients are L_{11}, L_{22}, R_{11}, and R_{22}. They are called (see Table I) mechanical conductance $(J_v/\Delta P)_{\Delta \psi = 0}$, electrical conductance $(I/\Delta \psi)_{\Delta P = 0}$, mechanical resistance $(\Delta P/J_v)_{I=0}$, and electrical resistance $(\Delta \psi / I)_{J_v = 0}$, respectively. The other eight relations containing the cross coefficients give the following four relations as a result of the Onsager law:

$$(I/J_v)_{\Delta \psi = 0} = -(\Delta P / \Delta \psi)_{J_v = 0}. \tag{71}$$

This is one of Saxen's relations connecting streaming current to electro-osmotic pressure (coupling flows to coupling forces). The other three relations are

$$(\Delta \psi / \Delta P)_{I=0} = -(J_v / I)_{\Delta P = 0} \tag{72}$$

(the relation between streaming potential and electro-osmotic flow)

$$(\Delta \psi / J_v)_{I=0} = (\Delta P / I)_{J_v = 0} \tag{73}$$

(the relation between second streaming potential and second electro-osmotic pressure)

$$(I/P)_{\Delta \psi = 0} = (J_v / \Delta \psi)_{\Delta P = 0} \tag{74}$$

(the relation between second streaming current and electro-osmotic flow). Each of the four equations (67)–(70) are useful in different experimental arrangements despite Eqs. (67) and (68) being mathematically equivalent to Eqs. (69) and (70). For example, when $\Delta \psi = 0$, Eq. (67) gives

$$(J_v / \Delta P)_{\Delta \psi = 0} = L_{11}. \tag{75}$$

L_{11} is the filtration coefficient at zero $\Delta \psi$. Because $\Delta \psi$ is zero, the current does not disappear. The flow of ions coupled to volume flow gives a current

TABLE I

Electrokinetic Relations

	$\Delta\psi$	ΔP	I	J_v
$\Delta\psi$	1	$\left(\dfrac{\Delta P}{\Delta\psi}\right)_{J_v=0} = \dfrac{R_{12}}{R_{22}}$ Electro-osmotic pressure	$\left(\dfrac{I}{\Delta\psi}\right)_{\Delta P=0} = L_{22}$ Electrical conductance	$\left(\dfrac{J_v}{\Delta\psi}\right)_{\Delta P=0} = L_{12}$ Second electro-osmotic flow
ΔP	$\left(\dfrac{\Delta\psi}{\Delta P}\right)_{I=0} = -\dfrac{L_{21}}{L_{22}}$ Streaming potential	1	$\left(\dfrac{I}{\Delta P}\right)_{\Delta\psi=0} = L_{21}$ Second streaming current	$\left(\dfrac{J_v}{\Delta P}\right)_{\Delta\psi=0} = L_{11}$ Mechanical conductance
I	$\left(\dfrac{\Delta\psi}{I}\right)_{J_v=0} = R_{22}$ Electric resistance	$\left(\dfrac{\Delta P}{I}\right)_{J_v=0} = R_{12}$ Second electro-osmotic pressure	1	$\left(\dfrac{J_v}{I}\right)_{\Delta P=0} = \dfrac{L_{12}}{L_{22}}$ Electro-osmotic flow
J_v	$\left(\dfrac{\Delta\psi}{J_v}\right)_{I=0} = R_{21}$ Second streaming potential	$\left(\dfrac{\Delta P}{J_v}\right)_{I=0} = R_{11}$ Mechanical resistance	$\left(\dfrac{I}{J_v}\right)_{\Delta\psi=0} = \dfrac{R_{21}}{R_{22}}$ Streaming current	1
	One of the flows is zero		One of the forces is zero	

according to Eq. (68). This current given by $(I)_{\Delta\psi=0} = L_{21}\Delta P$ can be measured by using a galvanometer and reversible electrodes in the two chambers separated by the membrane. If the filtration coefficient at $I = 0$ is to be measured, then the current must be made zero by connecting the electrodes to a potentiometer. The volume flow J_v measured when $I = 0$ is denoted in the literature by the symbol L_p. For this condition, Eq. (69) gives

$$(J_v/\Delta P)_{I=0} = 1/R_{11} = L_p. \tag{76}$$

The relation between L_p and L_{11} can be derived by rearranging Eqs. (67) and (68). Rearrangement of Eq. (68) gives

$$\Delta\psi = (I - L_{21}\Delta P)/L_{22}.$$

Introducing this expression into Eq. (67) gives on rearrangement

$$J_v = \left[L_{11} - \frac{L_{12}L_{21}}{L_{22}}\right]\Delta P + \frac{L_{12}}{L_{22}} I. \tag{77}$$

For the condition $I = 0$, Eq. (77) becomes

$$\left(\frac{J_v}{\Delta P}\right)_{I=0} = L_{11} - \frac{L_{12}^2}{L_{22}} = L_p. \tag{78}$$

It follows therefore that $L_{11} = L_p$ only if $L_{12} = 0$. This means that the filtration coefficient measured when $\Delta\psi = 0$ will be equal to that measured when $I = 0$ only if $L_{12} = 0$; i.e., there should be no coupling between the volume flow and the ion flow. So L_p is generally measured using only water on either side of the membrane in the case of artificial membranes.

Another example of coupling is electroconvection. In an ion-selective membrane, because of the presence of charged groups, there will be more counterions than coions in the membrane phase. Application of an electric field across the membrane allows counterions to impart more momentum to the solvent molecules surrounding them in the membrane than do coions. Consequently, in the case of a negatively charged membrane, cations migrate to the cathode, into which solvent will also flow. The velocity of this solvent transfer will be determined by the strength of the electric field and the solvent (water) permeability of the membrane. The counterions in their movement will experience less resistance to flow than do coions. Therefore with reference to the membrane matrix, the counterions move faster than they would otherwise if the liquid were standing still. Consequently, the migration or convection of the pore liquid will increase the flux of counterions. This will increase the electrical conductance of the membrane, which may be written

$$\bar{K}_{sp} = (I/\Delta\psi)_{\Delta P=0} = L_{22}. \tag{79}$$

Under these conditions, volume flow takes place [see Eq. (67)]. If this is prevented by application of pressure, then

$$\bar{k}_{sp} = (I/\Delta\psi)_{J_v=0} = 1/R_{22}. \tag{80}$$

Now the relation between $\bar{k}'_{sp}$ and $\bar{k}_{sp}$ may be derived again by using Eqs. (67) and (68). Equation (67) gives on rearrangement $\Delta P = (J_v - L_{12}\Delta\psi)/L_{11}$. Substitution of this relation into Eq. (68) gives

$$I = \frac{L_{21}}{L_{11}} J_v + \left[L_{22} - \frac{L_{21}L_{12}}{L_{11}} \right] \Delta\psi. \tag{81}$$

When $J_v = 0$, Eq. (81) becomes

$$(I/\Delta\psi)_{J_v=0} = L_{22} - L_{12}^2/L_{11}. \tag{82}$$

In view of Eqs. (79) and (80), Eq. (82) becomes

$$\bar{k}_{sp} = \bar{k}'_{sp} - L_{12}^2/L_{11}. \tag{83}$$

Only when $L_{12} = 0$ does $\bar{k}_{sp}$ become equal to $\bar{k}'_{sp}$. The term L_{12}^2/L_{11} is the electroconvective contribution to the total measured conductance.

In addition to $\bar{k}_{sp}$ and $\bar{k}'_{sp}$, the other practical electrokinetic parameter that can be measured is the electro-osmotic permeability β. This is defined by the relation

$$\beta = (J_v/I)_{\Delta P=0}. \tag{84}$$

Dividing Eq. (67) by Eq. (68) for the condition $\Delta P = 0$ gives, according to Eq. (84),

$$\beta = L_{12}/L_{22}. \tag{85}$$

Equation (85) in view of Eq. (79) gives

$$\beta = L_{12}/\bar{k}'_{sp}. \tag{86}$$

From the thermodynamic equations (67) and (68), a set of practical equations for the description of electrokinetic phenomena can be derived. Substituting Eq. (78) for L_{11} and Eq. (85) for L_{12} into Eq. (67) and then using Eqs. (85) and (68) gives

$$J_v = L_p \Delta P + \beta I. \tag{87}$$

Similarly, substitution of Eq. (86) for L_{21} and Eq. (79) for L_{22} into Eq. (68) gives

$$I = \bar{k}'_{sp}\beta \Delta P + \bar{k}'_{sp} \Delta\psi. \tag{88}$$

Alternatively, Eq. (88) can be transformed into the form

$$I = (\bar{k}'_{sp}\beta/L_p)J_v + \bar{k}_{sp} \Delta\psi. \tag{89}$$

The practical parameters are related to the thermodynamic electrokinetic parameters by the following relations:

$$\bar{K}'_{sp} = L_{22}, \quad \bar{K}'_{sp}\beta = L_{12}, \quad \bar{K}'_{sp}\beta^2 + L_p = L_{11},$$
$$L_p = 1/R_{11}, \quad -\beta/L_p = R_{12}, \quad \bar{K}_{sp} = 1/R_{22}.$$

IV. Transport of a Solution of Nonelectrolyte across a Simple Membrane

The dissipation function for this case is

$$\phi = J_s \Delta\mu_s + J_w \Delta\mu_w, \tag{90}$$

where s stands for the solute and w for water. The chemical potential μ_w is given by

$$\mu_w = \mu_w^0 + \bar{V}_w P + \mu_w^c, \tag{91}$$

where μ_w^c is the concentration-dependent part of μ_w. The gradient across a slab Δx of the membrane is given by

$$\Delta\mu_w = (\mu_w)_0 - (\mu_w)_{\Delta x} = \bar{V}_w(P_0 - P_{\Delta x}) + (\mu_w^c)_0 - (\mu_w^c)_{\Delta x}$$
$$= \bar{V}_w \Delta P + \Delta\mu_w^c. \tag{92}$$

Employing similar arguments gives for the solute the relation

$$\Delta\mu_s = \bar{V}_s \Delta P + \Delta\mu_s^c. \tag{93}$$

Integration of Eq. (28) of Chapter 2 gives

$$\mu_i = \bar{V}_i P + \mu_i^c, \tag{94}$$

where μ_i^c is the integration constant. Differentiation of Eq. (94) for the case $i = w$ gives

$$d\mu_w = \bar{V}_w\, dP + d\mu_w^c. \tag{95}$$

If a semipermeable membrane separates two phases (phase 1 = solution and phase 2 = solvent), the pressure that would just prevent the flow of solvent from phase 2 to phase 1 is equal to the osmotic pressure $\Delta\Pi$ (where $\Delta\Pi = \Delta\Pi_{im} + \Delta\Pi_s$, im = impermeable solute, s = permeable solute). At osmotic equilibrium $d\mu_w = 0$. Thus Eq. (95) gives

$$d\mu_w^c = -\bar{V}_w\, dP$$

or

$$\Delta\mu_w^c = -\bar{V}_w \Delta\Pi. \tag{96}$$

Substituting Eq. (96) into Eq. (92) gives

$$\Delta\mu_w = \bar{V}_w(\Delta P - \Delta\Pi). \tag{97}$$

The Gibbs–Duhem equation [Eq. (25) of Chapter 2] can be written for the solution

$$\sum_{j=1}^{k-1} n_j \, d\mu_j^c = -n_w \, d\mu_w^c. \tag{98}$$

Substituting Eq. (96) into Eq. (98) gives

$$\sum_{j=1}^{k-1} n_j \, d\mu_j^c = n_w \bar{V}_w \, d\Pi. \tag{99}$$

The solutions under consideration in general are so dilute that the volume contribution of the solvent to the total volume V is very large compared to the contributions of the other components of the solution. Thus $V \approx n_w \bar{V}_w$, and so Eq. (99) becomes

$$d\Pi = \sum_{j=1}^{k-1} \frac{n_j}{V} \, d\mu_j^c. \tag{100}$$

But n_j/V is the molar concentration of j, and so Eq. (100) becomes

$$d\Pi = \sum_{j=1}^{k-1} C_j \, d\mu_j^c. \tag{101}$$

When there is only one permeable solute s in the osmotic cell, Eq. (101) simplifies to

$$d\Pi_s = C_s \, d\mu_s^c. \tag{102}$$

In the osmotic cell, the concentration of the solute on either side of the membrane would be different unless the membrane is ideally semipermeable. In practice the membrane will not be perfectly semipermeable, and so an average concentration must be introduced into Eq. (102), which now can be written

$$\Delta\Pi_s = C_s^a \, \Delta\mu_s^c. \tag{103}$$

Substituting Eq. (103) into Eq. (93) gives

$$\Delta\mu_s = \bar{V}_s \Delta P + \Delta\Pi_s/C_s^a. \tag{104}$$

The average concentration for ideal solutions can be evaluated:

$$\Delta\mu_s^c = RT(\ln C_s^0 - \ln C_s^{\Delta x}), \tag{105}$$

and according to van't Hoff's law [Eq. (182) of Chapter 3]

$$\Delta\Pi = RT(C_s^0 - C_s^{\Delta x}). \tag{106}$$

Dividing Eq. (106) by Eq. (105) gives C_s^a according to Eq. (103). Thus

$$C_s^a = \frac{C_s^0 - C_s^{\Delta x}}{\ln(C_s^0/C_s^{\Delta x})}. \tag{107}$$

IV. Transport of a Solution of Nonelectrolyte

Expanding the logarithm

$$\left[\ln x = 2\left(\frac{x-1}{x+1}\right) + \frac{1}{3}\left(\frac{x-1}{x+1}\right)^3 + \cdots\right]$$

and ignoring the higher terms gives

$$C_s^a = (C_s^0 + C_s^{\Delta x})/2. \tag{108}$$

Substituting Eqs. (97) and (104) into Eq. (90) when $\Delta\Pi_{im} = 0$ gives on rearrangement

$$\phi = (J_w \bar{V}_w + J_s \bar{V}_s)\Delta P + (J_s/C_s^a - J_w \bar{V}_w)\Delta\Pi_s. \tag{109}$$

The first term of this equation is the volume flow J_v and so

$$J_v = J_w \bar{V}_w + J_s \bar{V}_s. \tag{110a}$$

Since $C_w \bar{V}_w \approx 1$, the second term may be written

$$\frac{J_s}{C_s^a} - J_w \bar{V}_w \approx \frac{J_s}{C_s^a} - \frac{J_w}{C_w} = \frac{C_s^a v_s}{C_s^a} - \frac{C_w v_w}{C_w} = v_s - v_w. \tag{110b}$$

Thus the flow conjugate to the force $\Delta\Pi$ is the velocity of the solute relative to that of the solvent. This may be denoted by J_D in view of the flow being similar to the diffusional flow. Thus the dissipation function is given by

$$\phi = J_v \Delta P + J_D \Delta\Pi, \tag{111}$$

where the new forces are the hydrostatic pressure difference and the osmotic pressure difference across the membrane.

The phenomenological equations corresponding to Eq. (111) are

$$J_v = L_p \Delta P + L_{PD} \Delta\Pi,$$
$$J_D = L_{DP} \Delta P + L_D \Delta\Pi. \tag{112}$$

The Onsager reciprocal relation gives $L_{PD} = L_{DP}$. Equation (112) shows that the hydrostatic pressure with $\Delta\Pi = 0$ produces not only J_v but also diffusional flow J_D:

$$(J_D)_{\Delta\Pi=0} = L_{DP} \Delta P. \tag{113}$$

L_{DP}, the coupling coefficient, indicates the extent of hyperfiltration properties of the membrane. L_{PD} on the other hand

$$(J_v)_{\Delta P=0} = L_{PD} \Delta\Pi \tag{114}$$

gives the osmotic flow. The condition for an ideally semipermeable membrane is that $J_s = 0$ for all values of the forces. Adding J_v and J_D [Eqs. (110a) and (110b)] gives

$$J_v + J_D = J_w\bar{V}_w + J_s\bar{V}_s + \frac{J_s}{C_s^a} - J_w\bar{V}_w = \frac{J_s}{C_s^a}(1 + C_s^a \bar{V}_s)$$

$$= \frac{J_s}{C_s^a}(1 + Q_s), \tag{115}$$

where Q_s is the volume fraction of solute and is negligible compared to Q_w (i.e., $Q_s \ll 1$). Substituting for J_v and J_D from Eq. (112) into Eq. (115) gives

$$\frac{J_s}{C_s^a} = (L_p + L_{DP})\Delta P + (L_{PD} + L_D)\Delta\Pi. \tag{116}$$

For $J_s = 0$, both terms in Eq. (116) must be zero. So for a perfectly semipermeable membrane

$$L_p = L_D = -L_{PD}. \tag{117}$$

Under these conditions, when $J_v = 0$, Eq. (112) becomes

$$(\Delta P)_{J_v=0} = -(L_{PD}/L_p)\Delta\Pi$$

and so for perfect semipermeability of the membrane $(\Delta P)_{J_v=0} = \Delta\Pi$. If the membrane allows solute to go through, then $-L_{PD}/L_p < 1$. Staverman introduced the term reflection coefficient σ to indicate the membrane selectivity. Thus

$$\sigma = -L_{PD}/L_p. \tag{118}$$

When $\sigma < 1$, the solute permeates the membrane, and when $\sigma = 1$, the solute does not go through the membrane but is "reflected." The meaning of σ becomes clearer when it is expressed in terms of solute and solvent velocities. Substituting for J_v and J_D in terms of velocities in the equation

$$[J_D/J_v]_{\Delta\Pi=0} = L_{PD}/L_p = -\sigma$$

gives

$$(v_s - v_w)/(v_s + v_w) = -\sigma.$$

But $v_w \gg v_s$ and so

$$\left(\frac{v_s - v_w}{v_w}\right)_{\Delta\Pi=0} = -\sigma$$

or

$$v_s/v_w = 1 - \sigma. \tag{119}$$

This means that when $v_s = 0$, $\sigma = 1$; that is, the membrane becomes ideally semipermeable when no solute goes through the membrane. When $\sigma = 0$,

$v_s = v_w$, and so both solute and solvent move through with equal velocities and no ultrafiltration takes place. When v_s becomes greater than v_w, σ becomes negative. This phenomenon when transfer of solute is greater than that of the solvent is called negative anomalous osmosis.

The phenomenological coefficient L_D may be replaced by ω, the solute permeability coefficient since it can be determined experimentally. Rearranging Eq. (112) gives $\Delta P = (J_v - L_{PD}\Delta\Pi)/L_p$. Substituting this into Eq. (116) and using Eq. (118) gives

$$J_s = C_s^a(1 - \sigma)J_v + \omega\,\Delta\Pi, \tag{120}$$

where

$$\omega = C_s^a(L_p L_D - L_{PD}^2)/L_p. \tag{121}$$

When $J_v = 0$, ω becomes the solute permeability. Thus $\omega = (J_s/\Delta\Pi)_{J_v=0}$. This is an important parameter of the membrane. When $L_{PD} = 0$, $\omega = C_s^a L_D$. Thus the set of experimentally determinable characteristic parameters of the membrane are L_p, σ, and ω.

The influence of volume flow J_v on solute flow J_s, i.e., the so-called "solvent drag," can be evaluated when $\Delta\Pi = 0$. Under this condition Eq. (120) becomes

$$J_s = C_s^a(1 - \sigma)J_v. \tag{122}$$

This solvent drag effect has been demonstrated in toad skin by using the same concentration of solute on either side of the skin and inducing volume flow by osmosis with the help of a concentration gradient by an impermeant solute. Thus measurements of both J_s and J_v enable determination of σ. In addition values for both L_p and ω may be derived from the experimental data. Furthermore, other methods have been used to measure σ. Substituting σ into Eq. (112) gives

$$J_v = L_p(\Delta P - \sigma\,\Delta\Pi). \tag{123}$$

When there are any number of solutes n present in the system, Eq. (123) becomes

$$J_v = L_p\left(\Delta P - \sum_{i=1}^{n} \sigma_i\,\Delta\Pi_i\right). \tag{124}$$

For a permeant solute when $\Delta P = 0$, Eq. (123) becomes

$$J_v = -L_p\sigma\,\Delta\Pi_s, \tag{125}$$

and for an impermeant solute i

$$J_v = -L_p\,\Delta\Pi_{im}. \tag{126}$$

Thus by observing the initial volume flow caused by a known solute concentration difference, σ can be estimated if L_p is known.

Another way is to compare the effects of concentration differences of permeant and impermeant solutes. Thus if both solutes are present simultaneously, the volume flow is given by Eqs. (125) and (126). That is, $J_v = -L_p(\Delta\Pi_{im} + \sigma\Delta\Pi_s)$. If $J_v = 0$, σ is given by

$$-\sigma = (\Delta\Pi_{im}/\Delta\Pi_s)_{J_v=0}.$$

Hence determination of that combination of $\Delta\Pi_{im}$ and $\Delta\Pi_s$ that gives an initial zero volume flow gives the value for σ.

The phenomenological coefficients (Ls) of Eq. (5) may be translated into friction coefficients (fs) by balancing the thermodynamic forces (X_i) by the algebraic sum of the frictional forces ($\mathscr{F}_{ij}$). The thermodynamic force X acting on the solute (X_s) existing with water in the membrane is counterbalanced by the sum of the frictional forces between solute and water and between solute and membrane, i.e.,

$$X_s = -\mathscr{F}_{sw} - \mathscr{F}_{sm}. \tag{127}$$

Similarly, the thermodynamic force on water is given by

$$X_w = -\mathscr{F}_{ws} - \mathscr{F}_{wm}. \tag{128}$$

The simple law of friction is that the frictional force $\mathscr{F}_{ik}$ which slows down the motion of an object (i) moving or gliding on another object (k) is proportional to the relative velocity of i with respect to k, i.e.,

$$\mathscr{F}_{ik} = -f_{ik}(v_i - v_k). \tag{129}$$

It is assumed that the interfaces are well lubricated. The values of the proportionality constant f_{ik} being determined by the difference in the values of v_i and v_k (velocities of i and k) are independent, unlike L values, of the frame of reference to which the velocities are related. Also, they will be concentration independent and thus bring out the specific interaction of i with k. Thus $\mathscr{F}_{sw} = -f_{sw}(\bar{v}_s - \bar{v}_w)$ and $\mathscr{F}_{sm} = -f_{sm}(\bar{v}_s - \bar{v}_m)$. If the membrane is chosen as the frame of reference, then $\bar{v}_m = 0$ and $\mathscr{F}_{sm} = -f_{sm}\bar{v}_s$. Introducing these values into Eqs. (127) and (128) gives

$$X_s = -\frac{d\mu_s}{dx} = f_{sw}(\bar{v}_s - \bar{v}_w) + f_{sm}\bar{v}_s = \bar{v}_s(f_{sw} + f_{sm}) - \bar{v}_w f_{sw},$$

$$X_w = -\frac{d\mu_w}{dx} = f_{ws}(\bar{v}_w - \bar{v}_s) + f_{wm}\bar{v}_w = -\bar{v}_s f_{ws} + \bar{v}_w(f_{ws} + f_{wm}). \tag{130}$$

These equations can be transformed into relations between forces and flows since J_i at a point x is given by $J_i = \bar{C}_{i(x)}\bar{v}_i$. Thus

$$X_s = -\frac{d\mu_s}{dx} = (f_{sw} + f_{sm})\frac{J_s}{\bar{C}_{s(x)}} - f_{sw}\frac{J_w}{\bar{C}_{w(x)}}, \tag{131}$$

$$X_w = -\frac{d\mu_w}{dx} = -f_{ws}\frac{J_s}{\bar{C}_{s(x)}} + (f_{ws} + f_{wm})\frac{J_w}{\bar{C}_{w(x)}}, \tag{132}$$

IV. Transport of a Solution of Nonelectrolyte 289

where $\bar{C}_{s(x)}$ and $\bar{C}_{w(x)}$ are the concentrations of solute and water respectively at a point x in the membrane. Equations (131) and (132) have the forms of Eqs. (8), i.e.,

$$X_s = R_{ss}J_s + R_{sw}J_w,$$
$$X_w = R_{ws}J_s + R_{ww}J_w. \qquad (133)$$

Thus equating the coefficients gives

$$R_{ss} = \frac{f_{sw} + f_{sm}}{\bar{C}_{s(x)}}, \qquad R_{sw} = -\frac{f_{sw}}{\bar{C}_{w(x)}},$$
$$R_{ws} = -\frac{f_{ws}}{\bar{C}_{s(x)}}, \qquad R_{ww} = \frac{f_{ws} + f_{wm}}{\bar{C}_{w(x)}}. \qquad (134)$$

As $R_{sw} = R_{ws}$, it follows that

$$f_{ws} = f_{sw}\bar{C}_{s(x)}/\bar{C}_{w(x)}. \qquad (135)$$

The local concentrations are unknown. Consequently, they are related to the external concentration by a partition coefficient. Thus

$$\bar{C}_{s(x)}/C_{s(x)} = K_s. \qquad (136)$$

The thermodynamic argument for writing Eq. (136) is that in the steady state the solution in an imaginary hole at a point x in the membrane has the same concentration as the external solution. The chemical potential of solute in the hole should be equal to that in the membrane. Similarly for water one can write

$$\bar{C}_w/C_w = K_w. \qquad (137)$$

Substituting Eq. (136) into Eq. (131) and imposing the condition that $J_w = 0$ gives

$$\left(-\frac{d\mu_s}{dx}\right)_{J_w=0} = \frac{(f_{sw} + f_{sm})J_s}{K_s C_{s(x)}}. \qquad (138)$$

But $d\mu_s = d\Pi/C_s$ [see Eq. (102)] and so

$$-\frac{d\Pi}{dx} = \frac{J_s(f_{sw} + f_{sm})}{K_s}. \qquad (139)$$

Equation (139) may be integrated across the membrane. Thus

$$[\Pi]_{\Pi_1 \text{ at } x=0}^{\Pi_2 \text{ at } x=d} = \frac{J_s}{K_s}(f_{sw} + f_{sm})[x]_0^d,$$

that is,

$$(\Delta\Pi)_{J_w=0} = \frac{J_s}{K_s}(f_{sw} + f_{sm})d. \qquad (140)$$

The solute permeability ω has been defined as $\omega = (J_s/\Delta\Pi)_{J_v=0}$. In most cases $J_v = 0$ is nearly equivalent to $J_w = 0$. Thus in view of Eq. (140) ω may be expressed as

$$\omega = \left(\frac{J_s}{\Delta\Pi}\right)_{J_v=0} \approx \left(\frac{J_s}{\Delta\Pi}\right)_{J_w=0} = \frac{K_s}{d(f_{sw} + f_{sm})}. \qquad (141)$$

Equation (141) shows that the solute permeability is directly proportional to the distribution coefficient of the solute and inversely proportional to the overall frictional interactions of the solute with the water and the membrane and to the membrane thickness.

The other membrane parameters to be evaluated in terms of the friction coefficients are the filtration or mechanical permeability coefficient L_P and the reflection coefficient σ.

Equation (137) may be written

$$\frac{\bar{C}_w \bar{V}_w}{C_w \bar{V}_w} = \frac{\bar{C}_w \bar{V}_w}{1} = Q_w = K_w, \qquad (142)$$

where $\bar{V}_w$ is the partial molar volume of water and Q_w is the volume fraction of water in the membrane. When only water exists on either side of the membrane, the driving force X_w is given by

$$X_w = -\frac{d\mu_w}{dx} = f_{wm}(\bar{v}_w - \bar{v}_m). \qquad (143)$$

When $\bar{v}_m = 0$, Eq. (143) becomes

$$-\frac{d\mu_w}{dx} = f_{wm}\bar{v}_w. \qquad (144)$$

Equation (144) may be rewritten

$$-\frac{d\mu_w}{dx} = f_{wm}\frac{\bar{v}_w \bar{C}_w}{\bar{C}_w} = \frac{f_{wm}J_w}{\bar{C}_w}.$$

Substituting for $\bar{C}_w$ from Eq. (142) gives

$$-\frac{d\mu_w}{dx} = \frac{J_w \bar{V}_w}{Q_w} f_{wm}. \qquad (145)$$

Integrating the left-hand side of Eq. (145) gives

$$\int_0^d -\frac{d\mu_w}{dx} dx = \mu_{w(0)} - \mu_{w(d)} = \Delta\mu_w.$$

But

$$\Delta\mu_w = \bar{V}_w \Delta P \qquad (146)$$

IV. Transport of a Solution of Nonelectrolyte

Integrating the right-hand side of Eq. (145) gives

$$\int_0^d \frac{J_w \bar{V}_w}{Q_w} f_{wm} \, dx = \frac{J_v}{Q_w} f_{wm} d, \qquad (147)$$

where f_{wm} and Q_w are assumed to be constant and $J_w \bar{V}_w = J_v$. Equating Eqs. (146) and (147) gives on rearrangement

$$J_v = \frac{Q_w \bar{V}_w}{f_{wm} d} \Delta P. \qquad (148)$$

Comparing Eq. (148) to Eq. (112) under the condition $\Delta \Pi = 0$ gives

$$L_p = Q_w \bar{V}_w / f_{wm} d. \qquad (149)$$

This equation shows that the filtration coefficient L_p is directly proportional to the water content of the membrane and inversely proportional to the thickness of the membrane and to the friction coefficient between water and the membrane matrix.

When $J_w = 0$, Eq. (132) becomes

$$-\frac{d\mu_w}{dx} = -f_{ws} \frac{J_s}{\bar{C}_s}. \qquad (150)$$

Substituting Eq. (135) and the relation $\bar{C}_w \bar{V}_w = Q_w$ into Eq. (150) gives on simplification

$$-\frac{d\mu_w}{dx} = -\frac{\bar{V}_w f_{sw}}{Q_w} J_s. \qquad (151)$$

Eliminating J_s between Eqs. (138) and (151) gives

$$-\frac{d\mu_s}{dx} \bar{C}_s = -\frac{f_{sw} + f_{sm}}{K_s f_{sw} \bar{V}_w} Q_w \left(-\frac{d\mu_w}{dx}\right). \qquad (152)$$

Introducing the relation

$$\frac{d\mu_s}{dx} \bar{C}_s = \frac{d\Pi}{dx}$$

[see Eq. (102)] into Eq. (152) and integrating it across the membrane gives

$$\int_0^d -\frac{d\Pi}{dx} dx = -\frac{f_{sw} + f_{sm}}{K_s f_{sw} \bar{V}_w} Q_w \int_0^d -\frac{d\mu_w}{dx} dx,$$

$$\Delta\Pi = -Q_w \frac{f_{sw} + f_{sm}}{K_s f_{sw} \bar{V}_w} \Delta\mu_w. \qquad (153)$$

Substituting Eq. (97) into Eq. (153) gives

$$\Delta\Pi = -Q_w \frac{f_{sw} + f_{sm}}{K_s f_{sw}} (\Delta P - \Delta\Pi). \tag{154}$$

Rearrangement of Eq. (154) yields

$$\left(\frac{\Delta P}{\Delta\Pi}\right)_{J_w = 0} = \sigma' = 1 - \frac{K_s f_{sw}}{(f_{sw} + f_{sm})Q_w}. \tag{155}$$

Although the difference between $J_w = 0$ and $J_v = 0$ is negligible and $(\Delta P/\Delta\Pi)_{J_w=0}$ may be approximated to $(\Delta P/\Delta\Pi)_{J_v=0}$, Kedem and Katchalsky (1961) evaluated the small difference between the two. Series expansion of ΔP in the vicinity of $J_v = 0$ as a function of J_v at constant $\Delta\Pi$ can be written

$$(\Delta P)_{J_w=0} = (\Delta P)_{J_v=0} + J_s \bar{V}_s \left(\frac{\partial \Delta P}{\partial J_v}\right)_{J_v=0} + \tfrac{1}{2}(J_s \bar{V}_s)^2 \left(\frac{\partial^2 \Delta P}{\partial J_v^2}\right)_{J_v=0} + \cdots.$$

Ignoring the higher-order terms gives

$$(\Delta P)_{J_w=0} = (\Delta P)_{J_v=0} + J_s \bar{V}_s \left(\frac{\partial \Delta P}{\partial J_v}\right)_{J_v=0}. \tag{156}$$

But at constant $\Delta\Pi$

$$\frac{\partial \Delta P}{\partial J_v} = \frac{1}{L_p}. \tag{157}$$

Substituting Eq. (157) into Eq. (156) and dividing by $\Delta\Pi$ gives

$$\left(\frac{\Delta P}{\Delta\Pi}\right)_{J_w=0} = \left(\frac{\Delta P}{\Delta\Pi}\right)_{J_v=0} + \frac{J_s \bar{V}_s}{L_p \Delta\Pi}, \tag{158}$$

$$\sigma' = \sigma + \omega \bar{V}_s / L_p.$$

Thus combination of Eqs. (155) and (158) gives

$$\sigma = 1 - \frac{\omega \bar{V}_s}{L_p} - \frac{K_s f_{sw}}{(f_{sw} + f_{sm})Q_w}. \tag{159}$$

Substituting Eq. (141) in the last term of Eq. (159) yields

$$\sigma = 1 - \frac{\omega \bar{V}_s}{L_p} - \frac{\omega f_{sw} d}{Q_w}. \tag{160}$$

On the basis of these equations the following conclusions can be reached. (1) When $\omega = 0$, $\sigma = 1$ and the membrane is semipermeable. (2) When ω decreases, σ decreases, and when ω becomes larger, σ can attain negative values (negative anomalous osmosis). (3) When the solute and the solvent take different pathways in the membrane, i.e., water entering the pores in the membrane and solute dissolving in the lipids of the biological membrane so that $f_{sw} = 0$, Eq. (160) becomes

$$\sigma = 1 - \omega \bar{V}_s/L_p. \tag{161}$$

Equation (161) therefore serves to identify the existence of independent pathways for the solute and the solvent.

From the membrane permeation parameters ω, L_p, and σ, which can be measured, the friction coefficients f_{sw}, f_{sm}, and f_{wm} may be calculated from the following equations:

$$f_{sw} = \frac{(1 - \sigma - \omega \bar{V}_s/L_p)Q_w}{\omega d}, \tag{162}$$

$$f_{sm} = \left[\frac{\sigma + \omega \bar{V}_s/L_p}{1 - \sigma - \omega \bar{V}_s/L_p}\right] f_{sw}, \tag{163}$$

$$f_{wm} = \frac{Q_w \bar{V}_w}{d}\left[\frac{1}{L_p} - \frac{(1-\sigma)(\sigma + \omega \bar{V}_s/L_p)C_s}{\omega}\right]. \tag{164}$$

The measured membrane parameters and the calculated friction coefficients for a membrane system (urea and Visking dialysis tubing) are given in Table II.

It is now known that nonelectrolyte solute flux interactions occur in membranes and result in the phenomenon known as solute drag. This is seen as an increased flux of tracer solute in the direction of the diffusion gradient of a second solute as well as a decreased tracer flux in the opposite direction. These solute flux interactions may be expressed by extending Eq. (120) thus:

$$J_i = C_i^a(1 - \sigma_i)J_v + P_{ii}\Delta C_i + P_{ij}\Delta C_j, \tag{165}$$

where J_i is the flux of solute i and is written as the sum of volume flow, diffusional flow and solute interaction. $C_i^a = [C_{i(1)} + C_{i(2)}]/2$, where (1) and (2) represent the two sides, σ_i is the reflection coefficient for solute i, P_{ii} is the self-permeability coefficient ($=\omega_i RT \Delta C_i$) and ΔC_i is the concentration gradient of solute i across the membrane, P_{ij} is the cross coefficient between the solutes i and j, and ΔC_j is the concentration gradient of solute j across the membrane.

Writing Eq. (165) for two solutes, d (driver) and t (tracer, driven) for the flow in the direction $2 \to 1$ gives

$$J_d^{2 \to 1} = C_d^a(1 - \sigma_d)J_v^{2 \to 1} + \omega_d RT[C_{d(2)} - C_{d(1)}] + f_{td}C_d^a J_t^{2 \to 1} \tag{166}$$

$$J_t^{2 \to 1} = C_t^a(1 - \sigma_t)J_v^{2 \to 1} + \omega_t RT[C_{t(2)} - C_{t(1)}] + f_{dt}C_t^a J_d^{2 \to 1} \tag{167}$$

TABLE II

Measured Membrane Parameters and the Calculated Friction Coefficients for the Urea–Visking Dialysis Tubing System[a]

Parameters (measured)	Urea concentration used 0.5 M (C_s)
Partial molar volume of solute $\bar{V}_s$ (ml)	~44.9
Membrane thickness d (cm)	5.5×10^{-3}
Mole fraction of membrane water Q_w	0.68
$L_p \left[\dfrac{cm^3}{dyn\ s} \right]$	3.2×10^{-11}
$\omega \left[\dfrac{mole}{dyn\ s} \right]$	20.8×10^{-15}
σ	0.013
Friction coefficients (calculated) $\left[\dfrac{dyn\ s}{mol\ cm} \right]$ [Eqs. (162)–(164)]	
f_{sw}	0.66×10^{16}
f_{sm}	0.065×10^{16}
f_{wm}	8.3×10^{13}

[a] From Ginzburg and Katchalsky (1963).

where the third term in Eqs. (166) and (167) is modified from the general Eq. (165). f_{td} and f_{dt} are the interaction or cross coupling coefficients between d and t (driver and the driven).

Experiments may be performed so that the osmotic flow due to ΔC_d may be counterbalanced by the application of hydrostatic pressure to maintain $J_v = 0$. Consequently the first term in Eqs. (166) and (167) drops out. Because of the use of low concentration of t on side 2, the last term in Eq. (166) becomes negligible. So Eq. (166) becomes

$$J_d^{2 \to 1} = \omega_d RT[C_{d(2)} - C_{d(1)}], \tag{168}$$

and the Eq. (167) for the tracer becomes

$$J_t^{2 \to 1} = \omega_t RT[C_{t(2)} - C_{t(1)}] + f_{dt} C_t^a J_d^{2 \to 1}. \tag{169}$$

For the condition when the solutes d and t are present on side 2 only (i.e., $C_{d(1)} = 0$ and $C_{t(1)} = 0$ and dropping subscript 2), substitution of Eq. (168) into Eq. (169) gives

$$J_t^{2 \to 1} = \omega_t RT C_t + f_{dt} C_t^a \omega_d RT C_d. \tag{170}$$

IV. Transport of a Solution of Nonelectrolyte

For the flow in the opposite direction the relation is

$$J_t^{1 \to 2} = \omega_t RTC_t - f_{dt} C_t^a \omega_d RTC_d. \tag{171}$$

Evaluation of the net flow,

$$J_t = J_t^{2 \to 1} - J_t^{1 \to 2},$$

on rearrangement gives the value for the solute interaction coefficient. Thus

$$f_{dt} = \frac{J_t^{2 \to 1} - J_t^{1 \to 2}}{C_t \omega_d RTC_d} \tag{172}$$

where $(2C_t^a = C_t)$. Expressing Eq. (172) in permeability coefficients gives

$$f_{dt} = \frac{P_t^{2 \to 1} - P_t^{1 \to 2}}{P_d C_d}, \tag{173}$$

where $P_t = J_t/C_t$ and $P_d = \omega_d RT$.

Van Bruggen and colleagues (1974) studied this solute flux coupling in a homopore membrane having a pore radius of 150 Å. Some of their data related to permeability coefficients, permeability of the tracer in the two directions, and the values for the solute interaction coefficient f_{dt} are given in Tables III–V.

The data of Table III show that an increase in sucrose concentration caused a decrease in permeability of the tracer solute. The data of Table IV show that in the case of sucrose tracer there is solute–solute interaction at a sucrose concentration of 1.5 M, in the case of trisaccharide tracer interaction occurs at concentrations of 1.0 and 1.5 M and in the other two cases at concentrations of 0.5, 1, and 1.5 M. The data of Table V indicate that solute–solute interaction is determined by (1) transmembrane flow of both the driver and the driven solute, (2) concentration of the driver solute, (3) size of the driver solute, and (4) membrane pore size.

TABLE III

Effect of Sucrose Concentration with Sucrose on Both Sides of a Homopore Membrane on the Permeability of a Tracer Solute[a]

Concentration of sucrose (M)	Permeability coefficient $\times 10^5$ (cm s^{-1})			
	Sucrose	Maltotriose	Inulin	Dextran
0	3.1	2.3	0.5	0.6
0.5	2.1	1.5	0.37	0.41
1.0	1.1	0.7	0.23	0.26
1.5	0.6	0.4	0.1	0.08

[a] From van Bruggen et al. (1974).

TABLE IV

Values of Permeability of Tracer $P_t \times 10^5$ (cm s^{-1}) in the two Directions[a]

Concentration of sucrose (M)		0	0.5	1.0	1.5
Sucrose	$p^{2\to1}$	2.94	1.06	1.11	1.10
	$p^{1\to2}$	3.27	1.83	0.95	0.63
Maltotriose	$p^{2\to1}$	2.19	1.25	1.06	1.42
	$p^{1\to2}$	2.39	1.23	0.80	0.53
Inulin	$p^{2\to1}$	0.45	0.41	0.47	0.61
	$p^{1\to2}$	0.45	0.13	0.05	0.06
Dextran	$p^{2\to1}$	0.55	0.43	0.53	0.55
	$p^{1\to2}$	0.55	0.24	0.13	0.09

[a] From van Bruggen et al. (1974).

V. Permeation of Electrolyte Solution through a Membrane

A simple case of two ions, one cation represented by 1 and the other anion represented by 2 and water represented by w, is chosen. The dissipation function is

$$\phi = J_1 \Delta\bar{\mu}_1 + J_2 \Delta\bar{\mu}_2 + J_w \mu_w. \tag{174}$$

It is convenient to transform Eq. (174) into one involving the flow of salt J_s and electric current I. For the salt the electroneutrality condition is

$$v_1 z_1 + v_2 z_2 = 0, \tag{175}$$

and the relation

$$\Delta\mu_s = v_1 \Delta\bar{\mu}_1 + v_2 \Delta\bar{\mu}_2 \tag{176}$$

TABLE V

Values of Solute Interaction Coefficient f_{dt} (cm^3 mol^{-1}) for Sucrose Driver in a Homopore Membrane[a]

Concentration of sucrose (M)	Sucrose	Maltotriose	Inulin	Dextran
0.5	<25	<25	520	360
1.0	<100	235	375	360
1.5	281	533	330	275

[a] From van Bruggen et al. (1974).

also holds. As the electrodes used for measurements are generally reversible to anions, the emf E acting in the system is given by

$$\Delta\bar{\mu}_2 = -z_2 EF, \tag{177}$$

and the current is given by

$$I = F(z_1 J_1 + z_2 J_2). \tag{178}$$

Identifying the flow of ion 1 with that of the salt since ion 2 is involved with electrode reactions, one can write

$$J_1 = v_1 J_s. \tag{179}$$

Introducing Eqs. (179) and (175) into Eq. (178) gives on rearrangement

$$J_2 = I/z_2 F + v_2 J_s. \tag{180}$$

Introducing Eqs. (177), (179), (180), and (176) into Eq. (174) gives on rearrangement

$$\phi = J_w \Delta\mu_w + J_s \Delta\mu_s + IE. \tag{181}$$

When $I = 0$, Eq. (181) reduces to Eq. (90). As $\Delta\mu_w$ and $\Delta\mu_s$ are given by Eqs. (97) and (104), Eq. (181) becomes

$$\phi = J_w \bar{V}_w(\Delta P - \Delta\Pi) + J_s\left(\bar{V}_s \Delta P + \frac{\Delta\Pi_s}{C_s^a}\right) + IE. \tag{182}$$

Combining Eqs. (110) and (182) and substituting for $\bar{V}_s = Q_s/C_s^a$ gives on simplification

$$\phi = J_v(\Delta P - \Delta\Pi) + J_s \frac{\Delta\Pi_s}{C_s^a}\left(1 + Q_s \frac{\Delta\Pi}{\Delta\Pi_s}\right) + IE. \tag{183}$$

As $Q_s \Delta\Pi/\Delta\Pi_s \ll 1$, Eq. (183) simplifies to

$$\phi = J_v(\Delta P - \Delta\Pi) + J_s \Delta\Pi_s/C_s^a + IE. \tag{184}$$

The set of phenomenological equations relating the three flows and forces are

$$\begin{aligned}
J_v &= L_{11}(\Delta P - \Delta\Pi) + L_{12} \Delta\Pi_s/C_s^a + L_{13} E, \\
J_s &= L_{21}(\Delta P - \Delta\Pi) + L_{22} \Delta\Pi_s/C_s^a + L_{23} E, \\
I &= L_{31}(\Delta P - \Delta\Pi) + L_{31} \Delta\Pi_s/C_s^a + L_{33} E.
\end{aligned} \tag{185}$$

According to the Onsager law, $L_{ij} = L_{ji}$. The inverse matrix of coefficients is given by

$$\begin{aligned}
\Delta P - \Delta\Pi &= R_{11} J_v + R_{12} J_s + R_{13} I, \\
\Delta\Pi_s/C_s^a &= R_{21} J_v + R_{22} J_s + R_{23} I, \\
E &= R_{31} J_v + R_{32} J_s + R_{33} I,
\end{aligned} \tag{186}$$

where $R_{ij} = R_{ji}$.

TABLE VI

Practical Transport Coefficients

Straight coefficients	Set I	Set II
Filtration coefficient	$L_p = \left(\dfrac{J_v}{\Delta P - \Delta \Pi}\right)_{\Delta\Pi_s, I=0}$	$L_p = \left(\dfrac{J_v}{\Delta P - \Delta \Pi}\right)_{\Delta\Pi_s, I=0}$
Solute permeability	$\omega = \left(\dfrac{J_s}{\Delta\Pi_s}\right)_{J_v, I=0}$	$\omega' = \left(\dfrac{J_s}{\Delta\Pi_s}\right)_{(\Delta P - \Delta\Pi), I=0}$
Electrical conductance	$k_{sp} = \left(\dfrac{I}{E}\right)_{J_v, \Delta\Pi_s=0}$	$k'_{sp} = \left(\dfrac{I}{E}\right)_{\Delta P - \Delta\Pi, \Delta\Pi_s=0}$
Coupling coefficients		
Reflection coefficient	$C_s^a(1-\sigma) = -\left(\dfrac{\Delta P - \Delta\Pi}{\Delta\Pi_s/C_s^a}\right)_{J_v, I=0}$	$C_s^a(1-\sigma) = \left(\dfrac{J_s}{J_v}\right)_{\Delta\Pi_s, I=0}$
Electron-osmotic pressure P_E Electro-osmotic permeability β	$P_E = \left(\dfrac{\Delta P - \Delta E}{E}\right)_{J_v, \Delta\Pi_s=0}$	$\beta = \left(\dfrac{J_v}{I}\right)_{\Delta P - \Delta\Pi, \Delta\Pi_s=0}$
	$= -\left(\dfrac{I}{J_v}\right)_{\Delta\Pi_s, E=0}$	$= -\left(\dfrac{E}{\Delta P - \Delta\Pi}\right)_{\Delta\Pi_s, I=0}$
	(streaming current)	(streaming potential)
Transport number	$t_1 = v_1 z_1 F \left(\dfrac{J_s}{I}\right)_{J_v, \Delta\Pi_s=0}$	$t'_1 = v_1 z_1 F \left(\dfrac{J_s}{I}\right)_{\Delta P - \Delta\Pi, \Delta\Pi_s=0}$
	$= -v_1 z_1 F \left(\dfrac{E}{\Delta\Pi_s/C_s^a}\right)_{J_v, I=0}$	$= -v_1 z_1 F \left(\dfrac{E}{\Delta\Pi_s/C_s^a}\right)_{\Delta P - \Delta\Pi, I=}$
	(membrane potential)	(membrane potential)

Equation (185) or (186) contains nine phenomenological coefficients. Because of the Onsager law, only six are independent. Therefore complete characterization of the system requires six independent experimental measurements. If the system contains more components, the experimental problem becomes formidable even though in theory the system can be described. Only the relevant equations of the three flow system as described by Kedem and Katchalsy (1961, 1963) are outlined. For details, their papers may be consulted. They chose two sets of practical transport coefficients, and these are given in Table VI. These practical coefficients follow from the following equations, which may be derived from the original phenomenological equations.

V. Permeation of Electrolyte Solution through a Membrane

Coefficients of set I:

$$J_v = L_p(\Delta P - \Delta \Pi) + C_s^a L_p(1 - \sigma)\Delta \Pi_s/C_s^a - (P_E L_p/k_{sp})I,$$
$$J_s = C_s^a(1 - \sigma)J_v + C_s^a \omega \, \Delta \Pi_s/C_s^a + (t_1/v_1 z_1 F)I, \quad (187)$$
$$E = (P_E/k_{sp})J_v - (t_1/v_1 z_1 F)(\Delta \Pi_s/C_s^a) + (1/k_{sp})I.$$

Coefficients of set II:

$$J_v = L_p(\Delta P - \Delta \Pi) + C_s^a L_p(1 - \sigma)\Delta \Pi_s/C_s^a + \beta I,$$
$$J_s = C_s^a(1 - \sigma)L_p(\Delta P - \Delta \Pi) + C_s^a \omega'(\Delta \Pi_s/C_s^a) + (t_1'/v_1 z_1 F)I, \quad (188)$$
$$E = -\beta(\Delta P - \Delta \Pi) - (t_1'/v_1 z_1 F)\Delta \Pi_s/C_s^a + (1/k_{sp}')I.$$

Comparison of Eqs. (187) and (188) gives the relations between the several transport coefficients of the two sets. Thus

$$\beta = -P_E L_p/k_{sp}, \quad (189)$$

$$k_{sp}' = \frac{k_{sp}}{1 - P_E^2 L_p/k_{sp}} = \frac{k_{sp}}{1 + P_E \beta}, \quad (190)$$

$$t_1'/v_1 z_1 F = t_1/v_1 z_1 F + C_s^a(1 - \sigma)\beta, \quad (191)$$

$$\omega' = \omega + C_s^a(1 - \sigma)L_p. \quad (192)$$

In the case of nonelectrolytes, three transport coefficients L_p, ω, and σ were considered in terms of three friction coefficients f_{sw}, f_{sm}, and f_{wm}. In the case of electrolytes also, similar analysis can be carried out. But a large number of friction parameters are required. In the case of simple (1:1) electrolyte and water, six parameters, f_{1w}, f_{1m}, f_{12}, f_{2w}, f_{2m}, and f_{wm} are involved, and this is too large a number. However, this can be simplified considerably if a charged membrane whose concentration of fixed charge ($\bar{X}$) is larger than the concentration of the simple electrolyte is considered. Let the membrane be negatively charged. Then one can write

$$\bar{C}_1 = \bar{C}_2 + \bar{X} = \bar{C}_s + \bar{X}, \quad (193)$$

where $\bar{C}_1$, $\bar{C}_2$, and $\bar{C}_s$ are the counterion, coion, and salt concentrations in the membrane. For a highly charged membrane, $\bar{X} \gg \bar{C}_s$ and $\bar{C}_1 = \bar{X}$. Using the general expression

$$X_i = f_{iw}(v_i - v_w) + f_{im}(v_i - v_m),$$

one can write

$$X_1 = -\frac{d\bar{\mu}_1}{dx} = f_{1w}(v_1 - v_w) + f_{1m}v_1, \quad (v_m = 0),$$

$$X_2 = -\frac{d\bar{\mu}_2}{dx} = f_{2w}(v_2 - v_w) + f_{2m}v_2,$$

since $\bar{C}_2 \approx 0$ and f_{12} is negligible.

The flows are given by $J_1 = \bar{C}_1 v_1$ and $J_2 = \bar{C}_2 v_2$. Thus

$$-\frac{d\mu_1}{dx} = \frac{J_1(f_{1w} + f_{1m})}{\bar{C}_1} - \frac{f_{1w} J_w}{\bar{C}_w}, \tag{194}$$

$$-\frac{d\mu_2}{dx} = \frac{J_2(f_{2w} + f_{2m})}{\bar{C}_2} - \frac{f_{2w} J_w}{\bar{C}_w}. \tag{195}$$

The force acting on a salt is given by

$$X_s = -\frac{d\mu_s}{dx} = -\frac{d\bar{\mu}_1}{dx} - \frac{d\bar{\mu}_2}{dx}. \tag{196}$$

Substituting Eqs. (194) and (195) into Eq. (196) gives for the condition when $I = 0$ (i.e., $J_1 = J_2 = J_s$)

$$-\frac{d\mu_s}{dx} = J_s\left[\frac{f_{1w} + f_{1m}}{\bar{C}_1} + \frac{f_{2w} + f_{2m}}{\bar{C}_2}\right] - \frac{J_w}{\bar{C}_w}(f_{1w} + f_{2w}). \tag{197}$$

As $\bar{C}_1 \gg \bar{C}_2$, the first term on the right-hand side of Eq. (197) becomes negligible in comparison with the second and $\bar{C}_2 = \bar{C}_s$. So Eq. (197) simplifies to

$$-\frac{d\mu_s}{dx} = \frac{J_s(f_{2w} + f_{2m})}{\bar{C}_s} - \frac{J_w(f_{1w} + f_{2w})}{\bar{C}_w}. \tag{198}$$

When $J_w = 0$, Eq. (198) reduces to

$$-\frac{d\mu_s}{dx} = \frac{J_s}{\bar{C}_s}(f_{2w} + f_{2m}). \tag{199}$$

But $d\mu_s = d\Pi_s/C_s$, and so

$$-\frac{d\Pi_s}{dx} = \frac{J_s(f_{2w} + f_{2m})}{K}, \tag{200}$$

where K is the distribution coefficient of the salt between the membrane and the aqueous phase. Integration of Eq. (200) gives

$$\Delta\Pi_s K = J_s(f_{2w} + f_{2m})d,$$

and so

$$\omega = \left(\frac{J_s}{\Delta\Pi_s}\right)_{J_w = 0} = \frac{K}{d(f_{2w} + f_{2m})}. \tag{201}$$

The reflection coefficient of the salt $\sigma[=(\Delta P/\Delta\Pi)_{J_v = 0}]$ may be obtained by following the steps similar to those used in Eqs. (151)–(160). The resulting relation is

$$\sigma = 1 - \frac{\omega \bar{V}_s}{L_p} - \frac{K}{Q_w}\frac{f_{1w} + f_{2w}}{f_{2w} + f_{2m}} = 1 - \frac{\omega \bar{V}_s}{L_p} - \frac{\omega(f_{1w} + f_{2w})d}{Q_w}. \tag{202}$$

Equations (201) and (202) may be applied to a charged membrane that contains a large quantity of water (loose membrane). The Donnan distribution of electrolyte in the membrane gives $-\bar{C}_1\bar{C}_2 = C_s^2$, or

$$\bar{C}_s(\bar{C}_s + \bar{X})/Q_w^2 = C_s^2, \tag{203}$$

where $\bar{C}_1/Q_w$, $\bar{C}_2/Q_w$, $\bar{C}_s/Q_w$, and $\bar{X}/Q_w = \bar{X}'$ are concentrations in moles per milliliter of pore water and the distribution coefficient is given by

$$K = \frac{\bar{C}_s}{C_s} = \frac{C_s Q_w^2}{\bar{C}_s + \bar{X}} \approx \frac{C_s Q_w^2}{\bar{X}} = \frac{C_s Q_w}{\bar{X}'}. \tag{204}$$

In ion-exchange membranes of high charge density, Q_w is high and $\bar{C}_2 \approx 0$. For such membranes, $f_{2m} \ll f_{2w}$, and further friction coefficients between ions and water approximate their values in free solution divided by a tortuosity factor θ. That is,

$$f_{1w} = f_{1w}^0/\theta \quad \text{and} \quad f_{2w} = f_{2w}^0/\theta. \tag{205}$$

Tortuosity θ shows that the water-filled ion pathway in the membrane is longer than the thickness d of the membrane. In this case ω and σ become

$$\omega = \frac{K\theta}{df_{2w}^0} = \frac{C_s Q_w^2 \theta}{d\bar{X} f_{2w}^0}, \tag{206}$$

$$\sigma = 1 - \frac{\omega \bar{V}_s}{L_p} - \frac{C_s Q_w}{\bar{X}}\left(\frac{f_{1w}^0 + f_{2w}^0}{f_{2w}^0}\right). \tag{207}$$

The transport number of counterion in water is given by

$$t_1^0 = \frac{u_1^0}{u_1^0 + u_2^0} = \frac{1/f_{1w}^0}{1/f_{1w}^0 + 1/f_{2w}^0} = \frac{f_{2w}^0}{f_{1w}^0 + f_{2w}^0}.$$

Introducing this into Eq. (207) gives

$$\sigma = 1 - \frac{\omega \bar{V}_s}{L_p} - \frac{C_s Q_w}{\bar{X} t_1^0} \tag{208}$$

or

$$\sigma' = 1 - \frac{C_s Q_w}{\bar{X} t_1^0} = 1 - \frac{C_s}{\bar{X}' t_1^0}.$$

According to Eq. (208), when t_1^0 is low, σ decreases rapidly with increasing concentration. If the coion is more mobile than the counterion, σ may become negative, generating the phenomenon of anomalous osmosis.

Evaluation of some of the permeability coefficients (see Table VI) for a charged membrane in equilibrium with an electrolyte solution and subject to an electric field will provide an insight into some of the membrane phenomena.

For the condition $J_w = 0$, Eqs. (194) and (195) become

$$-\frac{d\bar{\mu}_1}{dx} = \frac{J_1(f_{1w} + f_{1m})}{\bar{C}_1}, \tag{209}$$

$$-\frac{d\bar{\mu}_2}{dx} = \frac{J_2(f_{2w} + f_{2m})}{\bar{C}_2}. \tag{210}$$

The Donnan equilibrium [see Eq. (203)] when $\bar{C}_1 \approx \bar{X}$ gives

$$\bar{C}_s = \bar{C}_2 = C_s^2 Q_w^2 / \bar{X},$$

and furthermore, $f_{1m} \ll f_{1w}$ and $f_{2m} \ll f_{2w}$. Therefore Eqs. (209) and (210) can be written

$$-\frac{d\bar{\mu}_1}{dx} = \frac{f_{1w} J_1}{\bar{X}}, \tag{211}$$

$$-\frac{d\bar{\mu}_2}{dx} = \frac{f_{2w} J_2 \bar{X}}{C_s^2 Q_w^2}. \tag{212}$$

These on integration give

$$\Delta\bar{\mu}_1 = \frac{f_{1w} J_1}{\bar{X}} d, \tag{213}$$

$$\Delta\bar{\mu}_2 = \frac{f_{2w} J_2 \bar{X}}{C_s^2 Q_w^2} d. \tag{214}$$

Furthermore, the relations

$$\Delta\mu_s = \Delta\bar{\mu}_1 + \Delta\bar{\mu}_2 = 0 \tag{215}$$

(same 1:1 electrolyte solution on either side) and

$$\Delta\bar{\mu}_2 = -EF \tag{177'}$$

hold, and so Eq. (214) may be written

$$J_2 = -C_s^2 Q_w^2 EF / f_{2w} \bar{X} d. \tag{216}$$

Substituting Eqs. (213) and (214) into Eq. (215) gives

$$\frac{J_1 f_{1w}}{\bar{X}} = -\frac{f_{2w} J_2 \bar{X}}{C_s^2 Q_w^2},$$

and further use of Eq. (216) gives

$$J_1 = EF\bar{X} / f_{1w} d. \tag{217}$$

V. Permeation of Electrolyte Solution through a Membrane

The electric current in the membrane is given by the use of Eqs. (216) and (217) as

$$I = F(J_1 - J_2) = \frac{EF^2}{d}\left[\frac{\bar{X}}{f_{1w}} + \frac{C_s^2 Q_w^2}{f_{2w}\bar{X}}\right].$$

At low salt concentration, $\bar{X} \gg (C_s^2 Q_w^2/\bar{X})$, and so

$$I = EF^2\bar{X}/f_{1w}d. \tag{218}$$

Thus conductance

$$\bar{k}_{sp} = \left(\frac{I}{E}\right)_{J_v,\Delta\Pi_s = 0} = \frac{\bar{X}F^2}{f_{1w}d}. \tag{219}$$

Substitution of Eq. (205) into Eq. (219) gives

$$\bar{k}_{sp} = \bar{X}F^2\theta/f_{1w}^0 d. \tag{220}$$

The transference numbers of counterions and coions in the membrane may be evaluated in the same manner. The coion transference number is given by

$$\bar{t}_2 = -(FJ_2/I)_{J_v = 0,\Delta\Pi_s = 0}.$$

Substituting for J_2 and I from Eqs. (216) and (218) gives

$$\bar{t}_2 = \left(\frac{C_s Q_w}{\bar{X}}\right)^2 \frac{f_{1w}}{f_{2w}} \approx \left(\frac{C_s}{\bar{X}'}\right)^2. \tag{221}$$

As $\bar{t}_1 + \bar{t}_2 = 1$,

$$\bar{t}_1 = 1 - \left(\frac{C_s}{\bar{X}'}\right)^2 \frac{f_{1w}}{f_{2w}}. \tag{222}$$

β also can be expressed in terms of friction coefficients. Applying the Gibbs–Duhem equation to the salt solution in the membrane gives

$$-\frac{d\bar{\mu}_w}{dx}\bar{C}_w = \bar{C}_1\frac{d\bar{\mu}_1}{dx} + \bar{C}_2\frac{d\bar{\mu}_2}{dx}.$$

But $d\bar{\mu}_1/dx = -d\bar{\mu}_2/dx$ since there is no salt gradient in the membrane. So it follows that

$$\frac{d\bar{\mu}_w}{dx} = \left(\frac{\bar{C}_1 - \bar{C}_2}{\bar{C}_w}\right)\frac{d\bar{\mu}_2}{dx}.$$

Integrating across the membrane and substituting $\bar{C}_1 - \bar{C}_2 = \bar{X}$ gives

$$\Delta\bar{\mu}_w = (\bar{X}/\bar{C}_w)\Delta\bar{\mu}_2.$$

6. Steady-State Thermodynamic Approach to Membrane Transport

But $\Delta\mu_w = \bar{V}_w \Delta P$ when $\Delta\Pi = 0$ [see Eq. (97)], and $\Delta\bar{\mu}_2$ is given by Eq. (177). Thus

$$\bar{V}_w \Delta P = -(\bar{X}/\bar{C}_w)EF$$

or

$$\left(\frac{\Delta P}{E}\right)_{J_v=0,\Delta\Pi=0} = -\frac{\bar{X}F}{Q_w}. \tag{223}$$

Equations (87) and (89), when $J_v = 0$ and $\Delta\Pi = 0$ (substituting $\Delta\psi = E$), give

$$\beta = -L_p \Delta P / \bar{k}_{sp} E$$

or

$$\left(\frac{\Delta P}{E}\right)_{J_v=0,\Delta\Pi=0} = -\frac{\beta \bar{k}_{sp}}{L_p} \tag{224}$$

Equating Eqs. (223) and (224) gives

$$\beta = (\bar{X}F/\bar{k}_{sp}Q_w)L_p. \tag{225}$$

Substituting the values for L_p [Eq. (149)] and $\bar{k}_{sp}$ [Eq. (220)] gives

$$\beta = \frac{\bar{V}_w}{F} \frac{f_{1w}}{f_{wm}}. \tag{226}$$

Choosing the capillary model for a porous membrane, the volume flow of water and the pressure difference across it are related by the Poiseuille law [see Eq. (156) of Chapter 3], which may be written

$$Q/\Pi r^2 = \theta r^2 \Delta P/8\eta d,$$

where Q is the volume of water per unit capillary area flowing through a capillary of radius r and length d. η is the viscosity and θ is the tortuosity increasing the effective length d to d/θ.

Volume flow through unit membrane area is therefore given by

$$J_v = \frac{Q_w r^2}{8\eta(d/\theta)} \Delta P = \frac{\theta Q_w r^2}{8\eta d} \Delta P,$$

where Q_w is the fraction of membrane area available for water permeation. d/θ is the effective length of capillaries of equivalent radius r. Thus it follows that

$$L_p = \theta Q_w r^2 / 8\eta d. \tag{227}$$

Comparing this with Eq. (149) gives the relation

$$f_{wm} = 8\eta \bar{V}_w / \theta r^2. \tag{228}$$

Substituting Eq. (228) in Eq. (226) gives

$$\beta = f^0_{1w} r^2 / 8\eta F. \qquad (229)$$

Thus one can calculate a value for β assuming the following values: $r = 10$ Å $= 10^{-7}$ cm; f^0_{1w} (for potassium ion) $= 1.3 \times 10^{15}$ dyn s cm^{-1} mol^{-1}; $\eta = 10^{-2}$ P. Thus

$$\beta = \frac{1.3 \times 10^{15} \times 10^{-14}}{8 \times 10^{-2} \times 96{,}500}$$

$$= 1.7 \times 10^{-3} \quad \text{cm}^3 \, \text{C}^{-1}$$

$$= 9.1 \quad \text{mol/faraday}.$$

Simple homogeneous membranes for which the transport coefficients are described may be combined to form a composite membrane either as a parallel array or as a series array of simple membranes. Kedem and Katchalsky (1963) have derived the transport equations for a composite membrane, and these are given in Tables VII (parallel elements) and VIII (series elements). They have discussed several transport phenomena that arise in the composite membranes in the light of the equations given in Tables VII and VIII, and their papers should be consulted for details.

Several studies of transport phenomena in ion-exchange membranes have been described by using the pore model and the frictional interpretation. The dilute-solution approach is probably simpler because the influence of coion may be ignored. Meares and his colleagues (1972) have taken this

TABLE VII

Transport Coefficients for a Composite Membrane Formed of Parallel Combination of Two Membranes a and b[a,b]

$k'_{sp} = \gamma_a k'_{sp(a)} + \gamma_b k'_{sp(b)}; \; L_p = \gamma_a L_p^a + \gamma_b L_p^b + \alpha_a \alpha_b k'_{sp}(\beta_a - \beta_b)^2$

$\beta = \alpha_a \beta_a + \alpha_b \beta_b; \; \omega' = \gamma_a \omega'_a + \gamma_b \omega'_b + \alpha_a \alpha_b k'_{sp} \dfrac{(t'_{1a} - t'_{1b})^2}{C_s(v_1 z_1 F)^2}$

$1 - \sigma = \gamma_a(1 - \sigma_a)(L_p^a/L_p) + \gamma_b(1 - \sigma_b)(L_p^b/L_p)$

$\qquad + \dfrac{\alpha_a \alpha_b k'_{sp}(\beta_a - \beta_b)(t'_{1a} - t'_{1b})}{v_1 z_1 F C_s L_p}$

$t'_1 = \alpha_a t'_{1a} + \alpha_b t'_{1b}$

---

[a] From Roy Caplan and Mikulecky (1966).
[b] γ_i is the fraction of area occupied by membrane i, $\alpha_j = \gamma_j k'_{sp(j)}/k'_{sp}$.

306 6. Steady-State Thermodynamic Approach to Membrane Transport

TABLE VIII

Transport Coefficients for a Composite Membrane
Formed of a Series Combination of Two Membranes
a and $b^{a,b}$

$$\frac{1}{L_p} = \frac{1}{L_p^a} + \frac{1}{L_p^b} + C_s^* \frac{(\sigma_a - \sigma_b)^2}{\omega_a + \omega_b}, \quad \frac{1}{\omega} = \frac{1}{\omega_a} + \frac{1}{\omega_b}$$

$$\frac{1}{k_{sp}} = \frac{1}{k_{sp(a)}} + \frac{1}{k_{sp(b)}} + \frac{(t_{1a} - t_{1b})^2}{C_s^*(v_1 z_1 F)^2(\omega_a + \omega_b)}, \quad \sigma = \sigma_a \frac{\omega}{\omega_a} + \sigma_b \frac{\omega}{\omega_b}$$

$$\beta = \beta_a \frac{L_p}{L_p^a} + \beta_b \frac{L_p}{L_p^b} + \frac{(\sigma_a - \sigma_b)(t_{1a} - t_{1b})L_p}{v_1 z_1 F(\omega_a + \omega_b)}$$

$$t_1 = t_{1a} \frac{\omega}{\omega_a} + t_{1b} \frac{\omega}{\omega_b}$$

$$P_E = P_{Ea} \frac{k_{sp}}{k_{sp(a)}} + P_{Eb} \frac{k'_{sp}}{k_{sp(b)}} - \frac{(\sigma_a - \sigma_b)(t_{1a} - t_{1b})k_{sp}}{v_1 z_1 F(\omega_a + \omega_b)}$$

---

[a] From Roy Caplan and Mikulecky (1966).
[b] C^* is the corresponding average concentration in the membrane, $C^* = (C_s - C_s^{in})/\ln(C_s/C_s^{in})$, where C_s^{in} is the inner concentration in the layer between membranes a and b.

coion effect into consideration, and their review must be consulted for details regarding the several strategies that may be employed to determine and to interpret the several frictional parameters. A brief account of the dilute solution approach due to Spiegler (1958) and to Scattergood and Lightfoot (1968) is outlined in the following.

From the definition of β (see Table VI), it follows that $\beta = V/I$ (ml s^{-1} A^{-1}), where V is the volume ($J_w \approx J_v$) of water flowing through the membrane per second at one ampere of current I ($I \times t$ = coulomb). In terms of transport number of water $\bar{t}_w$ (mol/faraday), the relation becomes $\bar{t}_w = \beta F/18$. Another definition of electro-osmotic permeability, which may be denoted by β_E, is that it is the volume of water flowing per second through 1 cm^2 membrane area subject to an electric force of 1 V cm^{-1} of membrane thickness (ml cm^{-1} s^{-1} V^{-1}). The two are related by

$$\beta_E = \beta \bar{k}_{sp}. \tag{230}$$

Spiegler (1958), in his treatment of membrane transport phenomena based on the capillary model, defined electro-osmotic water transport in terms of moles of transport, i.e., $\Omega = \beta_E/18$, and derived the relation

$$\Omega = \frac{\bar{C}_3 F}{g}[\bar{C}_1 \bar{X}_{13}(\bar{X}_{23} + \bar{X}_{24}) - \bar{C}_2 \bar{X}_{23}(\bar{X}_{13} + \bar{X}_{14})], \tag{231}$$

where $\bar{C}$s are the concentrations in the membrane phase, $\bar{X}_{ij}$ are friction coefficients ($J\,s\,cm^{-2}\,mol^{-1}$). 1, 2, 3, and 4 refer to counterion, coion, water, and membrane matrix, respectively. F (faraday) is a generalized force acting across the membrane, and g is given by

$$g = \bar{C}_1\bar{X}_{13}\bar{X}_{14}(\bar{X}_{23} + \bar{X}_{24}) + \bar{C}_2\bar{X}_{23}\bar{X}_{24}(\bar{X}_{13} + \bar{X}_{14})$$
$$+ \bar{C}_3\bar{X}_{34}(\bar{X}_{13} + \bar{X}_{14})(\bar{X}_{23} + \bar{X}_{24}). \qquad (232)$$

The flux of the counterion J_1 under the influence of the generalized force $\mathscr{F}$ is given by

$$J_1 = \frac{\bar{C}_1\mathscr{F}}{g}[(\bar{C}_1\bar{X}_{13} + \bar{C}_3\bar{X}_{34})(\bar{X}_{23} + \bar{X}_{24}) + \bar{C}_2\bar{X}_{23}\bar{X}_{24}]. \qquad (233)$$

Equations (232) and (233) simplify for an ion-exchange membrane in dynamic equilibrium with dilute solution (i.e., $\bar{C}_2 \approx 0$), and the terms $\bar{X}_{23}$ and $\bar{X}_{24}$ become negligible. Then, the flux of water is given by

$$J_3 = \frac{\bar{C}_3\mathscr{F}}{g'}\bar{C}_1\bar{X}_{13}, \qquad (234)$$

$$J_1 = \frac{\bar{C}_1\mathscr{F}}{g'}(\bar{C}_1\bar{X}_{13} + \bar{C}_3\bar{X}_{34}), \qquad (235)$$

and

$$g' = \bar{C}_1\bar{X}_{13}\bar{X}_{14} + \bar{C}_3\bar{X}_{34}(\bar{X}_{13} + \bar{X}_{14}). \qquad (236)$$

Hence water flux to counterion flux is given by

$$\frac{J_3}{J_1} = \bar{C}_3 \bigg/ \left[\bar{C}_1 + \bar{C}_3\frac{\bar{X}_{34}}{\bar{X}_{13}}\right]. \qquad (237)$$

For the ideal case when $\bar{X}_{13}$ and $\bar{X}_{34}$ are zero, Eq. (237) becomes

$$J_3/J_1 = \bar{C}_3/\bar{C}_1. \qquad (238)$$

For the passage of a faraday of current $J_3 = \Omega F$ and $J_1 = 1$ mol for univalent counterion. Thus Eq. (238) may be written as $\Omega F = 55.56/\bar{m}_1$, where $\bar{m}_1$ is the moles of counterion associated with $\frac{1000}{18}$ mol of water in the membrane. This equation in terms of the other parameters defined, i.e., β, β_E, and $\bar{t}_w$, can be written

$$\Omega F = \beta_E F/18 = \beta k_{sp}F/18 = \bar{t}_w k_{sp} = 55.56/\bar{m}_1. \qquad (239)$$

Experimental data deviate considerably from these ideal relations (see Lakshminarayanaiah and Subrahmanyan, 1968).

In addition to Eqs. (231)–(237) Spiegler (1958) derived the following relations:

$$a = RT/\bar{D}_1 = \bar{X}_{13} + \bar{X}_{14}, \qquad (240)$$

$$H = \Omega/\bar{C}_3 F = \bar{k}_{sp}\bar{t}_w/\bar{C}_3 F^2$$
$$= \bar{C}_1 \bar{X}_{13}/[\bar{C}_1 \bar{X}_{13}\bar{X}_{14} + \bar{C}_3 \bar{X}_{34}(\bar{X}_{13} + \bar{X}_{14})], \qquad (241)$$

$$L = \bar{t}_1 \bar{k}_{sp}/\bar{C}_1 F^2, \qquad (242)$$

$$\bar{X}_{13} = (aL - 1)/H. \qquad (243)$$

Furthermore, Mackay and Meares (1959) included the coion interactions thus:

$$b = RT/\bar{D}_2 = \bar{X}_{23} + \bar{X}_{24}, \qquad (244)$$

$$\bar{X}_{23} = (\bar{C}_1 b/\bar{C}_2 H)(L - A/\bar{C}_1 + \bar{C}_2/\bar{C}_1 b), \qquad (245)$$

where $A = \bar{k}_{sp}/F^2$. The experimental data for the different membrane parameters determined by Mackay and Meares (1959) for a cation exchange membrane (Zeokarb 315) in 0.05-M NaCl solution and the values derived by the use of the experimental data in the above equations for the several friction coefficients are given in Table IX. In the treatment given above, $\bar{X}_{12}$ was neglected. It could be significant in strong solutions. Furthermore, the iso-

TABLE IX

Experimental Values for the Different Parameters of Zeokarb 315 Membranes in 0.05-M NaCl solution and the Calculated Friction Coefficients (J s cm^{-2} mol^{-1})[a]

Membrane parameters	Experimental values
Counterion diffusion coefficient $\bar{D}_1$ (cm^2 s^{-1})	3.5×10^{-6}
Coion diffusion coefficient $\bar{D}_2$ (cm^2 s^{-1})	7.42×10^{-6}
Counterion concentration $\bar{C}_1$ (mol cm^{-3})	0.468×10^{-3}
Coion concentration $\bar{C}_2$ (mol cm^{-3})	0.124×10^{-4}
Water concentration $\bar{C}_3$ (mol cm^{-3})	0.407×10^{-1}
Specific conductance $\bar{k}_{sp}$ (1/Ω cm)	0.01007
Counterion transport number $\bar{t}_1$	0.9795
Electro-osmotic permeability β_E (mol cm^{-1}s^{-1}V^{-1})	4.77×10^{-6}
Friction coefficients	Calculated [Eqs. (240)–(245)]
$\bar{X}_{13}$	4.93×10^8
$\bar{X}_{14}$	2.13×10^8
$\bar{X}_{34}$	4.65×10^6
$\bar{X}_{23}$	3.32×10^8
$\bar{X}_{24}$	0.02×10^8

[a] From Mackay and Meares (1959).

tope interaction term $\overline{X}_{i*i}$ (i.e., interaction of the tracer quantities of the isotope i^* with its abundant species i) has been omitted. This interaction term could be considerable and important. The work of Essig and colleagues (see Kedem and Essig, 1965; Li, et al., 1974, 1977) has shown that the isotope interaction influences the unidirectional fluxes measured and leads to abnormal values for the flux ratio. They demonstrated this by using an anion-exchange membrane across which fluxes were measured by using an isotope in the absence and in the presence of a driving force whose directions were reversed. When the membrane has the same solution (e.g., NaCl) on either side, the flux of ^{36}Cl (J^*) across the membrane is given by $J^*/\Delta\rho$, where $\Delta\rho$ is the difference in specific activity across the membrane. The fluxes in the two directions were equal. The exchange resistance is given by

$$R^* = \frac{\text{driving force}}{\text{flux}} = \frac{\overline{\Delta\mu^*}}{J^*} = \frac{RT\Delta\rho}{J^*}. \tag{246}$$

When a gradient was imposed, the tracer fluxes in the two directions were again measured. With ΔE imposed (hot side positive), $J^*/\Delta\rho$ was measured and gave the unidirectional flux $\vec{J}$. The cold side was drained, rinsed, and refilled, and the field was reversed making the hot side negative. This measurement gave $\overleftarrow{J}$ and enabled calculation of the flux ratio f.

In addition to exchange resistance, two other related parameters are the electrical resistance R_{el} and the phenomenological resistance to flow R_{ph}, which are given by

$$R_{el} = \Delta E/I, \tag{247}$$

$$R_{ph} = \frac{\text{driving force}}{\text{flux}} = \frac{\overline{\Delta\mu}}{J} = \frac{z\Delta EF}{J}. \tag{248}$$

Since $I = zFJ$, Eqs. (247) and (248) give

$$R_{ph} = (zF)^2 R_{el}. \tag{249}$$

When there is no interaction of the isotope (*) with its abundant species $R^* = R_{ph}$ and the flux ratio [see Eq. (107) of Chapter 4] is given by

$$f = \exp(\Delta\bar{\mu}/RT). \tag{250}$$

When there is isotope interaction, $R^* \neq R_{ph}$ and the flux ratio is given by

$$f = \exp(\Delta\bar{\mu}^*/RT).$$

Assuming that J^* represented the flux of the abundant species J, substituting from Eqs. (246) and (248) gives

$$f = \exp\left(\frac{R^*J}{RT}\right) = \exp\left(\frac{R^*}{R_{ph}}\frac{\Delta\bar{\mu}}{RT}\right). \tag{251}$$

TABLE X

Observed and Calculated Flux Ratio for an Anion-Exchange
Membrane in 0.03-M KCl ($\Delta C = 0$ and $\Delta \bar{\mu} = zF\,\Delta E$)[a]

Membrane	R^*/R_{ph}	ΔE (mV)	$\exp\left(\dfrac{\Delta\bar{\mu}}{RT}\right)$ [Eq. (250)]	$\exp\left[\dfrac{R^*}{R_{\text{ph}}}\dfrac{\Delta\bar{\mu}}{RT}\right]$ [Eq. (251)]	f Expt.
1	0.71	41	4.94	3.11	3.53
2	0.49	32	3.48	1.84	1.92
3	0.51	41.5	5.04	2.29	2.17
4	0.49	41.8	5.10	2.21	2.24

[a] From Li et al. (1974).

The data of Li et al. (1974) given in Table X show that Eq. (251) reproduced the experimental flux ratio values better than Eq. (250). Even when the driving force was ΔC ($\Delta E = 0$), Eq. (251) predicted the flux ratio better than Eq. (250). Here an average value of concentration of the electrolyte $C^a = (C_1 + C_2)/2$ was used to evaluate R^* and R_{ph}, and the results are shown in Table XI.

Scattergood and Lightfoot (1968) took this isotope interaction into account in their studies of ion transport across membranes and evaluated the diffusivity coefficient $\bar{D}^*_{1*1}$. This involved a fourth additional measurement of hydrodynamic permeability (L_p) of the membrane under the condition of zero current, the other three transport measurements being $\bar{D}_1$, $\bar{k}_{\text{sp}}$, and $\bar{t}_3$. They used the following relations:

$$\bar{t}_3 = \left(\frac{\bar{x}_3}{\bar{D}_{13}}\right) \bigg/ \left[\frac{\bar{x}_1}{\bar{D}_{13}} + \frac{\bar{x}_4}{\bar{D}_{34}}\right], \qquad (252)$$

TABLE XI

Observed and Calculated Flux Ratio for an Anion-Exchange
Membrane with $\Delta E = 0$ and $\Delta\bar{\mu} = RT \ln(C_1/C_2)$[a]

Membrane	R^*/R_{ph}	C_1, C_2 (M)		$\exp\left(\dfrac{\Delta\bar{\mu}}{RT}\right)$ [Eq. (250)]	$\exp\left[\dfrac{R^*}{R_{\text{ph}}}\dfrac{\Delta\bar{\mu}}{RT}\right]$ [Eq. (251)]	f Expt.
1	0.36	0.1	0.05	2.0	1.28	1.22
2	0.71	0.05	0.01	5.0	3.14	3.30
3	0.64	0.05	0.01	5.0	2.81	2.33
4	0.51	0.05	0.01	5.0	2.28	1.94

[a] From Li et al. (1974).

V. Permeation of Electrolyte Solution through a Membrane

$$L_p = (\bar{x}_3 \bar{C} \bar{V}_3^2) \bigg/ \left[\left\{ \frac{\bar{x}_1}{\bar{D}_{13}} + \frac{\bar{x}_4}{\bar{D}_{34}} \right\} RTd \right], \tag{253}$$

$$\bar{k}_{sp} = \left(\frac{\bar{C}_1 F^2}{RT} \right) \bigg/ \left[\frac{\bar{x}_3}{\bar{D}_{13}} + \frac{\bar{x}_4}{\bar{D}_{14}} - \left(\bar{t}_3 \frac{\bar{x}_1}{\bar{D}_{13}} \right) \right], \tag{254}$$

$$\frac{1}{\bar{D}_1} = \frac{\bar{x}_1 + \bar{x}_{1*}}{\bar{D}_{1*1}} + \frac{\bar{x}_3}{\bar{D}_{13}} + \frac{\bar{x}_4}{\bar{D}_{14}}, \tag{255}$$

where the $\bar{x}_i$ are the mole fractions, the $\bar{D}_{ij}$ are the Stefan–Maxwell diffusivity coefficients, $\bar{C}$ is the total concentration, and $\bar{V}_3$ is the partial molar volume of water. This treatment and the friction-coefficient treatment are equivalent in that the interaction coefficients of the two treatments are related by

$$\bar{X}_{ij} = (RT/\bar{D}_{ij})\bar{x}_j, \qquad \bar{D}_{ij} = \bar{D}_{ji}, \quad \text{and} \quad \bar{x}_i \bar{X}_{ij} = \bar{x}_j \bar{X}_{ji}.$$

The experimental data related to a cation- and an anion-exchange membrane together with the calculated values for the several parameters are given in Table XII.

There are several important points that follow from the values of the $\bar{D}_{ij}$ of Table XII. The isotope interaction is considerable, and $\bar{X}_{1*1}$ is the highest compared to the other friction coefficients. The experimental data of Table XII in comparison of those of Table IX show that the water content of AMF

TABLE XII

Transport Data for Membranes in Contact with 0.01-M NaCl Solution[a]

Parameter	AMF C-103	AMF A-100
Membrane thickness d (cm)	0.0134	0.0205
Counterion concentration $\bar{C}_1$ (mol/cm^3) × 10^3	0.92	1.223
Fixed group concentration $\bar{C}_4$ (mol/cm^3) × 10^3	0.92	1.223
Water concentration $\bar{C}_3$ (mol/cm^3) × 10^3	9.45	10.30
Total concentration $\bar{C}$ (mol/cm^3) × 10^3	11.29	12.75
$\bar{x}_1$	0.0814	0.096
$\bar{x}_4$	0.0814	0.096
$\bar{x}_3$	0.837	0.808
Counterion self-diffusion $\bar{D}_1$ (cm^2/s) × 10^8	5.74	9.51
Specific conductance $\bar{k}_{sp}$ (1/Ω cm) × 10^3	1.793	2.28
Water transport number $\bar{t}_3$ (mol/F)	7.0	3.0
L_p (cm^4/J s) × 10^6	2.1	2.8
$\bar{D}_{ij}$ (cm^2/s) × 10^8		
$\bar{D}_{1*1}$	0.5	1.14
$\bar{D}_{14}$	4.4	4.86
$\bar{D}_{13}$	270	1150
$\bar{D}_{34}$	575	634

[a] From Lakshminarayanaiah (1970, 1971).

C-103 is small compared to that of Zeokarb 315. On this basis one might expect the Spiegler model to be inapplicable to AMF C-103 membrane. That this is so is confirmed by the fact that Eqs. (240)–(243) give negative values for $\bar{X}_{14}$.

Electrokinetic effects are described by Eqs. (67)–(70). By making appropriate measurements, values for the first-order coefficients of these equations can be derived. Similarly, values for the second-order phenomenological coefficients can be derived, provided linearity was observed between the flows and forces. The phenomenological equations can be written as a power series (see Rastogi and Jha, 1966a,b) thus:

$$J_v = L_{11}\frac{\Delta P}{T} + L_{12}\frac{\Delta\psi}{T} + L_{111}\left(\frac{\Delta P}{T}\right)^2 + L_{112}\frac{\Delta P\,\Delta\psi}{T^2} + L_{122}\left(\frac{\Delta\psi}{T}\right)^2 + \cdots, \tag{256}$$

$$I = L_{21}\frac{\Delta P}{T} + L_{22}\frac{\Delta\psi}{T} + L_{211}\left(\frac{\Delta P}{T}\right)^2 + L_{212}\frac{\Delta P\,\Delta\psi}{T^2} + L_{222}\left(\frac{\Delta\psi}{T}\right)^2 + \cdots, \tag{257}$$

$L_{ijk}(i, j, k = 1, 2 \ldots)$ are the second-order coefficients.

When $\Delta\psi = 0$, the volume flow is given by

$$(J_v)_{\Delta\psi=0} = L_{11}\frac{\Delta P}{T} + L_{111}\left(\frac{\Delta P}{T}\right)^2 + \cdots. \tag{258}$$

When $\Delta P = 0$, Eq. (256) becomes

$$(J_v)_{\Delta P=0} = L_{12}\frac{\Delta\psi}{T} + L_{122}\left(\frac{\Delta\psi}{T}\right)^2 + \cdots. \tag{259}$$

Equation (256) can be written in terms of Eqs. (258) and (259):

$$(J_v)_{\text{total}} = (J_v)_{\Delta\psi=0} + (J_v)_{\Delta P=0} + L_{112}\frac{\Delta\psi\,\Delta P}{T^2} + \cdots, \tag{260}$$

or

$$\left(\frac{\Delta P}{\Delta\psi}\right)_{J_v=0} = -\frac{L_{12}}{L_{11}} - \frac{L_{111}}{L_{11}}\frac{(\Delta P)^2}{T\Delta\psi} - \frac{L_{112}}{L_{11}}\frac{\Delta P}{T} - \frac{L_{122}}{L_{11}}\frac{\Delta\psi}{T}. \tag{261}$$

Similar information can be obtained from Eq. (257) when $I = 0$ [i.e., streaming potential $(\Delta\psi/\Delta P)_{I=0}$] and when $\Delta\psi = 0$ [i.e., streaming current $(I/\Delta P)_{\Delta\psi=0}$]. Thus

$$\left(\frac{\Delta\psi}{\Delta P}\right)_{I=0} = -\frac{L_{21}}{L_{22}} - \frac{L_{211}}{L_{22}}\frac{\Delta P}{T} - \frac{L_{212}}{L_{22}}\frac{\Delta\psi}{T} - \frac{L_{222}}{L_{22}}\frac{(\Delta\psi)^2}{T\Delta P}, \tag{262}$$

$$\left(\frac{I}{\Delta P}\right)_{\Delta\psi=0} = \frac{L_{21}}{T} + \frac{L_{211}}{T^2}\Delta P + \cdots. \tag{263}$$

V. Permeation of Electrolyte Solution through a Membrane

When the higher-power terms are negligible, the following relations are obtained:

$$(J_v)_{\Delta\psi=0} = L_{11}\frac{\Delta P}{T} \quad \text{and} \quad (J_v)_{\Delta P=0} = L_{12}\frac{\Delta\psi}{T}$$

follow from Eq. (256);

$$(I)_{\Delta\psi=0} = L_{21}\frac{\Delta P}{T} \quad \text{and} \quad (I)_{\Delta P=0} = L_{22}\frac{\Delta\psi}{T}$$

follow from Eq. (257);

$$\left(\frac{\Delta P}{\Delta\psi}\right)_{J_v=0} = -\frac{L_{12}}{L_{11}}$$

follows from Eq. (256); and

$$\left(\frac{\Delta\psi}{\Delta P}\right)_{I=0} = -\frac{L_{21}}{L_{22}}$$

follows from Eq. (257). Thus values for the first-order coefficients may be derived. Rastogi and Jha (1966b) evaluated the second-order coefficient L_{112} both from Eqs. (260) and (261) by making the appropriate measurements. It was found that the range of validity of Poiseulle's law was so wide that the coefficient L_{111} was zero. It was also found that up to 440 V, L_{122}/T^2 was zero. Under these conditions, Eqs. (260) and (261) become

$$(J_v)_{\text{total}} = L_{11}\frac{\Delta P}{T} + (J_v)_{\Delta P=0} + L_{112}\frac{\Delta\psi\,\Delta P}{T^2}, \quad (264)$$

$$(J_v)_{\text{total}} = (J_v)_{\Delta\psi=0} + L_{12}\frac{\Delta\psi}{T} + L_{112}\frac{\Delta\psi\,\Delta P}{T^2}, \quad (265)$$

$$\left(\frac{\Delta P}{\Delta\psi}\right)_{J_v=0} = -\frac{L_{12}}{L_{11}} - \frac{L_{112}}{L_{11}}\frac{\Delta P}{T}. \quad (266)$$

By plotting the different measured parameters, viz.,

$$\frac{(J_v)_{\text{total}} - (J_v)_{\Delta P=0}}{\Delta P} \quad \text{against} \quad \Delta\psi$$

or

$$\frac{(J_v)_{\text{total}} - (J_v)_{\Delta\psi=0}}{\Delta\psi} \quad \text{against} \quad \Delta P,$$

L_{112}/T^2 was evaluated. Similarly, it can also be evaluated from Eq. (266) by plotting $(\Delta P/\Delta\psi)_{J_v=0}$ against ΔP. The agreeing value derived by the two methods for L_{112} was 17×10^{-5} (cm^5 s^{-1} V^{-1} dyne^{-1}).

Even higher-order terms may be incorporated into Eq. (256) thus:

$$J_v = L_{11}\frac{\Delta P}{T} + L_{12}\frac{\Delta\psi}{T} + L_{111}\left(\frac{\Delta P}{T}\right)^2 + L_{112}\frac{\Delta P\,\Delta\psi}{T^2} + L_{122}\left(\frac{\Delta\psi}{T}\right)^2$$
$$+ L_{1111}\left(\frac{\Delta P}{T}\right)^3 + L_{1112}\left(\frac{\Delta P}{T}\right)^2\frac{\Delta\psi}{T} + L_{1122}\frac{\Delta P}{T}\left(\frac{\Delta\psi}{T}\right)^2$$
$$+ L_{1222}\left(\frac{\Delta\psi}{T}\right)^3 + \cdots. \qquad (267)$$

When $\Delta\psi = 0$ or $\Delta P = 0$, Eq. (267) becomes

$$(J_v)_{\Delta\psi=0} = L_{11}\frac{\Delta P}{T} + L_{111}\left(\frac{\Delta P}{T}\right)^2 + L_{1111}\left(\frac{\Delta P}{T}\right)^3 + \cdots, \qquad (268)$$

$$(J_v)_{\Delta P=0} = L_{12}\frac{\Delta\psi}{T} + L_{122}\left(\frac{\Delta\psi}{T}\right)^2 + L_{1222}\left(\frac{\Delta\psi}{T}\right)^3 + \cdots. \qquad (269)$$

For the high values of ΔP and $\Delta\psi$, if Ohm's and Poiseuille's laws become applicable, then L_{111}, L_{1111}, L_{122}, and L_{1222} may be negligible. Under these conditions, Eq. (267) can be transformed, using Eqs. (268) and (269), into

$$\frac{(J_v)_{\text{total}} - (J_v)_{\Delta\psi=0}}{\Delta\psi} = \frac{L_{12}}{T} + \frac{L_{112}}{T}\frac{\Delta P}{T} + \frac{L_{1112}}{T}\left(\frac{\Delta P}{T}\right)^2 + \frac{L_{1122}}{T}\frac{\Delta P}{T}\frac{\Delta\psi}{T} + \cdots$$
(270)

or

$$\frac{[(J_v)_{\text{total}} - \{(J_v)_{\Delta\psi=0} + (J_v)_{\Delta P=0}\}]}{\Delta P\,\Delta\psi} = \frac{L_{112}}{T^2} + \frac{L_{1112}}{T^2}\frac{\Delta P}{T} + \frac{L_{1122}}{T^2}\frac{\Delta\psi}{T} + \cdots. \qquad (271)$$

The numerator of the quantity on the left-hand side of Eq. (271) can be determined by experiments for any value of ΔP and $\Delta\psi$. At constant temperature and a fixed value of ΔP, the left-hand side may be evaluated and plotted as a function of $\Delta\psi$. According to Eq. (271), this plot should be a straight line with a slope equal to L_{1122}/T^3 and intercept equal to $L_{112}/T^2 + (L_{1112}/T^3)\Delta P$. So far, no studies involving these higher order coefficients have been reported.

VI. Nature of Water Flow across Membranes

The nature of water flow occurring across a membrane subject to a hydrostatic pressure difference depends on the size of the pores in the membrane, the size of the permeating species, and the degree of bonding the permeating species has for the membrane material. In general, when the size of the pore is large compared to the size of the permeating species, the flow would be mostly of a viscous nature, but as the size of the pore becomes comparable to the size of the permeating species, the diffusional component of the total flow (viscous plus diffusional) would increase, and ultimately when the sizes of both become comparable the flow would be mostly diffusional. Comparison of these two types of flows would give useful information about the structure of the membrane.

Solute flux J_s, following Fick's law, can be described by

$$J_s = D_s A_s \Delta C_s / \Delta x. \tag{272}$$

D_s is the solute diffusion coefficient in water (square centimeters per second). A_s is a parameter reflecting the constraints imposed by the membrane on solute diffusion. One such constraint among others is that the total area of the aqueous pores is less than the area of the membrane. The conventional permeability P_s is given by $J_s/\Delta C_s$ or

$$P_s = D_s A_s / \Delta x. \tag{273}$$

As D_s is known, measurement of P_s gives an estimate for the factor $A_s/\Delta x$, which must be interpreted in terms of solute diffusion in aqueous pores.

If A is the membrane area, Eq. (272) may be written

$$J_s = D'_s \frac{A_p}{A} \frac{\Delta C_s}{\Delta x}, \tag{274}$$

where D' is called the restricted diffusion coefficient and A_p is the total area of the pores and equal to $n\pi r^2$. Again A_p can be roughly estimated in artificial membranes, but it is impossible to do so in biological membranes. Since Eqs. (272) and (274) are equivalent, it follows that

$$D_s A_s = D'_s A_p / A. \tag{275}$$

Experiments with artificial membranes have shown that $A_s/\Delta x$ decreases with an increase in the molecular weight of the diffusing solute. This can be attributed to (i) a frictional force acting on the diffusing solute and (ii) steric hindrance to solute diffusion at the entrance to the pore.

The frictional restriction to solute diffusion has been corrected by the use of Faxen equation

$$\frac{f_s^0}{f_s} = 1 - 2.104 \left(\frac{a_s}{r}\right) + 2.09 \left(\frac{a_s}{r}\right)^3 - 0.95 \left(\frac{a_s}{r}\right)^5, \tag{276}$$

where f_s^0 and f_s are frictions on the solute molecule in free solution and at the wall of the pore, respectively, and a_s is the radius of the solute molecule. The steric hindrance is caused by the fact that the size of the solute molecule in relation to the size of the pore may not be negligible. The effective pore area is given by $\pi(r - a_s)^2$, and the available pore area is πr^2. The probability of the solute entering the pore is thus given by the ratio

$$\frac{\pi(r - a_s)^2}{\pi r^2} \left[= \left(1 - \frac{a_s}{r}\right)^2 \right].$$

In view of these two factors, the apparent area for diffusion A_{sd} ($=A_sA$) is less than A_p. Consequently, the ratio A_{sd}/A_p represents the correction for friction and steric hindrance. One can therefore write

$$\frac{A_{sd}}{A_p} = \left(1 - \frac{a_s}{r}\right)^2 \left[1 - 2.104\left(\frac{a_s}{r}\right) + 2.09\left(\frac{a_s}{r}\right)^3 - 0.95\left(\frac{a_s}{r}\right)^5\right]. \quad (277)$$

The measured value of $A_s/\Delta x$ includes the several interactions, and therefore can be written

$$\frac{A_s}{\Delta x} = \frac{A_{sd}}{A \Delta x} \frac{A_p}{A_p} = \frac{A_p}{A \Delta x} \left(\frac{A_{sd}}{A_p}\right).$$

Substituting from Eq. (277) for A_{sd}/A_p gives

$$\frac{A_s}{\Delta x} = \frac{A_p}{A \Delta x} \left(1 - \frac{a_s}{r}\right)^2 \left[1 - 2.104\left(\frac{a_s}{r}\right) + 2.09\left(\frac{a_s}{r}\right)^3 - 0.95\left(\frac{a_s}{r}\right)^5\right]. \quad (278)$$

In some artificial membranes, $A_p/A\Delta x$ can be estimated, and thus can be interpreted in terms of the pore model from which Eq. (278) is derived.

Application of a hydrostatic pressure across the membrane causes the solute to move as well (hyperfiltration). The steric hindrance factor $(1-a_s/r)^2$ has to be further corrected for the radial velocity profile in the pore. This correction can be effected by using the Ferry equation for steric hindrance,

$$2\left(1 - \frac{a_s}{r}\right)^2 - \left(1 - \frac{a_s}{r}\right)^4. \quad (279)$$

Thus the overall restriction to solute movement through the pores becomes

$$\frac{A_{sf}}{A_p} = \left[2\left(1 - \frac{a_s}{r}\right)^2 - \left(1 - \frac{a_s}{r}\right)^4\right]$$
$$\times \left[1 - 2.104\left(\frac{a_s}{r}\right) + 2.09\left(\frac{a_s}{r}\right)^3 - 0.95\left(\frac{a_s}{r}\right)^5\right], \quad (280)$$

where A_{sf} is the apparent area for solute movement during filtration. Thus one can compare the extent of constraint that solute molecules feel during diffusion and ultrafiltration. The first is expressed by Eq. (277), and the second by Eq. (280). Computations of A_{sd}/A_{sf} as a function of a_s/r using Eqs. (277) and (280) showed that a given molecule experienced less hindrance during filtration than during diffusion through an aqueous pore.

VI. Nature of Water Flow across Membranes

Information about the porosity of the membranes can be derived from studies of rates of diffusion and viscous flow across them.

Let the porous membrane separate two compartments, each filled with water containing different concentrations of labelled water ω^*. The rate of its diffusion is given by

$$J_{w^*} = D_w A_w \Delta C_{w^*}/\Delta x, \tag{281}$$

where J_{w^*} is the net flux of labeled water, A_w is the area of the pores, D_w is the diffusion coefficient of the isotope used in water, and Δx is the pore length (approximately the membrane thickness when there is no tortuosity in the path). Measurements of J_{w^*} and ΔC_{w^*} evaluate the diffusional permeability P_d (cm/sec) of the membrane to water, and thus

$$P_d = D_w A_w/\Delta x. \tag{282}$$

Water flux when a hydrostatic pressure difference is maintained across the membrane is given by

$$J_w^d = \frac{D_w A_w}{RT} C_w \frac{d\mu_w}{dx}, \tag{283}$$

where C_w is the concentration of water at a point x in the membrane, and $d\mu_w/dx$ the gradient of chemical potential for water at x. But $d\mu_w = \bar{V}_w dP$, and since $C_w \bar{V}_w \approx 1$, Eq. (283) becomes

$$J_w^d = \frac{D_w A_w}{RT} \frac{dP}{dx}. \tag{284}$$

Assuming that $dP/dx \approx \Delta P/\Delta x$ gives

$$J_w^d = \frac{D_w A_w}{RT \Delta x} \Delta P. \tag{285}$$

Equation (285) can be equated to $J_v = L_p \Delta P$ provided the diffusional flux is the sole constituent of the total flux. Under this condition,

$$J_v = J_w^d \bar{V}_w$$

and

$$L_p = \frac{D_w A_w \bar{V}_w}{RT \Delta x}. \tag{286}$$

Inserting Eq. (282) into Eq. (286) gives

$$L_p RT/\bar{V}_w = P_d. \tag{287}$$

This relation permits L_p (cm s^{-1} atm^{-1}) to be expressed in the same units as P_d (cm/sec). Equation (286) holds only when the pressure-driven water flow is all diffusional. In practice, it is generally observed that $(L_p RT/\bar{V}_w) > P_d$, suggesting that the flow of water due to a difference in chemical potential is composed of both diffusional and viscous components. Since the presence of viscous flow across a membrane points to the existence of pores or aqueous channels in the membrane, several studies exist comparing values of P_d with P_f (filtration permeability $= L_p RT/\bar{V}_w$).

When there is a pressure gradient acting across the porous membrane containing n pores of radius r, it may be assumed that both the diffusional and viscous flows go through A_p only; then $A_w = A_p$. The diffusional flux of water according to Eqs. (285) and (282) is given by

$$J_w^d = \frac{D_w A_p \Delta P}{RT \Delta x} = \frac{P_d}{RT} \Delta P, \qquad (288)$$

where $P_d = n\pi r^2 D_w/\Delta x$.

In addition to the diffusional flux, there is the viscous flux through the pores, and according to Poiseuille's law, the volume flow $J_w^v \bar{V}_w$ due to ΔP is given by

$$J_w^v = \frac{n\pi r^4 \Delta P}{8\eta \bar{V}_w \Delta x}, \qquad (289)$$

where η is the viscosity of the fluid in the pores. If Eq. (289) is compared to $J_v = L_p \Delta P$, it can be written

$$J_w^v = L_p' \Delta P/\bar{V}_w, \qquad (290)$$

where the parameter L_p' is analogous to L_p.

Dividing Eq. (290) by Eq. (289) and inserting $P_d = n\pi r^2 D_w/\Delta x$ give

$$r^2 = 8\eta D_w L_p'/P_d. \qquad (291)$$

L_p' cannot be obtained directly, and so can be removed from Eq. (291) as follows.

The total flux is given by $J_w^d + J_w^v$, and so $J_v = (J_w^d + J_w^v)\bar{V}_w$. This equation on substitution of the relation $J_v/\bar{V}_w = L_p \Delta P/\bar{V}_w$ and Eqs. (288) and (290) becomes

$$\frac{L_p \Delta P}{\bar{V}_w} = \frac{P_d}{RT} \Delta P + \frac{L_p' \Delta P}{\bar{V}_w}. \qquad (292)$$

Combining Eqs. (291) and (292) to eliminate L_p' yields on rearrangement

$$r = \left\{ \frac{8\eta D_w \bar{V}_w}{RT} \left[\frac{RT L_p}{\bar{V}_w P_d} - 1 \right] \right\}^{1/2}, \qquad (293)$$

where η is the viscosity of water. Use of appropriate values for D_w, η, $\bar{V}_w$, and RT at 25°C in Eq. (293) simplifies it to

$$r = 3.8\sqrt{L_p RT/\bar{V}_w P_d - 1} \quad \text{Å}$$

or
$$r = 3.8\sqrt{P_f/P_d - 1} \quad \text{Å}. \tag{294}$$

Thus a value for the equivalent pore radius can be derived by reliable measurements of L_p and P_d.

When this treatment is applied to barriers where the pore radius is likely to be small, say $r = 2a_w$, consideration must be given to the effective areas available for diffusion and filtration of water in the membrane. This was done by Paganelli and Solomon (1957) by equating A_{sd} and A_{sf} to A_{w*d} and A_{w*f} and incorporating the ratio into Eq. (293) thus:

$$r^2 = \lambda\left(\frac{A_{w*d}}{A_{w*f}}\right), \tag{295}$$

where

$$\lambda = \frac{8\eta D_w \bar{V}_w}{RT}\left[\frac{L_p RT}{\bar{V}_w P_d} - 1\right].$$

Equation (295) can be obtained as usual by replacing A_p in Eq. (288) by A_{w*d} and $n\pi r^2$ in Eq. (289) by A_{w*f}. Dividing Eq. (277) by Eq. (280) and replacing solute s by solute w gives

$$\frac{A_{w*d}}{A_{w*f}} = \frac{1}{2 - (1 - a_w/r)^2}. \tag{296}$$

Substituting Eq. (296) into Eq. (295) gives on rearrangement

$$r^2 + 2a_w r - (a_w^2 + \lambda) = 0. \tag{297}$$

Solving this quadratic equation in r gives

$$r = -a_w + \sqrt{2a_w^2 + \lambda}. \tag{298}$$

Thus a value for the equivalent radius of the pore can be evaluated.

Equation (293) indicates that a pressure-driven flow across the membrane will have a completely diffusional flow in character when $P_f/P_d = 1$, i.e., $r = 0$. The realistic expectation is that at some critical value of r, transport of water becomes obligatorily diffusional in nature due to the inability of one water molecule to overtake another water molecule in the pore (single-file transport). Single-file transport has been theoretically analyzed by several investigators, and the results are controversial. According to Manning (1975), the ratio of $P_f/P_d = 1$; but according to Dick (1966), Levitt (1974), and Rosenberg and Finkelstein (1978) $P_f/P_d = N_w$, the number of water molecules in the channel. The derivation of this result due to Finkelstein and Rosenberg (1979) is outlined below.

The osmotic pressure difference $\Delta\Pi$ (equivalent to hydrostatic pressure ΔP) across a pore (length l containing N_w water molecules) in which single-file transport occurs is given by

$$\Delta\Pi = \kappa T \Delta n_s, \quad (299)$$

where Δn_s is the difference in the number of impermeant solute molecules (molecules per cubic centimeter) in the two solutions on either side of the pore. The force acting on the water molecules in the pore is given by

$$\mathscr{F}_\Pi = V_{n_w} N_w \Delta\Pi/l, \quad (300)$$

where V_{n_w} is the volume per water molecule in the bulk solution. The frictional force $\mathscr{F}_f$ acting on the molecules when they are moving at a velocity v can be written

$$\mathscr{F}_f = N_w f v, \quad (301)$$

where f is the friction coefficient per water molecule. Equating Eqs. (300) and (301) and substituting Eq. (299) for $\Delta\Pi$ simplifies to

$$v = V_{n_w} \kappa T \Delta n_s / l f. \quad (302)$$

Thus the flux of water is given by

$$J_w = \frac{N_w v}{l} = \frac{N_w V_{n_w} \kappa T}{l^2 f} \Delta n_s. \quad (303)$$

P_f for a single pore is given by $J_w = P_f \Delta n_s$, and so substituting Eq. (303) gives

$$P_f = N_w V_{n_w} \kappa T / f l^2. \quad (304)$$

When a tracer is used, application of an external force $\mathscr{F}$ acts not only on the tracer but also on the abundant species, and so all the molecules in the pore move. This force is balanced by the frictional force, and so Eq. (301) becomes applicable. If N^* is the number of tracer molecules in the pore, its flux is given by

$$J_{w^*}^F = N^* v / l. \quad (305)$$

Substituting for v from Eq. (301) gives

$$J_{w^*}^F = N^* \mathscr{F}_f / N_w f l. \quad (306)$$

Equation (306) described the flow of the tracer when it is subject to an external force ($\mathscr{F} = \mathscr{F}_f$). Imagine now that this force is balanced by a concentration gradient of the tracer. That is, the flux due to diffusion $J_{w^*}^d$ balances the flux $J_{w^*}^F$ caused by the external force establishing a state of equilibrium. If n_l^* and n_r^* are the number of tracer molecules on the left and right side of the pore, then they are related by the Boltzmann distribution, and so

$$n_l^* / n_r^* = \exp(-\mathscr{F}_f l / \kappa T). \quad (307)$$

If the initial concentration of the tracer on either side of the pore is n^*, then for the steady state, conservation of the tracer mass gives

$$n_l^* = n^* - \Delta n^*, \quad n_r^* = n^* + \Delta n^*, \tag{308}$$

where Δn^* are the tracer molecules lost from one side and gained by the other side. Combining Eqs. (307) and (308) gives

$$\frac{n^* - \Delta n^*}{n^* + \Delta n^*} = \exp\left(\frac{-\mathscr{F}_f l}{\kappa T}\right). \tag{309}$$

For a small force, Eq. (309) approximates to

$$1 - \Delta n^*/n^* = 1 - \mathscr{F}_f l/\kappa T. \tag{310}$$

Thus

$$\Delta n^* = (\mathscr{F}_f l/\kappa T) n^*, \tag{311}$$

and as $J_{w^*}^d = P_d \Delta n^*$, substituting from Eq. (311) gives

$$J_{w^*}^d = P_d(\mathscr{F}_f l/\kappa T) n^*. \tag{312}$$

Since $J_{w^*}^d = J_{w^*}^F$, equating Eqs. (312) and (306) gives

$$P_f = N_w V_{n_w} \kappa T/f l^2. \tag{313}$$

N^*/N_w is the fraction of molecules labeled in the pore. This will be equal to $n^* V_{n_w}$, the fraction of molecules labeled in the bulk solution. It is assumed that there is little isotope effect. Consequently Eq. (313) becomes

$$P_d = \kappa T V_{n_w}/f l^2. \tag{314}$$

Dividing Eq. (304) by Eq. (314) gives

$$P_f/P_d = N_w. \tag{315}$$

If pores in a membrane are identical in all respects, allowing only single-file transport to occur, then N_w will be proportional to l. According to Eq. (304), $P_f \propto 1/l \propto 1/N_w$. This is in agreement with Poiseuille's law. But Eq. (314) shows that $P_d \propto 1/l^2 \propto 1/N_w^2$. This proportionality is quite unusual and is confined to single-file transport.

Structural considerations indicate that the pore created in a lipid bilayer membrane by gramicidin A is narrow (the radius of the channel is about 2 Å), and so transport occurs by a single-file process. The measurements on a macroscopic level gave values of 34.2×10^{-3} and 6.5×10^{-3} cm/s for P_f and P_d, respectively. Thus $P_f/P_d \approx 5$. This value according to Eq. (315) gives the number of water molecules in the gramicidin pore.

In most situations, membranes would be generally surrounded by aqueous electrolyte solutions. In such a case a pore in which transport occurs by single filing will be filled with both ions and water. The relation of water flux to ion flux in such a pore can be approached by the study of electro-kinetic phenomena such as electro-osmosis and streaming potential, which according to Eq. (72) are equivalent. The relation has been derived by Levitt *et al.* (1978), and Rosenberg *et al.* (1978), and Finkelstein and Andersen (1981). The two approaches are outlined below.

The electrochemical potential difference across a membrane for the cation (+) is given by

$$\Delta\bar{\mu}_+ = RT\ln\frac{x_{+(1)}}{x_{+(2)}} + \bar{V}_+ \Delta P + F\Delta\psi, \qquad (316)$$

where $x_{+(1)}$ and $x_{+(2)}$ are the mole fractions of the cation on the two sides of the membrane. If n_+, n_w, and n_{im} are the number of moles of cation, water, and impermeable ion, then

$$\frac{x_{+(1)}}{x_{+(2)}} = \left[\frac{n_{+(1)}}{n_{+(1)} + n_{w(1)} + n_{im(1)}}\right] \bigg/ \left[\frac{n_{+(2)}}{n_{+(2)} + n_{w(2)} + n_{im(2)}}\right]$$

$$\approx \left(\frac{n_{+(1)}}{n_{w(1)}}\right)\left(\frac{n_{w(2)}}{n_{+(2)}}\right)\left[1 + \frac{n_{+(2)}}{n_{w(2)}} - \frac{n_{+(1)}}{n_{w(1)}} + \frac{n_{im(2)}}{n_{w(2)}} - \frac{n_{im(1)}}{n_{w(1)}}\right]. \qquad (317)$$

The molality of the electrolyte on the two sides of the membrane being the same, $n_{+(1)}/n_{w(1)} = n_{+(2)}/n_{w(2)}$, and so Eq. (317) becomes

$$\frac{x_{+(1)}}{x_{+(2)}} = 1 + \frac{n_{im(2)}}{n_{w(2)}} - \frac{n_{im(1)}}{n_{w(1)}} \approx 1 + \bar{V}_w \Delta m_{im}, \qquad (318)$$

where Δm_{im} is the impermeant ion molality difference across the membrane. Substituting Eq. (318) into Eq. (316) gives

$$\Delta\bar{\mu}_+ = RT\ln(1 + \bar{V}_w \Delta m_{im}) + \bar{V}_+ \Delta P + F\Delta\psi.$$

This equation when $\bar{V}_w \Delta m_{im} \ll 1$,

$$\ln(1 + \bar{V}_w \Delta m_{im}) \approx \bar{V}_w \Delta m_{im}$$

approximates

$$\Delta\bar{\mu}_+ \approx RT\bar{V}_w \Delta m_{im} + \bar{V}_+ \Delta P + \Delta\psi F$$
$$= \bar{V}_w \Delta\Pi + \bar{V}_+ \Delta P + F\Delta\psi. \qquad (319)$$

A similar procedure gives for $\Delta\mu_w$

$$\Delta\mu_w = \bar{V}_w(\Delta\Pi + \Delta P). \qquad (320)$$

The dissipation function can be written

$$\phi = J_w \Delta\mu_w + J_+ \Delta\bar{\mu}_+. \qquad (321)$$

Substituting Eqs. (319) and (320) into Eq. (321) gives, when $\Delta P = 0$,

$$\phi = \bar{V}_w(J_w + J_+)\Delta\Pi + J_+ F \Delta\psi. \tag{322}$$

Equation (322) can be written

$$\phi = J_v \alpha \Delta\Pi + I \Delta\psi, \tag{323}$$

where $J_v \alpha = \bar{V}_w(J_w + J_+)$ and $I = J_+ F$. Thus the linear equations become

$$J_v = L_{11}\alpha \Delta\Pi + L_{12}\Delta\psi, \tag{324}$$

$$I = L_{21}\alpha \Delta\Pi + L_{22}\Delta\psi. \tag{325}$$

Electro-osmotic permeability β is given by

$$\left(\frac{J_v}{I}\right)_{\Delta\Pi=0} = \frac{L_{12}}{L_{22}} = \frac{J_w \bar{V}_w + J_+ \bar{V}_+}{F J_+}$$

$$= \frac{\bar{V}_w}{F}\left[\frac{J_w}{J_+} + \frac{\bar{V}_+}{\bar{V}_w}\right]. \tag{326}$$

Since $J_w/J_+ = N_w$, the number of moles of water associated with 1 mol of cation, Eq. (326) becomes

$$\beta = \frac{L_{12}}{L_{22}} = \frac{\bar{V}_w}{F}\left[N_w + \frac{\bar{V}_+}{\bar{V}_w}\right]. \tag{327}$$

The streaming potential is given by Eq. (325) when $I = 0$. Thus

$$\left(\frac{\Delta\psi}{\Delta\Pi}\right)_{I=0} = -\frac{L_{21}\alpha}{L_{22}} = -\frac{L_{12}\alpha}{L_{22}}. \tag{328}$$

But it can be shown from the definition of α and N_w that

$$\alpha = \frac{1 + N_w}{N_w + \bar{V}_+/\bar{V}_w}.$$

Substituting this value of α and the ratio L_{12}/L_{22} from Eq. (327) into Eq. (328) gives

$$\left(\frac{\Delta\psi}{\Delta\Pi}\right)_{I=0} = -\frac{\bar{V}_w}{F}(1 + N_w). \tag{329}$$

In Eq. (327), $\bar{V}_+$ is the volume associated with the transport of a mole of the cation across the membrane. Determination of this involves use of reversible electrodes and applying the necessary correction for the electrode reactions. If Ag/AgCl electrodes are used, $\bar{V}_+$ is given by

$$\bar{V}_+ = \bar{V}_{MCl} + \bar{V}_{Ag} - \bar{V}_{AgCl}.$$

$\bar{V}_{Ag} = 10.28$, $\bar{V}_{AgCl} = 25.77$, and so $\bar{V}_+ = \bar{V}_{MCl} - 15.15$. $\bar{V}_{MCl}$ values are concentration dependent, and particular values can be obtained from standard texts. $\bar{V}_{KCl}$ value at 0.15 M is 27.5 ml; at 3 M, it is 30.2; and for NaCl, the corresponding values are $\bar{V}_{NaCl} = 17.5$ ml at 0.15 M and 19.8 at 3 M. Compared to the $\bar{V}_w$ value (18 ml), $\bar{V}_+$ values are smaller (at 0.15 M, KCl = 12 ml, NaCl = 2 ml). In view of this, Eq. (327) becomes

$$(\beta F/\bar{V}_w) = N_w \qquad \text{(electro-osmosis)},$$

$$\left(\frac{\Delta\psi}{\Delta\Pi}\right)_{I=0} \frac{F}{\bar{V}_w} = 1 + N_w \qquad \text{(streaming potential)}.$$

Thus by measuring streaming potential (easy to measure with bilayers) or electro-osmotic flow (difficult to measure with bilayers), values for N_w can be derived. For the gramicidin channel, $N_w = 6$.

If the correction introduced into Eqs. (324) and (325) is dropped (negligible error introduced), one can write

$$L_{11} = \left(\frac{J_v}{\Delta\Pi}\right)_{\Delta\psi=0} = L_p = \frac{\bar{V}_w}{RT} P_f \qquad (330)$$

and

$$L_{22} = \left(\frac{I}{\Delta\psi}\right)_{\Delta\Pi=0} = G_{M+}. \qquad (331)$$

Furthermore, Eq. (325) gives

$$L_{12} = -\left(\frac{\Delta\psi}{\Delta\Pi}\right)_{I=0} L_{22} = L_e. \qquad (332)$$

$(\Delta\psi/\Delta\Pi)_{I=0}$ is the streaming potential per one osmol pressure difference across the pore. A quantity η' having values between 0 and 1 indicating the extent of coupling between ion flow and water flow can be defined and is given by $L_{12}L_{21}/L_{11}L_{22}$ (see Finkelstein and Andersen, 1981), that is

$$\eta' = (L_e)^2/L_{11}L_{22}. \qquad (333)$$

This definition is equivalent to

$$\left(\frac{\Delta\psi}{\Delta\Pi}\right)_{I=0} \left(\frac{\Delta\Pi}{\Delta\psi}\right)_{J_v=0} = 1.$$
$$\quad\text{streaming} \qquad\quad \text{electro-osmotic}$$
$$\quad\text{potential} \qquad\qquad \text{pressure}$$

The first term arises from the movement of ion through the pore, and the second arises from the movement of water through the pore. Substituting

Eq. (332) into Eq. (333) gives

$$\eta' = \left(-\frac{\Delta\psi}{\Delta\Pi}\right)^2_{I=0} \frac{L_{22}}{L_{11}}. \tag{334}$$

But combining Eqs. (327) and (328) gives $-(\Delta\psi/\Delta\Pi)_{I=0} = (\bar{V}_w/F)N_w$, and so Eq. (334) becomes

$$\eta' = \frac{L_{22}}{L_{11}} \left(\frac{\bar{V}_w}{F}\right)^2 N_w^2. \tag{335}$$

When there is perfect coupling between ion and water flows (i.e., $\eta' = 1$), Eq. (335) becomes

$$L_{22}/L_{11} = F^2/\bar{V}_w^2 N_w^2. \tag{336}$$

Substituting Eqs. (330) and (331) into Eq. (336) gives

$$G_{M^+}/P_f = F^2/\bar{V}_w RT N_w^2. \tag{337}$$

Use of Eq. (115) of Chapter 4 in Eq. (337) gives

$$J_{M^+} = P_f/N_w^2 \bar{V}_w. \tag{338}$$

Water flux is given by $J_w = P_d C_w$. Thus the ratio (water flux to ion flux), on introducing the relations $P_f/P_d = N_w$ [Eq. (315)] and $C_w \bar{V}_w \approx 1$, becomes

$$J_w/J_{M^+} = N_w. \tag{339}$$

Equation (339) is valid only when there is perfect coupling of ion and water movements in the pore. This may occur in a single-file transport channel (e.g., gramicidin A channel) at high concentrations of the electrolyte when the channel always contains an ion. At low concentrations of the electrolyte, the channel may not contain an ion, and so water transport will not be coupled to ion transport.

References

de Groot, S. R. (1963). "Thermodynamics of Irreversible Processes." North-Holland Publ., Amsterdam.
Denbigh, K. G. (1951). "The Thermodynamics of the Steady State." Methuen, London.
Dick, D. A. T. (1966). "Cell Water," p. 108. Butterworth, London.
Finkelstein, A., and Andersen, O. S. (1981). J. Membr. Biol. **59**, 155.
Finkelstein, A., and Rosenberg, P. A. (1979). In "Membrane Transport Processes" (C. F. Stevens and R. W. Tsien, eds.), Vol. 3, p. 73. Raven Press, New York.
Ginzburgh, B. Z., and Katchalsky, A. (1963). J. Gen. Physiol. **47**, 403.
House, C. R. (1974). "Water Transport in Cells and Tissues." Williams & Wilkins, Baltimore, Maryland.

Katchalsky, A., and Curran, P. F. (1965). "Nonequilibrium Thermodynamics in Biophysics," Harvard Univ. Press, Cambridge, Massachusetts.
Kedem, O., and Essig, A. (1965). *J. Gen. Physiol.* **48**, 1047.
Kedem, O., and Katchalsky, A. (1958) *Biochim. Biophys. Acta* **27**, 229.
Kedem, O., and Katchalsky, A. (1961). *J. Gen. Physiol.* **45**, 143.
Kedem, O., and Katchalsky, A. (1963). *Trans. Faraday Soc.* **59**, 1918, 1931, 1941.
Kirkwood, J. G. (1954). *In* "Ion Transport across Membranes" (H. T. Clarke, ed.), p. 119. Academic Press, New York.
Lakshminarayanaiah, N. (1970). *J. Phys. Chem.* **74**, 2385.
Lakshminarayanaiah, N. (1971). *J. Macromol. Sci.—Phys. B* **5**, 159.
Lakshminarayanaiah, N., and Subrahmanyan, V. (1968). *J. Phys. Chem.* **72**, 1253.
Levitt, D. G. (1974). *Biochim. Biophys. Acta* **373**, 115.
Levitt, D. G., Elias, S. R., and Hautman, J. M. (1978). *Biochim. Biophys. Acta* **512**, 436.
Li, J. H., and Essig, A. (1977). *Biochim. Biophys. Acta* **465**, 421.
Li, J. H., DeSousa, R. C., and Essig, A. (1974). *J. Membr. Biol.* **19**, 93.
Mackay, D., and Meares, P. (1959). *Trans. Faraday Soc.* **55**, 1221.
Manning, G. S. (1975). *Biophys. Chem.* **3**, 147.
Meares, P., Thain, J. F., Dawson, D. G. (1972). *In* "Membranes" (G. Eisenman, ed.), Vol. 1, p. 55. Dekker, New York.
Nernst, W. (1888). *Z. Phys. Chem.* **2**, 613.
Nernst, W. (1889). *Z. Phys. Chem.* **4**, 129.
Onsager, L. (1931). *Phys. Rev.* **37**, 405; **38**, 2265.
Paganelli, C. V., and Solomon, A. K. (1957). *J. Gen. Physiol.* **41**, 259.
Rastogi, R. P., and Jha, K. M. (1966a). *J. Phys. Chem.* **70**, 1017.
Rastogi, R. P., and Jha, K. M. (1966b). *Trans. Faraday Soc.* **62**, 585.
Rosenberg, P. A., and Finkelstein, A. (1978). *J. Gen. Physiol.* **72**, 327, 341.
Roy Caplan, S., and Mikulecky, D. C. (1966). *In* "Ion Exchange" (J. A. Marinsky, ed.), Vol. 1, p. 1. Dekker, New York.
Scattergood, E. M., and Lightfoot, E. N. (1968). *Trans. Faraday Soc.* **64**, 1135.
Spiegler, K. S. (1958). *Trans. Faraday Soc.* **54**, 1408.
Staverman, A. J. (1952). *Trans. Faraday Soc.* **48**, 176.
Van Bruggen, J. T., Boyett, J. D., Bueren, A. L. V., and Galey, W. R. (1974). *J. Gen. Physiol.* **63**, 639.

Chapter 7

IMPEDANCE, CABLE THEORY, AND HODGKIN–HUXLEY EQUATIONS

In biological cells the interior world is separated from the exterior surroundings by a membrane that is a bilayer of lipids with their polar groups oriented toward the interior and exterior aqueous phases of the cell. Protein is considered to exist according to the Danielli–Davson model (1935) at the two ends of the bilayer. According to the Singer–Nicolson model (1972), some proteins (peripheral) exist on the periphery of the lipid bilayer and others (integral proteins) exist in the membrane at discrete places (see Lakshminarayanaiah, 1979). Certain proteins of the membrane act as ion carriers or undergo conformation to form pores or channels through which ions can pass. Thus the membrane seems to be a mosaic of lipids acting as insulators (i.e., dielectric) and proteins serving as potential channels formers or conductors (i.e., resistor). So the membrane is composed of both resistive and capacitative elements, which in the passive (i.e., resting) and in the active (i.e., subject to a threshold stimulus) states of the membrane have definite values. In addition, the aqueous phases of the cell contain small ions, mostly potassium in the cell interior and mostly sodium in the cell exterior. In the resting state the membrane is more permeable to potassium ions than to sodium ions and relatively impermeable to calcium, magnesium, sulfate, etc., ions. Free exchange between the cell interior and its exterior occurs across the membrane during the course of cell life, and so any injury to the membrane will prove fatal to the cell.

In the resting state, the cell has a characteristic membrane potential (the value dependent on the nature of the cell; about -70 to -90 mV in nerve and muscle cells), which is due to the difference in the concentration of potassium ions across the cell membrane. When the cell is stimulated by a depolarizing pulse of sufficient strength, the resting potential becomes less negative, reaches zero, becomes positive, and finally reverts back to its original level. This is the all-or-none action potential, which is generally propagated (impulse) along the length of the nerve cell. Hodgkin and Huxley (1952a–d) proposed the ionic hypothesis to explain these several events that occur when the biological cell is electrically stimulated. According to this hypothesis, when the cell is stimulated some dipolar molecules (probably transport or channel proteins) move, giving rise to a current that is capacitative in nature (displacement or gating current), and create sodium channels through which sodium ions move down their electrochemical gradient, giving rise to the early Na current. Later, potassium channels open, and repolarization of the membrane occurs by the outward movement of potassium ions, giving rise to the outward K current. These events were demonstrated by the use of the voltage-clamp technique, without which it is difficult to follow the events that occur during an explosive all-or-none reaction, i.e., generation of the action potential.

In the resting state, the cell membrane resistance is relatively high. In the active state it is considerably reduced, with little change in the membrane capacitance. One technique that is used to derive quantitative values for the different elements of the cell, i.e., the electrical cell constants, is the measurement of the complex impedance of the cell system. With the help of appropriate models or equivalent circuits, definite values for the resistances and capacitances that might exist in the cell system can be derived. Measurement of complex impedance of biological samples involves use of platinized platinum electrodes for macroscopic samples and glass microelectrodes for isolated single cells. As some of the cylindrical cells in their conductance properties resemble a submarine cable (external and internal conductive media separated by a nonconductive phase or membrane) equations (cable equations) developed to describe the behavior of a cable are used in the derivation of values for the electrical constants of the biological cell preparations. The general concepts related to impedance, an introduction to cable equations, and an outline of the empirical equations of Hodgkin and Huxley (1952d) that are used to generate the action potential are given in this chapter.

I. Impedance

When a potential that is time dependent in a sinusoidal manner, i.e., $E \sin \omega t$, where E is small (less than 5 mV), is applied across a membrane (see Fig. 1), a sinusoidal current $i \sin(\omega t + \phi)$ will flow at angular frequencies 2ω, 3ω, etc., where $\omega = 2\pi f$ (f is sinusoidal frequency in hertz). Thus impedance Z is defined as having the magnitude given by

$$Z = E/i, \qquad (1)$$

and phase angle ϕ, which is given by the phase difference between the

I. Impedance

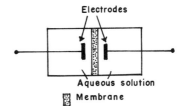

Fig. 1. Schematic representation of a two-electrode membrane system.

applied sinusoidal potential and the resulting sinusoidal current. Impedance is therefore a vector quantity having both magnitude and direction. Such two-component vectors are conveniently represented as a point in a plane (see Fig. 2). The vector is characterized by $|Z|$ and ϕ or by its real and imaginary components Z' and Z'' projected onto the X and Y axes. When a pure resistance of magnitude R ohms exists in the place of the membrane (see Fig. 1), then $Z = R$ and $\phi = 0$. In Fig. 3a the pure resistance is represented as a point on the X axis for any given frequency. When a pure capacitance of C farads is substituted for the membrane system, the situation becomes complex. For this case $\phi = 90°$ and the impedance Z is given by $-1/\omega C$. So as the frequency is changed, Z assumes different values, and the representative point varies in the manner shown in Fig. 3b. Figures 3a and b represent the simplest forms of impedance spectra that result as a function of frequency.

Invariably, in a majority of cases, the cell membrane is represented by a complex network of resistances and capacitances, and this is called the equivalent circuit. In general, these show a complicated behavior in the complex impedance plane. Examples are shown in Figs. 3c–e. Figure 3c represents an impedance spectrum due to series combination of a resistance and a capacitance. In this case $Z' = R_s$ and $Z'' = 1/\omega C_s$. Since $Z^2 = (Z')^2 + (Z'')^2$, it follows that

$$Z_s^2 = R_s^2 + (1/\omega C_s)^2. \tag{2}$$

ϕ can vary from $0°$ to $90°$, depending on the frequency used. But Z (vector) can be written

$$Z = R_s + jX_s, \tag{3}$$

where $j = \sqrt{-1}$ and X_s is the reactance equal to $-1/\omega C_s$.

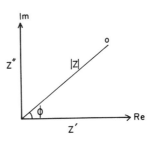

Fig. 2. Schematic illustration of impedance plotted in a complex plane.

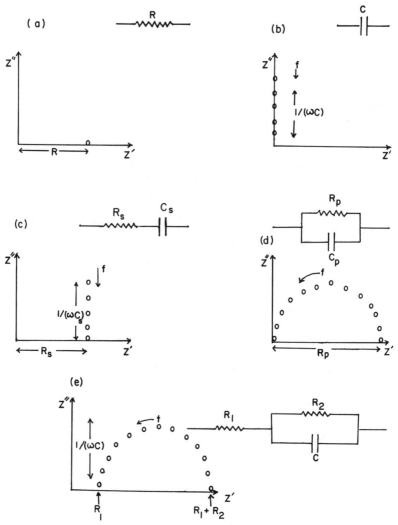

Fig. 3. Schematic representation of complex plane impedances and their associated equivalent circuits. (After Archer and Armstrong, 1980.)

For a parallel combination of resistance and capacitance, the spectrum is completely different and the impedance is given by (parallel combination of R_p and $1/\omega C_p$)

$$\frac{1}{Z} = \frac{1}{R_p} + j\omega C_p$$

I. Impedance

or

$$Z = \frac{R_p}{1 + j\omega R_p C_p}. \tag{4}$$

In this case the capacitance is additive. Sometimes it is more useful to consider the reciprocal of impedance, i.e., admittance Y, and it is given by $Y = G + jB$, where G is the conductance $(1/R)$ and B is the susceptance (ωC). Thus

$$Y = \frac{1}{Z} = \frac{1 + j\omega R_p C_p}{R_p}. \tag{5}$$

Equation (4) can be rearranged to

$$Z = \frac{R_p(1 - j\omega R_p C_p)}{1 + \omega^2 R_p^2 C_p^2} = \frac{R_p}{1 + \omega^2 R_p^2 C_p^2} - j\frac{\omega R_p^2 C_p}{1 + \omega^2 R_p^2 C_p^2}. \tag{6}$$

Comparing Eqs. (3) and (6) (i.e., equating the real and the imaginary parts) gives

$$R_s = \frac{R_p}{1 + \omega^2 R_p^2 C_p^2}, \tag{7}$$

$$C_s = \frac{1 + \omega^2 R_p^2 C_p^2}{\omega^2 R_p^2 C_p}. \tag{8}$$

If the circuits in Figs. 3c and d are expressed in terms of admittances, then they are given, respectively, by

$$Y = \frac{1}{R_s - j/\omega C_s}, \tag{9}$$

$$Y = \frac{1}{R_p} + j\omega C_p. \tag{10}$$

Equation (9) can be written

$$Y = \frac{\omega C_s(j + \omega R_s C_s)}{1 + \omega^2 R_s^2 C_s^2}$$

$$= \frac{\omega^2 R_s C_s^2}{1 + \omega^2 R_s^2 C_s^2} + j\frac{\omega C_s}{1 + \omega^2 R_s^2 C_s^2}. \tag{11}$$

Equating the real and the imaginary parts of Eqs. (10) and (11) gives

$$R_p = \frac{1 + \omega^2 R_s^2 C_s^2}{\omega^2 R_s C_s^2}, \tag{12}$$

$$C_p = \frac{C_s}{1 + \omega^2 R_s^2 C_s^2}. \tag{13}$$

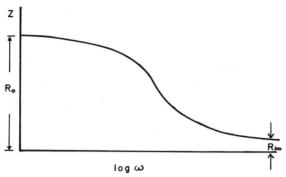

Fig. 4. Impedance Z as a function of angular frequency ω (schematic). (After Schanne and Ruiz P.-Ceretti, 1978.)

One of the most commonly used circuits to interpret an impedance measurement related to biological cell structure is that shown in Fig. 3e. Measurement of the impedance of this circuit as a function of frequency is shown in Fig. 4. It shows that there are two limiting resistances. At low frequencies, the capacitative reactance is large and the impedance is given by

$$R_0 = R_1 + R_2. \tag{14}$$

At high frequencies, the resistance becomes negligible, and so

$$R_\infty = R_1. \tag{15}$$

Measurement of biological-tissue impedance as a function of frequency enables one to determine values for R_0 and R_∞. The impedance in complex form can be written (see Fig. 3e)

$$Z = R_1 + \frac{R_2}{1 + j\omega CR_2} = R_1 + \frac{R_2(1 - j\omega CR_2)}{1 + \omega^2 C^2 R_2^2}.$$

Separating the real and the imaginary parts gives

$$Z = R_1 + \frac{R_2}{1 + \omega^2 C^2 R_2^2} - j\frac{\omega CR_2^2}{1 + \omega^2 C^2 R_2^2}. \tag{16}$$

Substituting the time constant $\tau = R_2 C$ and R_0 and R_∞ from Eqs. (14) and (15) into Eq. (16) gives

$$Z = \left[R_\infty + \frac{R_0 - R_\infty}{1 + \omega^2 \tau^2}\right] - j\left[\frac{\omega\tau(R_0 - R_\infty)}{1 + \omega^2 \tau^2}\right]. \tag{17}$$

I. Impedance

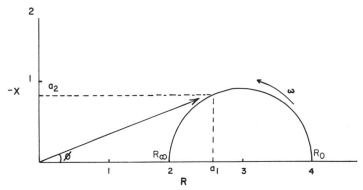

Fig. 5. Impedance locus for the circuit of Fig. 3e for arbitrary units of $R_0 = 4$ and $R_\infty = 2$. The angular frequency ω increases counterclockwise as shown. The impedance vector Z turns around its origin and its point moves along the impedance locus. At any given frequency, a_1 and a_2 are the values of resistance R and reactance $-X$. (After Schanne and Ruiz P.-Ceretti, 1978.)

The real component of the tissue impedance is given by

$$R = R_\infty + \frac{R_0 - R_\infty}{1 + \omega^2 \tau^2}. \tag{18}$$

The reactance X in the RX plane is given by

$$X = -\frac{\omega \tau (R_0 - R_\infty)}{1 + \omega^2 \tau^2}. \tag{19}$$

Both of these and hence the impedance Z are dependent on the angular frequency ω. Z will change continuously along a curve in the RX plane as the frequency is changed from 0 to ∞. This curve is called an impedance locus (see Fig. 5) and can be obtained analytically.

Equations (18) and (19) are related to Z by

$$Z = \sqrt{R^2 + X^2}. \tag{20}$$

The phase angle is given by

$$\tan \phi = X/R. \tag{21}$$

Equation (19) is a quadratic in $\omega \tau$, and its solution is

$$\omega \tau = -\frac{R_0 - R_\infty}{2X} + \sqrt{\left(\frac{R_0 - R_\infty}{4X}\right)^2 - 1}.$$

Dividing Eq. (18) by Eq. (19) gives $X = -\omega \tau (R - R_\infty)$ and eliminating $\omega \tau$ between these two equations gives on simplification

$$X^2 = (R - R_\infty)(R_0 - R)$$

or
$$X^2 + R^2 - R(R_0 + R_\infty) = -R_0 R_\infty. \tag{22}$$

Adding $(R_0 + R_\infty)^2/4$ to both sides of Eq. (22) gives on rearrangement

$$X^2 + [R - (R_0 + R_\infty)/2]^2 = [(R_0 - R_\infty)/2]^2. \tag{23}$$

This is the equation of a circle with the center on the R axis at $(R_0 + R_\infty)/2$ and a radius of $(R_0 - R_\infty)/2$, and represents the analytical expression for the impedance locus. As shown in Fig. 5, the impedance locus exists in the $-X$ region as a semicircle.

The characteristic frequency is defined as

$$\omega_c = 1/\tau, \tag{24}$$

and at this frequency, the reactance [see Eq. (19)] has a value

$$X_{w_c} = (R_0 - R_\infty)/2. \tag{25}$$

From the foregoing it follows that, provided the impedance or admittance spectrum is known, one can calculate the components of an equivalent circuit of resistances and capacitances generating such a spectrum (see Fig. 3e). So the general procedure is to measure first the impedance of the biological system and then try to find the appropriate equivalent circuit. This involves comparing the experimental data with a theoretical model. Since this type of work requires a knowledge of the cable theory, particularly where the experimental data are derived by the use of microelectrodes, an outline of the cable theory is presented before consideration of some models among several theoretical models used to fit the experimental data.

II. Elements of the Cable Theory

Several authors have considered the cable theory, and the papers by Hodgkin and Rushton (1946), Davis and Lorente de Nó (1947), Taylor (1963), and Rall (1977) and the text by Cole (1972) should be consulted since only an outline is provided here.

The cable-theory treatment of the distribution of current and potential in nerve, muscle, and other related preparations of the cylindrical type is based on the core-conductor model. A simple core conductor may be considered as a thin tube of membrane that is filled with electrically conducting fluid (e.g., axoplasm) and immersed in another electrically conducting medium (e.g., extracellular fluid). This membrane tube is such that its length is very much greater than its diameter, and the membrane resistance to current flow

II. Elements of the Cable Theory

TABLE I

Symbols Used in Cable Theory

Symbol	Meaning
V_e	Electric potential on the outside (extracellular) of the membrane (V)
V_i	Electric potential on the inside (intracellular) of the membrane (V)
V_m	Membrane potential difference ($= V_i - V_e$)
E_r	Resting membrane emf
$V = V_m - E_r$	Electrotonic potential (this is the extent of deviation from the resting membrane potential).
V_0	Value of V at $x = 0$ and $t = 0$
R_i	Specific resistivity of the intracellular medium (Ω cm)
R_m	Resistance of unit area of membrane (Ω cm^2)
C_m	Capacitance per unit area of membrane (F cm^{-2})
a	Radius of cylinder (cm)
r_i	Core resistance per unit length (Ω cm^{-1}): $r_i = R_i/\pi a^2$
r_e	Resistance per unit length of extracellular fluid (Ω cm^{-1}).
r_m	Resistance across a unit length of passive membrane (Ω cm); $r_m = R_m/2\pi a$
c_m	Capacitance of membrane per unit length of cylinder (F cm^{-1}): $c_m = C_m 2\pi a$
λ	Length constant of core conductor (cm); $\lambda \, (=\sqrt{r_m/(r_e + r_i)})$.
$\tau_m = r_m c_m = R_m C_m$	Passive membrane time constant (s)
I_i	Current flowing through the intracellular core (A)
I_e	Current, extracellular, parallel to cylinder axis (A)
i_m	Membrane current per unit length (A cm^{-1})
$I_m = i_m/2\pi a$	Membrane current density (A cm^{-2})

is large compared to that of a short length (i.e., small compared to the length constant λ; see p. 338 and Table I) of core fluid. Consequently, the current will flow inside the core conductor parallel to its axis for considerable length before the current leaks out across the membrane. The several symbols and definitions are given in Table I (see Rall, 1977).

The relation of core current to potential according to the core-conductor model (Fig. 6a) is given by

$$I_i r_i \Delta x = V_{i(1)} - V_{i(2)} = -\Delta V_i$$

or

$$I_i r_i = -\frac{\partial V_i}{\partial x}. \quad (26)$$

The relation of core current to membrane current (Fig. 6b) is given by

$$i_m \Delta x = I_{i(1)} - I_{i(2)} = -\Delta I_i$$

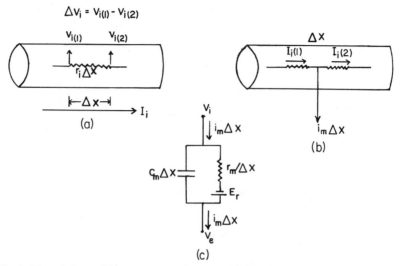

Fig. 6. The relations of (a) core current I_i to potential V; (b) core current I_i to membrane current i_m; (c) membrane current i_m divided into two parallel components, one capacitative and one resistive. (After Rall, 1977.)

or

$$i_m = -\frac{\partial I_i}{\partial x}. \tag{27}$$

Here I_i and i_m are functions of both x and t. For a uniform conductor, r_i is constant independent of x. Differentiation of Eq. (26) gives

$$r_i \frac{\partial I_i}{\partial x} = -\frac{\partial^2 V_i}{\partial x^2}. \tag{28}$$

This substituted into Eq. (27) yields

$$i_m r_i = \frac{\partial^2 V_i}{\partial x^2}. \tag{29}$$

Equation (29) does not depend either on the membrane model or on the distribution of potential in the extracellular phase. It depends only on the property of the core with r_i and without injection of any current into the core.

From the definitions of potentials given above, it follows that

$$V = V_i - V_e - E_r, \tag{30}$$

where E_r, the resting membrane emf, is independent of x, and so $\partial E_r/\partial x = 0$. Furthermore, the extracellular potential V_e is equipotential over a large extracellular volume compared to that of the intracellular space, and so

$\partial V_e/\partial x = 0$. This means that differentiation of Eq. (30) twice gives

$$\frac{\partial^2 V}{\partial x^2} = \frac{\partial^2 V_i}{\partial x^2}.$$

Using this relation in Eq. (29) gives

$$i_m = \frac{1}{r_i}\frac{\partial^2 V}{\partial x^2}. \tag{31}$$

Again no membrane model has been mentioned. This equation becomes complicated if extracellular equipotentiality breaks down, as, for example, when current is injected into the region of interest.

In Fig. 6c is given the equivalent circuit to represent the passive membrane properties. The total outward membrane current $i_m \Delta x$ is the sum of resistive and capacitative currents and can be expressed as

$$i_m \Delta x = (V_m - E_r)\frac{\Delta x}{r_m} + C_m \Delta x \frac{\partial V_m}{\partial t}. \tag{32}$$

Since $V = V_m - E_r$ and $\partial V/\partial t = \partial V_m/\partial t$, Eq. (32) becomes

$$i_m = \frac{V}{r_m} + C_m \frac{\partial V}{\partial t}. \tag{33}$$

For the simplest case, equating Eqs. (33) and (31) gives

$$V + r_m C_m \frac{\partial V}{\partial t} = \frac{r_m}{r_i}\frac{\partial^2 V}{\partial x^2}. \tag{34}$$

Substituting for $r_m C_m (= \tau_m)$ and $r_m/r_i (= \lambda^2)$, Eq. (34) becomes

$$\lambda^2 \frac{\partial^2 V}{\partial x^2} - V - \tau_m \frac{\partial V}{\partial t} = 0. \tag{35}$$

The physical meaning of the passive membrane time constant τ_m and the space constant λ can be assessed in the following manner. When a patch of the membrane is held at uniform potential and electrically isolated so that the membrane current is held at zero, Eq. (33) becomes

$$\frac{dV}{dt} = -\frac{V}{r_m C_m} = -\frac{V}{\tau_m}. \tag{36}$$

Equation (36) is a first-order kinetic equation, and its solution is

$$V = V_0 \exp(-t/\tau_m), \tag{37}$$

where V_0 is the initial value of V when $t = 0$. This means that the value of V will have fallen to $1/e$ of V_0 (i.e., $0.37 V_0$) in time $t = \tau_m$. If the potential is not space clamped, i.e., $V(x, t)$ is not uniform, then its decay is more complex than Eq. (37). The equations for such passive decay give the solutions for the cable equation for particular initial-boundary conditions. The terms on the right-hand side of Eq. (34) are independent of membrane capacity but are dependent on membrane resistance with the extracellular medium being equipotential.

The significance of λ as a length constant can be brought out by considering the steady-state distribution of V along a semi-infinite cable going from $x = 0$ to $x = \infty$. In the steady state, V depends on x but not on t, and so $\partial V/\partial t = 0$. The cable equation (35) reduces to

$$\lambda^2 \frac{\partial^2 V}{\partial x^2} - V = 0. \tag{38}$$

For the boundary conditions that $V = V_0$ at $x = 0$ and that no other applied voltage exists anywhere else, the solution of Eq. (38) is

$$V = V_0 \exp(-x/\lambda). \tag{39}$$

The voltage decreases exponentially with distance, resembling the voltage decrease with time. The length constant λ plays a role similar to that of the time constant. When the boundary conditions are different from those stated above, the steady-state decay of V with distance can be very different.

In deriving Eq. (31) it was assumed that the extracellular potential was equipotential over a large volume, and so $\partial V_e/\partial x = 0$. If it is otherwise, then

$$\frac{\partial V}{\partial x} = \frac{\partial V_i}{\partial x} - \frac{\partial V_e}{\partial x}. \tag{40}$$

Substituting Eq. (26) and the relation $I_e r_e = -\partial V_e/\partial x$ in Eq. (40) gives

$$\frac{\partial V}{\partial x} = -I_i r_i + I_e r_e. \tag{41}$$

Both r_i and r_e are independent of x. Differentiation of Eq. (41) gives

$$\frac{\partial^2 V}{\partial x^2} = -r_i \frac{\partial I_i}{\partial x} + r_e \frac{\partial I_e}{\partial x}. \tag{42}$$

When no current is applied through electrodes, conservation of current gives

$$i_m = -\frac{\partial I_i}{\partial x} = \frac{\partial I_e}{\partial x}. \tag{43}$$

Substituting Eq. (43) into Eq. (42) gives

$$\frac{\partial^2 V}{\partial x^2} = i_m(r_i + r_e). \tag{44}$$

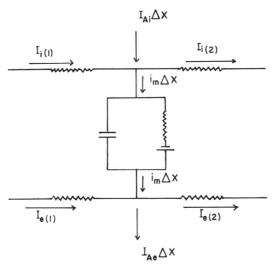

Fig. 7. Circuit diagram indicating the injection of current. This may be used to consider the application of Kirchhoff's law for conservation of current. (After Rall, 1977.)

Combining Eqs. (44) and (33) gives

$$V + r_m C_m \frac{\partial V}{\partial t} = \left(\frac{r_m}{r_i + r_e}\right) \frac{\partial^2 V}{\partial x^2}. \tag{45}$$

This equation is more general than Eq. (35), in which $\lambda^2 = r_m/(r_i + r_e)$ is restricted to r_m/r_i.

If the condition that no applied current is injected considered above is relaxed, then the circuit diagram of Fig. 7 in which injection of current through a microelectrode is indicated becomes applicable. An electrode (microelectrode or axial wire electrode) introduced into the cylindrical cell facilitates injection of current I_{Ai} per unit length of the core. $I_{Ai} \Delta x$ is the amount of current introduced. Applying Kirchhoff's law (see Chapter 2, Section XIV) to the node in the upper half of Fig. 7 gives

$$I_{i(1)} - I_{i(2)} + I_{Ai} \Delta x - i_m \Delta x = 0.$$

But $I_{i(1)} - I_{i(2)} = -\Delta I_i$, and so $[I_{Ai} - i_m] \Delta x = \Delta I_i$. This equation, in the limit $\Delta x \to 0$, may be written

$$\frac{\partial I_i}{\partial x} = I_{Ai} - i_m. \tag{46}$$

The current going out of Δx of external layer through the externally applied electrode is $I_{Ae} \Delta x$. Again applying Kirchhoff's law to the node in the lower half of Fig. 7 gives

$$I_{e(1)} - I_{e(2)} + i_m \Delta x - I_{Ae} \Delta x = 0.$$

This means

$$\frac{\partial I_e}{\partial x} = i_m - I_{Ae}. \tag{47}$$

Substituting Eqs. (46) and (47) into Eq. (42) gives

$$\frac{\partial^2 V}{\partial x^2} = i_m(r_i + r_e) - r_i I_{Ai} - r_e I_{Ae}. \tag{48}$$

When Eq. (48) is combined with Eq. (33) and substitutions of $\tau_m = r_m C_m$ and $\lambda^2 = r_m/(r_i + r_e)$ are made, the result is

$$\lambda^2 \frac{\partial^2 V}{\partial x^2} - V - \tau_m \frac{\partial V}{\partial t} = -\lambda^2 (r_i I_{Ai} + r_e I_{Ae}). \tag{49}$$

This is an augmented cable equation, which, when the function on the right-hand side becomes zero, yields Eq. (35). Similarly, if $r_i I_{Ai} = -r_e I_{Ae}$, again Eq. (35) results. But when $I_{Ai} = I_{Ae} = I_{Aie}$, Eq. (49) becomes

$$\lambda^2 \frac{\partial^2 V}{\partial x^2} - V - \tau_m \frac{\partial V}{\partial T} = -r_m I_{Aie}. \tag{50}$$

Equation (49) becomes important in studies related to applied fields, synaptic excitation, and active membrane properties.

For the dc steady state, the potential V depends on x but not on t, and so $\partial V/\partial t = 0$. The cable equation (35) or (45) becomes

$$\lambda^2 \frac{d^2 V}{dx^2} - V = 0. \tag{51}$$

Many useful results pertaining to the steady state follow from the solutions of Eq. (51) derived for several boundary conditions. The general solution to Eq. (51) is expressed in different but equivalent forms. The one familiar form is

$$V = A_1 \exp(x/\lambda) + A_2 \exp(-x/\lambda), \tag{52}$$

where A_1 and A_2 are arbitrary constants evaluated for the particular boundary conditions. These are (see Fig. 8) $x = 0$ to $x = +\infty$ for the semi-infinite case and $x = -\infty$ to $x = +\infty$ for the doubly infinite case. In the doubly infinite case, the steady subthreshold current to be applied so that $V = V_0$ at $x = 0$ is twice that necessary in the semi-infinite case, i.e., $G_{\pm\infty} = 2G_\infty$. In other words, the input resistance in the semi-infinite case is twice that of the doubly infinite case (i.e., $R_\infty = 2R_{\pm\infty}$).

II. Elements of the Cable Theory

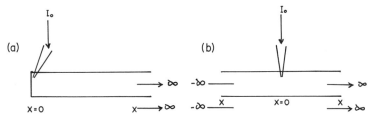

Fig. 8. Intracellular electrode applied current I_0, which is subthreshold at $X = 0$. (a) Steady state for semi-infinite length sealed at $X = 0$ and extending to $X = +\infty$. (b) Steady state for infinite length extending from $-\infty$ to $+\infty$. Extracellular medium is assumed to be equipotential.

The boundary conditions to evaluate the constants of Eq. (52) are $V = V_0$ at $x = 0$, and V remains bounded as $x \to \infty$. A_1 must be zero; otherwise V becomes unbounded and tends to reach ∞. $A_2 = V_0$ at $x = 0$ since $\exp(-x/\lambda) = 1$. Thus the solution of Eq. (51) for the specified boundary conditions is

$$V = V_0 \exp(-x/\lambda) \quad \text{for} \quad x \geq 0. \tag{53}$$

If a steady current I_0 is injected into the core through a microelectrode at $x = 0$ (see Fig. 8a) no current leaks to the left. This means, according to Eq. (26), that

$$I_0 = -\frac{1}{r_i}\left(\frac{dV_i}{dx}\right)_{x=0}. \tag{54}$$

Assuming the extracellular phase to be equipotential, $dV_i/dx = dV/dx$, and so using Eq. (53) gives $-dV/dx = dV_i/dx = -V_0/\lambda$. Substituting this value of dV_i/dx into Eq. (54) gives

$$I_0 = V_0/(\lambda r_i). \tag{55}$$

The input resistance by definition is V_0/I_0, and so

$$R_\infty = \lambda r_i = \sqrt{r_m r_i} = r_m/\lambda. \tag{56}$$

But for a cable where the extracellular volume is restricted, as, for example, placing a nonmyelinated axon in oil, the input resistance for a pair of electrodes, one inside and the other outside at $x = 0$, would be

$$R_\infty = \lambda(r_i + r_e) = \sqrt{r_m(r_i + r_e)} = r_m/\lambda. \tag{57}$$

For a doubly infinite case

$$R_{\pm\infty} = R_\infty/2 = \tfrac{1}{2}\sqrt{r_m(r_i + r_e)}. \tag{58}$$

When both the electrodes at $x = 0$ are placed outside, as used in the work of Hodgkin and Rushton (1946) and that of Katz (1948), the electrotonic

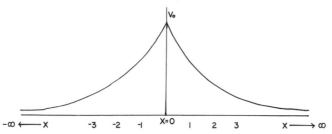

Fig. 9. Symmetric distribution of potential with distance X ($X = -\infty$ to $X = +\infty$) with injection of current I_0 at $X = 0$.

resistance $Y = V_A/I_0$, where V_A is the electrotonic potential at the polarizing electrode, is given by

$$Y = m\lambda r_e/2r_i,$$

$$m = r_i r_e/(r_i + r_e), \qquad (59)$$

and

$$m\lambda/2Y = r_i/r_e.$$

In view of these relations it follows that

$$\frac{R_\infty}{2} \Big/ Y = \left(\frac{r_e + r_i}{r_e}\right)^2. \qquad (60)$$

The solution of Eq. (51) for the doubly infinite case consists of two solutions, one for each region, requiring four boundary conditions. These are (1) $x = +\infty$, (2) $x = -\infty$, (3) $V = V_0$ at $x = 0$, and (4) continuity of V at $x = 0$, where the two regions join. The potential distribution profile is shown in Fig. 9.

Equation (53) applies for the region $x = 0$ to $x = +\infty$. But for the domain $x = 0$ to $x = -\infty$, A_2 should be zero; otherwise that term tends to ∞. At the origin for $x = 0$, $A_1 = V_0$, and so

$$V = V_0 \exp(x/\lambda) \qquad \text{for} \quad x \le 0. \qquad (61)$$

The solution to the cable equation (35) or (45) for the case of the transient response is complicated. However, Hodgkin and Rushton (1946) and Davis and Lorente de Nó (1947) have provided the details of the general solution. For the semi-infinite region ($0 \le X \le \infty$), the solution can be expressed as (see Rall, 1977)

$$\frac{V(X, T)}{V(0, \infty)} = \frac{1}{2}\left[\exp(-X)\,\text{erfc}\left(\frac{X}{2\sqrt{T}} - \sqrt{T}\right) - \exp(X)\,\text{erfc}\left(\frac{X}{2\sqrt{T}} + \sqrt{T}\right)\right], \qquad (62)$$

where $V(0, \infty) = I_0 R_\infty$ for semi-infinite length having a closed end at $X = 0$ (X and T are dimensionless variables, $X = x/\lambda$ and $T = t/\tau_m$) or $V(0, \infty) = I_0 R_{\pm\infty} = \frac{1}{2} I_0 R_\infty$ for a doubly infinite length extending from $-\infty$ to $+\infty$ and I_0 is the current applied at $x = 0$. The complementary error function is defined as $\text{erfc}(x) = 1 - \text{erf}(x)$, where the error function is defined as

$$\text{erf}(x) = \frac{2}{\sqrt{\pi}} \int_0^x \exp(-Y^2) dY$$

and is available as a tabulated function. Hodgkin and Rushton (1946) give a table for the function within the brackets of Eq. (62).

Equation (62) simplifies for either $T = \infty$ or $X = 0$ [note that $\text{erf}(-x) = -\text{erf}(x)$, $\text{erf}(0) = 0$, and $\text{erf}(\infty) = 1$]. When $T = \infty$ [i.e., steady state and $\text{erfc}(-\infty) = 2$, $\text{erfc}(\infty) = 0$], the right-hand side of Eq. (62) becomes $\exp(-X)$ [same as Eq. (53)]. When $X = 0$, Eq. (62) becomes

$$V(X, T) = I_0 R_\infty \text{erf} \sqrt{T}$$

or

$$V = I_0 R_\infty \text{erf} \sqrt{t/\tau_m}. \tag{63}$$

The effect of distance X on the rate of rise of V (i.e., dV/dt; see Fig. 10) can be shown by differentiating Eq. (62). Hence

$$\frac{dV/dT(X, T)}{V(0, \infty)} = (\pi T)^{-1/2} \exp\left[-\frac{X^2}{4T} - T\right]. \tag{64}$$

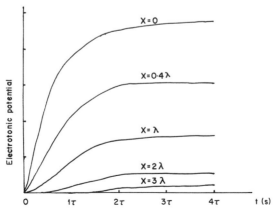

Fig. 10. Electrotonic potential plotted against time for different values of X (fraction or multiples of space constant λ). (After Rall, 1977.)

At a distance $X = 0$, the slope is given by

$$\frac{dV/dT(0, T)}{V(0, \infty)} = (\pi T)^{-1/2} \exp(-T).$$

Thus the slope dV/dt at $X = 0$ is greater than the slope at X by

$$\frac{dV/dT(0, T)}{dV/dt(X, T)} = \exp\left(\frac{X^2}{4T}\right). \tag{65}$$

On substituting for X and T and taking logarithms, Eq. (65) becomes

$$x^2 = 4\left(\frac{\lambda^2}{\tau_m}\right) t \ln\left[\frac{dV/dt \text{ at } (0, t)}{dV/dt \text{ at } (x, t)}\right]. \tag{66}$$

Thus a plot of x^2 versus $4t \ln[(dV/dt) \text{ ratio}]$ should give a straight line with a slope equal to (λ^2/τ_m). Thus if λ is shown, τ_m can be determined, or vice versa. In addition, other properties of Eq. (62) can be exploited to evaluate λ or τ_m. For example, transient response of V to I_0 at any point x is such that an inflection point exists. For that point, double differentiation of Eq. (62) or single differentiation of Eq. (64) with respect to time is zero. This operation gives the relation

$$x^2 = \lambda^2(4T^2 + 2T). \tag{67}$$

A plot of paired values of x^2 versus $(4T^2 + 2T)$ should give a straight line having a slope of λ^2.

Another property found by Hodgkin and Rushton (1946) was that the theoretical curves of Eq. (62) showed a linear relation between each value of X and its corresponding value of T for which the transient V at X reached half its steady-state value. This means that paired values of X and T give a slope of 2. That is, $\Delta x/\Delta T \approx 2$, or a plot of x and t (time to reach half maximum) would give

$$\Delta x/\Delta t = 2\lambda/\tau_m. \tag{68}$$

Application of some of these equations and others (given in Section III) are outlined in Section III, which deals with models that help in relating input impedance to the electrical constants of the cell.

III. Models to Relate Input Impedance to Electrical Cell Constants

In the measurement of biological impedances, because of the use of electrodes to make electrical contact with the biological preparation, several factors interfere with the impedance of interest. This point is illustrated for

III. Models to Relate Input Impedance to Electrical Cell Constants

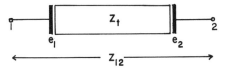

Fig. 11. Bioimpedance measurement. The two electrodes e_1 and e_2 make electrical contact with the tissue via appropriate saline. Z_t is the tissue impedance.

a simple case in Fig. 11. Terminals 1 and 2 are connected to two electrodes of equal surfaces. Between these two electrodes a piece of the tissue with tissue impedance Z_t is held. The contact between tissue and electrodes occurs through saline appropriate to the biological tissue. The impedance measured between the electrodes is Z_{12} and is higher than Z_t because of the contact impedances Z_{1t} and Z_{2t}. Thus the impedance of interest Z_t is given by

$$Z_t = Z_{12} - Z_{1t} - Z_{2t}. \tag{69}$$

Methods exist to separate Z_{1t} and Z_{2t} experimentally from the tissue impedance (see Schanne and Ruiz P.-Ceretti, 1978). But as Z_t is dependent on the geometry of the cell and the points of contact of the electrodes, interpretation of impedance measurements is difficult without mathematical models. At this point it becomes important to generalize Eq. (69) so that it becomes applicable to impedance measurements with the help of microelectrodes. So Eq. (69) is written

$$Z_{\text{app}} = Z_{\text{inp}} + Z_{\text{el}}^0 + Z_{\text{int}}, \tag{70}$$

where Z_{app} is called the apparent impedance (Z_{12}), Z_{inp} is the input impedance (Z_t), Z_{el}^0 is microelectrode impedance under specified conditions, and Z_{int} is the interaction impedance arising from all electrode interactions with its surroundings.

Schanne and Ruiz P.-Ceretti (1978) have listed several models used in the case of a variety of biological preparations, and some of the simpler ones are considered here.

The simplest model or the equivalent circuit pertains to that of a lipid bilayer membrane (see Fig. 12), where the surface of the membrane is exposed

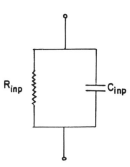

Fig. 12. Simplest equivalent circuit to represent a lipid bilayer membrane. (After Schanne and Ruiz P.-Ceretti, 1978.)

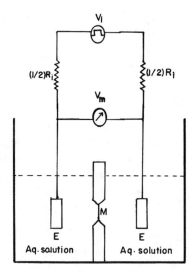

Fig. 13. Experimental arrangement for the measurement of bilayer membrane impedance. M is the bilayer membrane, Es, are two calomel electrodes, V_i is the known stimulating voltage, V_m is the recorded membrane potential, R_i is the series input resistance (10^9 Ω).

to current flow. The relation between input resistance and membrane resistance is simply governed by the area A of the membrane. Thus

$$R_m = \Delta V/I = AR_{inp}, \tag{71}$$

where ΔV is the voltage drop across the membrane divided by the current density. R_m is thus given as ohm square centimeters.

In the time domain (dc measurements) R_m and C_m of bilayer membrane are determined by using the simple setup shown in Fig. 13. When a known voltage V_i is applied to the membrane system of Fig. 13, V_m, the voltage developed across the membrane, is recorded with a high input impedance electrometer and is given by

$$V_m = i_m R_m, \tag{72}$$

where i_m and R_m are membrane current and resistance, respectively. But

$$i_m = V_i/(R_i + R_m). \tag{73}$$

Thus eliminating i_m between Eqs. (72) and (73) gives on rearrangement

$$R_m = V_m R_i/(V_i - V_m). \tag{74}$$

As the membrane area can be determined, R_m can be expressed in ohm square centimeters.

To measure the capacitance, a potential is applied to charge the membrane, and the applied potential is immediately removed. The capacitance

III. Models to Relate Input Impedance to Electrical Cell Constants

discharges and the potential decays with time according to

$$V_m = V_0 \exp(-t/RC_m),$$

where V_0 is the voltage at time $t = 0$ (i.e., when the external current is interrupted) and R the resistance through which the capacity discharges. If time $t = \tau_m$ is measured when $V_m/V_0 = 1/e (= 0.36)$, membrane capacitance is given by $C_m = [\tau_m/R]$. In the circuit of Fig. 13, capacitance discharges through R_m and R_i, which are parallel, and so $R = R_m R_i/[R_m + R_i]$. As the membrane area is known, C_m can be expressed as farads per square centimeter.

In the frequency domain (ac measurements), R_m and C_m can be evaluated by using impedance or admittance bridges for measurements. For the equivalent circuit of Fig. 12 (see also Fig. 3d), the equivalent series resistance and capacitative reactance are given by Eqs. (7) and (8). Bridge-balance measurements give values for R_s and $-X_s$. But $-X_s$ follows from Eq. (8), and so

$$-X_s = \frac{1}{\omega C_s} = \frac{\omega R_{inp}^2 C_{inp}}{1 + \omega^2 R_{inp}^2 C_{inp}^2}. \tag{75}$$

Dividing Eq. (7) by Eq. (75) gives

$$R_s/X_s = 1/\omega R_{inp} C_{inp}. \tag{76}$$

Substituting Eq. (76) into Eq. (7) gives the value for R_{inp} (R_p replaced by R_{inp}) thus:

$$R_{inp} = R_s [1 + (X_s/R_s)^2].$$

Using this value of R_{inp} in Eq. (76), C_{inp} can be calculated at the frequency used in the measurements (usually 10^3 Hz). As the area of the membrane can be determined, C_m and R_m can be calculated. The equivalent circuit 12 is probably the simplest. Other complex circuits can be drawn and analyzed in the way Hanai et al. (1964) have done.

The equivalent circuit shown in Fig. 3e is a general model that can be applied to a variety of biological tissues. The electrical property of such a circuit has been considered already, and its input impedance is given by

$$Z_{inp} = \left[\frac{R_0^2 + \omega^2 \tau^2 R_\infty^2}{1 + \omega^2 \tau_2} \right]^{1/2}.$$

Without additional information about the structure and geometry of the cell, this model will not give the values for the electrical constants of the cell. An example of the application of the model is the work of Tarr and Trank (1971), who used it to interpret the results of voltage clamp measurements on frog atrium. R_2 and C pertain to the cell membrane, whereas R_1 is assigned to a resistance in series with the parallel resistance–capacitance circuit

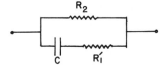

Fig. 14. Probable equivalent circuit for the frog atrium.

of the membrane. They found that a 20-mV depolarization step gave a capacitive current that rose rapidly to a maximum value I_0 and then decayed exponentially with a time constant of 1 ms to a steady-state value of I_s. The capacitative current continued to flow even when the clamp potential was steady. Two equivalent circuits to explain this current were proposed. One was Fig. 3e, and the other is shown in Fig. 14. But the results were found to conform to the circuit of Fig. 3e. From the equation

$$R_1 = \Delta V/I_0$$

a value of $R_1 = 7.7$ kΩ was determined; from

$$R_2 = \frac{\Delta V}{I_s} - R_1 = R_1 \left[\frac{I_0}{I_s} - 1\right],$$

$R_2 = 87.6$ kΩ; and from

$$C = \tau \bigg/ \left[\frac{R_1 R_2}{R_1 + R_2}\right] = \tau \bigg/ \left[R_1 \left(1 - \frac{I_s}{I_0}\right)\right],$$

$C = 0.17$ μF. The time constant τ was evaluated by measuring the current at a given time t and using the equation

$$\tau = -t/\ln[(I_t - I_0)/(I_0 - I_s)], \qquad \tau = 0.96 \quad \text{ms}.$$

The structure and the geometry of the fibers were as follows. The average diameter of a bundle of heart muscle fibers was 200 μm. This bundle was held in a gap between two streams of sucrose solution (double sucrose gap), the gap length was 100 μm. Each single fiber (i.e., cell) had a diameter of 4 μm. Thus the number of cells involved is $(100 \ \mu\text{m})^2/(2 \ \mu\text{m})^2 = 2500$. The surface area of 2500 cells is 3×10^{-2} cm^2. Thus the specific membrane resistance is 2600 Ω cm^2, and the specific membrane capacitance is 5.7 μF cm^{-2}.

The basic model that is used for the application of the theory of a linear cable to biological cells is the core-conductor model shown in Fig. 15. The symbols and the description have already been given.

Hodgkin and Rushton (1946) used single nerve fibers, and by measuring the response potential V to a subthreshold stimulus as a function of distance, evaluated λ with the help of Eq. (53). The electrotonic resistance Y was evaluated by measuring the voltage at the cathode for the passage of a known quantity of current. The parallel resistance of the core and external fluid, i.e., m, was determined by measuring the current and recording the voltage gradient at various points. τ_m can be evaluated by several methods, as

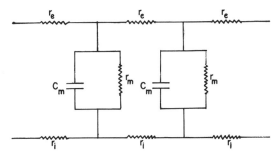

Fig. 15. Core conductor model: r_e is the resistance of the extracellular fluid; r_i is the resistance of the intracellular fluid; C_m and R_m are membrane capacitance and resistance respectively. (After Schanne and Ruiz P.-Ceretti, 1978.)

discussed already. Appropriate measurements relevant to the parameters of Eq. (66) or (68) yield a value for τ_m. With these four determinations, values for the cell constants can be computed from the following relations:

axoplasmic resistance (Ω cm): $R_i = \pi a^2 m(1 + m\lambda/2Y)$,

membrane resistance (Ω cm^2): $R_m = 2\pi a \lambda^2 m [2 + m\lambda/2Y + 2Y/m\lambda]$,

membrane capacitance (F cm^{-2}): $C_m = \tau_m/R_m$.

Katz (1948) used the procedure given above to determine the cell constants of the muscle cell. He used a bundle of muscle fibers. In this case a correction for the extracellular space is required.

When an intracellular microelectrode and a large outside reference electrode are used to follow changes in electrotonic potential following injection of current, the same procedure as above may be followed. In this case, since the outside medium is equipotential, r_e is eliminated from the cable equations.

Intracellular application of rectangular current (square-wave analysis) through a microelectrode and intracellular recording of electrotonic potential at a distance x from the site of current injection can be carried out on nerve or muscle fiber. Recall Eq. (53):

$$V = V_0 \exp(-x/\lambda). \tag{53}$$

The electrotonic potential V_0 at the site of current application and in a steady state (i.e., $x = 0$, $t = \infty$) is given by Eq. (55) for a semi-infinite case. For the doubly infinite case, it is given by

$$V_0 = \tfrac{1}{2} r_i \lambda I_0. \tag{77}$$

Substituting Eq. (77) into Eq. (53) gives

$$V/I_0 = \tfrac{1}{2}\sqrt{r_m r_i} \exp\left[-\frac{x}{\sqrt{r_m/r_i}}\right]. \tag{78}$$

Taking logarithms gives

$$\ln(V/I_0) = -x/\lambda + \ln(\tfrac{1}{2}\sqrt{r_m r_i}).$$

Measurement of V at several distances from the site of application of a known current I_0 and plotting $\ln(V/I_0)$ against x gives a straight line whose slope yields value for λ ($= -0.4343/\text{slope}$); the intercept gives the value for $\tfrac{1}{2}\sqrt{r_m r_i}$. From these values, r_m and r_i can be computed. τ_m can be evaluated as described above by measuring the time to reach half maximum at different distances.

The model that is used to interpret impedance measurements of cell suspensions is shown in Fig. 16. The magnitude of its input impedance is again given by combining Eqs. (18)–(20), i.e.,

$$Z_{\text{inp}} = \left[\frac{R_0^2 + \omega^2 \tau^2 R_\infty^2}{1 + \omega^2 \tau^2}\right]^{1/2}, \tag{79}$$

where R_0, R_∞, and τ are defined by

$$\begin{aligned} R_0 &= R_1, \\ R_\infty &= R_1 R_2/(R_1 + R_2), \\ \tau &= C(R_1 + R_2). \end{aligned} \tag{80}$$

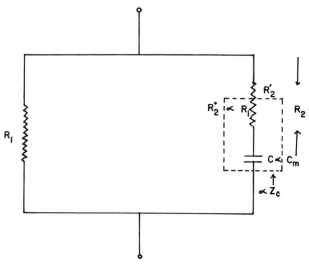

Fig. 16. An equivalent circuit for cell suspensions. To make this consistent with the circuit of Fig. 3e, the symbols R_1 and R_2 are given, Z_c is the impedance of the cell (Ω cm), R_i is the resistivity of the intracellular phase (Ω cm). R_2' is the part of R_2 that is extracellular, and R_2'' is the part of R_2 that is intracellular. (After Schanne and Ruiz P-Ceretti, 1978.)

III. Models to Relate Input Impedance to Electrical Cell Constants

The impedance locus is the same as that shown in Fig. 5. In the interpretation of an impedance measurement, R_1 and part of R_2 are assigned to the extracellular phase. The cell impedance Z_c is proportional to the elements in the box indicated by broken lines. The several relations (see Cole, 1972) between the cell components in the terminology of Schanne and Ruiz P.-Ceretti (1978) are

$$Z_{\text{inp}}^* = \frac{R_e[(\gamma + \rho')Z_c^* + (1 - \rho')R_e]}{\gamma(1 - \rho')Z_c^* + (1 + \gamma\rho')R_e} \tag{81}$$

(for cylindrical cells parallel to electric field $\gamma = 1$) and

$$Z_c^* = R_i - j/a\omega C_m \quad \text{for} \quad R_m \to \infty, \tag{82}$$

$$R_0^* = R_e[(1 + \rho')/(1 - \rho')], \tag{83}$$

$$R_\infty^* = \frac{R_e[(1 + \rho')R_i + (1 - \rho')R_e]}{(1 - \rho')R_i + (1 + \rho')R_e}, \tag{84}$$

$$R_1^* = R_e[(1 + \rho')/(1 - \rho')], \tag{85}$$

$$R_2^* = R_2' + R_2'' = R_e \frac{(1 - \rho')^2}{4\rho'} + R_i \frac{(1 + \rho')^2}{4\rho'}, \tag{86}$$

$$C^* = aC_m \frac{4\rho'}{(1 + \rho')^2}, \tag{87}$$

where Z_{inp}^* is the specific input impedance of suspension (Ω cm); Z_c^* the specific impedance of cell (Ω cm); R_e the resistivity of extracellular medium (Ω cm); γ the form factor of cells; ρ' the volume concentration of cells; R_0^* the resistivity of suspension for $\omega \to 0$; R_∞^* the resistivity of suspension for $\omega \to \infty$; R_1^*, R_2^*, and C^* the specific resistances and capacitance of symbols R_1, R_2, and C in Fig. 16; and a the cell radius (cm).

Equation (81) is applicable to suspensions. γ is a form factor that is 1 for cylindrical cells running parallel to the electric field and 2 for spherical cells. Z_c^* is the specific input impedance of a single cell and is represented by R_i in series with the specific impedance of the membrane capacity ($1/a\omega C_m$, a the cell radius). The shunt resistance of C_m (i.e., membrane resistance R_m) is assumed to be very high. Solving Eqs. (83) and (84) gives

$$R_i = R_0^* \left(\frac{1 - \rho'}{1 + \rho'}\right)^2 \left[\frac{R_0^*}{R_\infty^* - R_0^*} - \frac{R_\infty^*}{R_\infty^* - R_0^*}\left(\frac{1 + \rho'}{1 - \rho'}\right)^2\right]. \tag{88}$$

This equation is different from Eq. (1.20) given by Schanne and Ruiz P.-

Ceretti (1978). C_m is given by Eq. (87) as

$$C_m = \frac{(1 + \rho')^2}{4\rho'} \frac{C^*}{a}. \tag{89}$$

More elaborate circuits for Z_c can be used to represent the electrical properties of a single cell. Texts by Cole (1972) and Schanne and Ruiz P.-Ceretti (1978) must be consulted for details related to all aspects of impedance measurements in different biological preparations.

IV. Hodgkin–Huxley Equations

Some general principles relating to the passive electrical properties of biological membranes are recapitulated. The biological membrane can be considered in the electrical sense as equivalent to "leaky" capacitors separating two conducting phases containing mostly sodium ions in one phase and mostly potassium ions in the second phase. Existence of a steady voltage difference (E, in volts) across a capacitor (C, in farads) indicates charge (q, in coulombs) separation. The relation between these quantities is

$$q = CE. \tag{90}$$

Differentiation of Eq. (90) gives

$$\frac{dq}{dt} = C \frac{dE}{dt}.$$

Thus change in transmembrane voltage with time gives rise to a change in charge separation or displacement of charge with time. This is the capacitative current. In addition, the biological membranes are leaky to small charged particles (i.e., ions), giving rise to the ionic currents. Thus the membrane exhibits the property of conductance G, and so

$$GE = I. \tag{91}$$

When $I = 0$, there should be no voltage difference across the membrane. This will be so if the membrane is composed of only passive electrical components. It is common knowledge that a source of voltage (i.e., resting potential E_r equivalent to a battery) exists across biological membranes. So Eq. (91) can be modified to

$$G(E - E_r) = I. \tag{92}$$

Only when $E = E_r$ does I become zero. The value of voltage difference E across the membrane when $I = 0$ is called the equilibrium potential. These

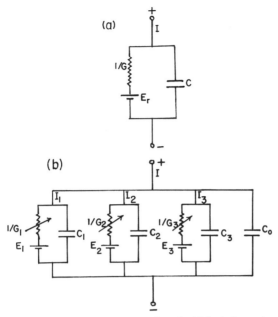

Fig. 17. Representation of the electrical properties of a biological membrane by equivalent circuits. (a) Simple circuit showing resting voltage E_r, conductance G, and capacitance C. (b) Complex circuit showing several conductive pathways. (After Horowicz et al., 1980.)

characteristics may be incorporated into a simple electrical circuit (Fig. 17a) or a complex one (Fig. 17b). If the conductances and capacitances shown in Fig. 17b do not vary with time and voltage, the circuit is electrically equivalent to that of Fig. 17a. The relations between the parameters of Fig. 17a and b become

$$C = C_0 + C_1 + C_2 + C_3,$$
$$G = G_1 + G_2 + G_3, \tag{93}$$
$$E_r = \frac{G_1 E_1 + G_2 E_2 + G_3 E_3}{G_1 + G_2 + G_3}.$$

The ideal nature of these equations breaks down in the case of biological membranes since the pathways or channels of current flow are both voltage and time dependent. Furthermore, the pathways are selective to ions, and they can be modified by changing the ionic compositions of solutions on both sides of the membrane in such a way that the current through one pathway can be maximized. In addition application of specific chemical compounds can eliminate current through certain pathways.

If a membrane-permeable ion is not in equilibrium across the membrane, then it will move down its electrochemical potential gradient. The rate of this particular ion (j) movement, i.e., I_j, is determined by the product of the driving force on the ion (i.e., $E - E_j$) and the conductance of the membrane to that ion (G_j). This may be written

$$I_j = G_j(E - E_j). \tag{94}$$

Although Eq. (94) is formally equivalent to Eq. (92), they convey different meanings. In Eq. (92) I is the total current due to movement of all ions (i.e., $I = \sum_j I_j$) when $E \neq E_r$. In Eq. (94) I_j is the current carried by the particular ion j when $E \neq E_j$. When $I = 0$, $E_r = E$ is the internal potential, whereas E_j is the internal potential when the current I_j is zero. This simply means that E_r is the resting membrane potential, whereas E_j is the equilibrium potential for that ion j.

The total current I for two ions, j and k, is given by

$$I = I_j + I_k. \tag{95}$$

Substituting Eq. (94) gives

$$I = G_j(E - E_j) + G_k(E - E_k). \tag{96}$$

Equation (92) gives $E = E_r$ when $I = 0$. Therefore substituting E_r for E in Eq. (96) and setting $I = 0$ give on rearrangement

$$E_r = \frac{G_j}{G_j + G_k} E_j + \frac{G_k}{G_j + G_k} E_k. \tag{97}$$

The fraction of the current carried by j ion is its transport number, and so $t_j = G_j/(G_j + G_k)$ and $t_j + t_k = 1$. Equation (97) therefore can be written

$$E_r = t_j E_j + t_k E_k. \tag{98}$$

If the ions j and k are identified with extracellular sodium ions and intracellular potassium ions, E_r will be close to the equilibrium potential E_k for the potassium ions because t_{Na} is relatively low in the resting state of the membrane. Similarly, when an action potential is generated, the membrane potential at the peak of the spike will be close to E_{Na} since t_K is relatively low. Values for some of these parameters pertaining to the squid giant axon are given in Table II. The relative permeabilities were as $P_K : P_{Na} : P_{Cl}$ equal to 1:0.04:0.45 in the resting state and equal to 1:20:0.45 in the active state (see Hodgkin and Katz, 1949). In keeping with this, the membrane resistance fell from its resting value of 1000 Ω cm^2 to a value of 25 Ω cm^2 corresponding to the active state. There was, however, no appreciable change in either the electrical capacitance of the membrane or the resistance of the axoplasm.

IV. Hodgkin–Huxley Equations

TABLE II

Membrane and Equilibrium Potentials

Tissue	Resting potential E_r (mV)	Action potential E_a (mV)	Equilibrium potentials		
			E_{Na} (mV)	E_K (mV)	E_{Cl} (mV)
Loligo axon	−61	+35	+49	−91	−54
Sepia axon	−62	+60	+52	−89	
Carcinus axon	−82	+52		−85	

The electrical events occurring during an action potential were examined by Hodgkin and Huxley (1952a–d) using the voltage-clamp technique (Cole, 1972).

The current through the axon membrane is assumed to consist of two components, a capacitative current due to changes in the charge density at the inner and outer surfaces of the membrane and an ionic current due to the flow of ions through the pathways in the membrane. Hence the total membrane current I is given by

$$I = C_m \frac{dE}{dt} + \sum_i I_i. \tag{99}$$

When the voltage across the membrane is held constant ("clamped") by electronic feedback $dE/dt = 0$, and so the current flow observed or recorded gives a measure of the total ionic flow. Usually at any given voltage step the current record shows three features. First, there is a brief "blip" of outward current (see Fig. 18) due to the discharge of the membrane capacity. After

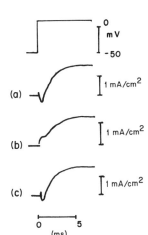

Fig. 18. Membrane currents observed during voltage clamp experiments. In (a) and (c) the axon was in seawater, while in (b) it was in sodium-free choline chloride solution. (After Hodgkin, 1958.)

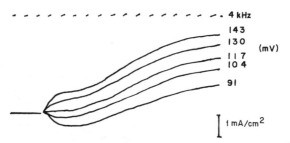

Fig. 19. Membrane currents recorded at large depolarizations which are shown at the right of each record. (After Hodgkin, 1958.)

this, the current is inward for about 1 ms, and then the current becomes outward and reaches a steady level at which it stays as long as the clamp lasts. According to the ionic hypothesis, the initial inward current is due to inward flow of sodium ions. If this is so, it should disappear if the axon is bathed in saline containing no sodium ions (choline chloride is used in place of sodium chloride). This was found to be so (see Fig. 18b), and only outward current is recorded. Now, pharmacologically, sodium current can be eliminated by using tetrodotoxin in the external saline. The direction of sodium ion flow is dependent on its electrochemical potential gradient and the membrane potential. When the membrane potential E is equal to the sodium equilibrium potential E_{Na}, net flow of sodium ions will not occur, and so if the membrane potential is held at E_{Na} (this brought about by a depolarizing pulse of magnitude V_{Na} where $V_{Na} = E_{Na} - E_r$), inward current will cease. If the magnitude of depolarizing pulse is less than V_{Na}, sodium current will be inward, and if it is greater than V_{Na}, the current will be outward. This is shown in Fig. 19, from which it is seen that V_{Na} corresponds to approximately 117 mV. The outward current remains unaffected by changes in the external sodium concentration, and so that component of membrane current is due to some other ion, possibly potassium. Direct evidence for this was obtained by the study of efflux of ^{42}K, which varied linearly with the current density giving a slope equal to Faraday's constant.

From the foregoing it follows that the ionic current following depolarization has two components, one due to transient inward flow of sodium ions and the other due to late outward flow of potassium ions. The maximum values of the early transient current and the steady-state current are shown in Fig. 20 as functions of voltage. The current–voltage curve for the transient peak currents shows a negative electrical resistance in the voltage region in which the threshold for the generation of action potential lies.

The kinetics of the conductance change following changes in membrane potential were described by Hodgkin and Huxley (1952d) by using a set of

IV. Hodgkin–Huxley Equations

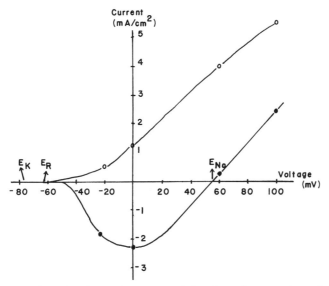

Fig. 20. Peak values of early transient current (●) and steady-state values of the delayed current (○) plotted as a function of membrane potential. (After Ehrenstein and Lecar, 1972.)

empirical equations. The total current I across the membrane of unit area is given by Eq. (99), and $\sum_i I_i = I_K + I_{Na} + I_L$, where I_L is an ohmic current usually referred to as leakage current. The ionic permeabilities of the membrane are expressed in terms of their conductances [see Eq. (94)]. Hence

$$I_K = G_K(E - E_K) = G_K(V - V_K), \tag{100}$$

$$I_{Na} = G_{Na}(E - E_{Na}) = G_{Na}(V - V_{Na}), \tag{101}$$

$$I_L = G_L(E - E_L) = G_L(V - V_L), \tag{102}$$

where $V = E - E_r$, $V_K = E_K - E_r$, $V_{Na} = E_{Na} - E_r$, and $V_L = E_L - E_r$; Vs are measured directly as displacements from E_r. E_L is the potential at which the leakage current due to chloride and other ions is zero. The electrical behavior of the membrane may be represented by the equivalent circuit shown in Fig. 21. The currents I_{Na}, I_K, and I_L run in the arms of the circuit in parallel with the capacitance C_m. Experiments indicate that G_{Na} and G_K vary with time, and the other parameters are assumed constant.

From an examination of the experimental voltage-clamp data, Hodgkin and Huxley (1952b) devised empirical equations to evaluate G_K and G_{Na} as functions of voltage and time. To guide them in this process, they assumed that it required four particles to occupy specific sites in the membrane to cause a change in potassium conductance. If n is the probability for a single

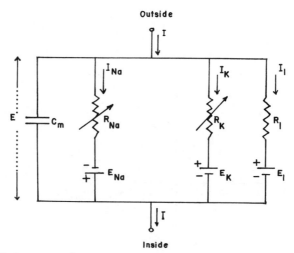

Fig. 21. Equivalent electrical circuit representing the membrane. C_m is the membrane capacitance (constant) $R_{Na} = 1/G_{Na}$, $R_K = 1/G_K$, $R_L = 1/G_L$. R_{Na} and R_K vary with time and membrane potential. (After Hodgkin and Huxley, 1952d.)

particle to be in the right place, then

$$G_K = \bar{G}_K n^4, \tag{103}$$

where $\bar{G}_K$ is the maximum potassium conductance. The value of n is given by the solution to the equation

$$\frac{dn}{dt} = \alpha_n(1 - n) - \beta_n n, \tag{104}$$

where α_n and β_n are the rate constants (t^{-1}), which depend only on potential. The voltage dependency of α_n and β_n as determined by Hodgkin and Huxley (1952d) from their experimental curves is given by

$$\alpha_n = 0.01(V + 10) \bigg/ \left[\exp\left(\frac{V + 10}{10}\right) - 1\right], \tag{105}$$

$$\beta_n = 0.125 \exp(V/80). \tag{106}$$

With the help of Eqs. (105) and (106), values for α_n and β_n can be calculated for any potential.

Rearranging Eq. (104) gives

$$\frac{dn}{dt} = \alpha_n - (\alpha_n + \beta_n)n. \tag{107}$$

IV. Hodgkin–Huxley Equations

If $y = \alpha_n - (\alpha_n + \beta_n)n$, then $dy/dt = -(\alpha_n + \beta_n) \, dn/dt$, and so Eq. (107) can be written

$$\frac{dy}{y} = -(\alpha_n + \beta_n) \, dt. \tag{108}$$

In the resting state, $V = 0$, and n has a resting value given by $n_0 = \alpha_{n_0}/(\alpha_{n_0} + \beta_{n_0})$.

When the membrane is held at a constant voltage (voltage clamp), the rate constants remain constant, and so Eq. (108) can be integrated. The result is

$$\ln y = -(\alpha_n + \beta_n)t + \text{const.} \tag{109}$$

At $t = 0$, $n = n_0$, and so the constant of integration, on substituting for y, gives $\ln[\alpha_n - (\alpha_n + \beta_n)n_0]$. Thus Eq. (109) becomes

$$\ln \frac{\alpha_n - (\alpha_n + \beta_n)n}{\alpha_n - (\alpha_n + \beta_n)n_0} = -(\alpha_n + \beta_n)t. \tag{110}$$

Substituting

$$n_\infty = \alpha_n/(\alpha_n + \beta_n) \tag{111}$$

and

$$\tau_n = 1/(\alpha_n + \beta_n) \tag{112}$$

in Eq. (110) gives

$$(n_\infty - n)/(n_\infty - n_0) = \exp(-t/\tau_n)$$

or

$$n = n_\infty - (n_\infty - n_0)\exp(-t/\tau_n). \tag{113}$$

The experimental potassium current generated (voltage-clamp experiment in the absence of NaCl or in the presence of tetrodotoxin in the saline) when the membrane potential is stepped up from a resting value of -60 mV to about zero is illustrated in Fig. 22. The onset of current is relatively slow, increases with time, and reaches a plateau in 5 ms. At the end of the depolarizing pulse, the current decays exponentially. At different potentials I_K turns on and off at different rates, reaching different steady-state values. These currents can be described by Eqs. (103) and (113). In an experiment in which the potential is stepped up, G_K changes from an initial value of G_{K_0} to one of G_{K_∞}; the value at any time t is given by

$$G_K = [(G_{K_\infty})^{1/4} - [(G_{K_\infty})^{1/4} - (G_{K_0})^{1/4}]\exp(-t/\tau_n)]^4, \tag{114}$$

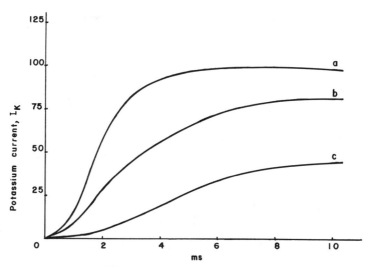

Fig. 22. Potassium current I_K (expressed as the percentage of maximally obtainable I_K) plotted as a function of time. Data obtained from axons bathed in choline seawater (Na free). Before application of depolarizing step, membrane potential was held at the resting level. In (a) depolarizing step was 109 mV, in (b) 63mV, and in (c) 32 mV. (After Palti, 1971.)

where τ_n is given by Eq. (112). Equation (114) can be fitted to experimental data such as shown in Fig. 22 by choosing specific values for G_{K_∞} and G_{K_0} and a value for τ_n to give the best fit. As $n_\infty = (G_{K_\infty}/\bar{G}_K)^{1/4}$, $\alpha_n = n_\infty/\tau_n$, and $\beta_n = (1 - n_\infty)/\tau_n$, the values for α_n and β_n can be derived. These values derived as a function of potential are described by the empirical equations (105) and (106).

The voltage dependence of n for the steady state (n_∞) can be determined analytically with the help of Eqs. (105)–(107). The results are shown in Fig. 23. Near the resting potential, $n \approx 0.3$. It tends to reach zero in the hyperpolarizing direction and unity in the depolarizing direction. These changes in n_∞ probably reflect the changes in the number of sites or channels available at any given membrane potential.

The several values of τ_n evaluated as a function of voltage are shown in Fig. 24.

Relating the time dependence of sodium conductance to membrane potential is similar to that of G_K. But because of its transient nature, it is more complex. It turns on rapidly and is quickly turned off while the depolarization continues. Membrane depolarization exerts two effects, which are called sodium activation and sodium inactivation. To describe the experimental results, two parameters m and h were introduced thus:

$$G_{Na} = \bar{G}_{Na} m^3 h. \tag{115}$$

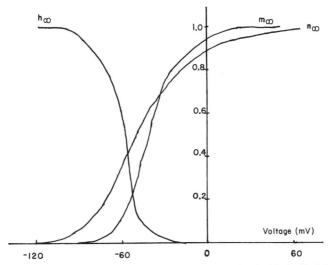

Fig. 23. The steady-state values of m, n, and h parameters associated with the Hodgkin–Huxley equations as a function of membrane potential. The parameters are normalized. Representative values of maximum conductance for squid axon membrane are $G_{Na} = 120$ mS/cm^2, $G_K = 36$ mS/cm^2. (After Palti, 1971.)

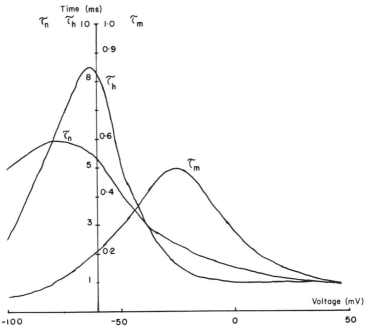

Fig. 24. Values of time constants τ_m, τ_h, and τ_n related to conductance parameters as a function of membrane potential. All time constants peak close to the resting membrane potential. The scale of τ_m is 10-fold larger than that of τ_n and τ_h. (After Palti, 1971.)

The activation parameter m is similar to n, being both potential and time dependent and having values from 0 to 1. $\bar{G}_{Na}$ is a constant equal to the maximum value of G_{Na}. Equation (115) is based on the idea that the sodium channel can be opened by the movement of three particles, each with a probability m being in the correct place and inactivated by a process of probability $1 - h$, m and h are given by

$$\frac{dm}{dt} = \alpha_m(1 - m) - \beta_m m, \tag{116}$$

$$\frac{dh}{dt} = \alpha_h(1 - h) - \beta_h h. \tag{117}$$

The solutions for these equations satisfying the boundary conditions $m = m_0$ and $h = h_0$ at $t = 0$ are

$$m = m_\infty - (m_\infty - m_0)\exp(-t/\tau_m), \tag{118}$$

$$h = h_\infty - (h_\infty - h_0)\exp(-t/\tau_h), \tag{119}$$

where

$$m_\infty = \frac{\alpha_m}{\alpha_m + \beta_m}, \quad h_\infty = \frac{\alpha_h}{\alpha_h + \beta_h},$$

$$\tau_m = \frac{1}{\alpha_m + \beta_m}, \quad \tau_h = \frac{1}{\alpha_h + \beta_h},$$

In the resting state G_{Na} is very small compared to its value for a large depolarization, and also m_0 can be neglected as a good approximation. Under these conditions inactivation is complete at $t \to \infty$, and hence $h_\infty = 0$. Thus equation for sodium conductance becomes

$$G_{Na} = \bar{G}_{Na} m_\infty^3 h_0 [1 - \exp(-t/\tau_m)]^3 \exp(-t/\tau_h)$$

or

$$G_{Na} = G'_{Na}[1 - \exp(-t/\tau_m)]^3 \exp(-t/\tau_h), \tag{120}$$

where $G'_{Na} = \bar{G}_{Na} m_\infty h_0$ and so is a constant. By plotting curves of G_{Na} versus time for different values of voltage and by fitting the experimental data with curves predicted by Eq. (120), it is possible to obtain values for G'_{Na}, τ_m, and τ_h corresponding to each clamped potential.

If there were no inactivation (i.e., h remained at its resting value h_0), then the steady-state value of sodium conductance would approach G'_{Na} according to Eq. (120). If sodium conductance is normalized by choosing an appropriate $\bar{G}_{Na}$ so that $m_\infty = 1$ for large depolarizations, then

$$m_\infty = \left(\frac{G'_{Na}}{G'_{Na(max)}}\right)^{1/3}, \tag{121}$$

IV. Hodgkin–Huxley Equations 363

where $G'_{Na(max)}$ is the asymptotic value of G'_{Na} at large depolarization. Hence from voltage-clamp data values of m_∞, τ_m, and τ_h can be obtained as functions of voltage, and so it is possible to compute α_m, β_m, α_h, and β_h as functions of voltage by using the relations

$$\alpha_m = \frac{m_\infty}{\tau_m}, \quad \beta_m = \frac{1 - m_\infty}{\tau_m}, \quad \alpha_h = \frac{h_\infty}{\tau_h}, \quad \beta_h = \frac{1 - h_\infty}{\tau_h}.$$

But the latter two relations require values for h_∞. In the foregoing considerations, the condition set was that $h_\infty = 0$ for $t \to \infty$ at large depolarizations. Hodgkin and Huxley (1952c) used a double-pulse technique to evaluate h_∞ as a function of voltage. In this technique, the first step involves application of a conditioning voltage V_1, and in the second step a test voltage V_2 is applied. V_1 is subthreshold, depolarizing or hyperpolarizing, whereas V_2 is suprathreshold and always depolarizing. The membrane current response I depends on V_1, V_2, and the duration of V_1. In general, larger and longer depolarizing V_1 produced greater inactivation in that the peak sodium current observed on application of test pulse V_2 after conditioning was smaller. On the other hand, the larger the hyperpolarizing V_1, the greater the sodium current in response to test pulse V_2. If I_{max} is the maximum sodium current that can be evoked in a nerve in its ideal state and if I is the actual Na current at a given extent of depolarization, then $I/I_{max} = h$ will be less than unity in an isolated axon. The experimental points determined by this technique conformed to the relation

$$h_\infty = [1 + \exp\{(V_h - V)/7\}]^{-1}, \quad (122)$$

where V_h is the potential at which $h_\infty = 0.5$ (V_h was about 2.5 mV). With the help of Eq. (122), values of h_∞ can be calculated for different voltages. Thus values for m_∞ and τ_m for small depolarizations can now be derived since τ_h and h_∞ are known for this condition. The potential dependence of both m_∞ and h_∞ is shown in Fig. 23, while the potential dependence of τ_m and τ_h is shown in Fig. 24.

m_∞ (see Fig. 23) increases with depolarization, but hyperpolarization decreases it. Under resting conditions it is close to zero (0.05), while n_∞ has a value of 0.3. This is the reason that in the resting state G_K is greater than G_{Na}. The potential dependence of the h parameter is the inverse of the potential dependence of either the m or n parameter. Depolarization causes a decrease in the value of h while the m and n parameters increase. All the time constants (Fig. 24) have similar shapes in their behavior toward voltage, but the magnitudes are very different $[\tau_n > \tau_m < \tau_h]$ under normal conditions, with the m process being the fastest.

The rate constants α_m, β_m, α_h, and β_h, like the other constants α_n and β_n derived as functions of potential [see Eqs. (105) and (106)], fit the following equations:

$$\alpha_m = 0.1(V + 25)/[\exp\{(V + 25)/10\} - 1],$$
$$\beta_m = 4\exp(V/18),$$
$$\alpha_h = 0.07\exp(V/20),$$
$$\beta_h = 1/[\exp\{(V + 30)/10\} + 1].$$

αs and βs are per millisecond, and Vs are in millivolts.

The complete expression for the membrane current density I is

$$I = C_m \frac{dV}{dt} + \bar{G}_K n^4 (V - V_K) + \bar{G}_{Na} m^3 h (V - V_{Na}) + \bar{G}_L (V - V_L). \quad (123)$$

If I is known, V can be determined by numerical integration. But the simplest case occurs when a patch of the membrane is excited simultaneously by means of an internal metal electrode. In this case there are no local circuit currents and the net current through the membrane is zero, the ionic current being equal and opposite to the capacitative current. In the case of propagated action potential, the situation is complex. From the cable theory the membrane current is given by

$$i_m = \frac{1}{r_i} \frac{\partial^2 V}{\partial x^2}$$

or

$$I_m = \frac{a}{2R_i} \frac{\partial^2 V}{\partial x^2}, \quad (29')$$

where $I_m = i_m/2\pi a$ and $r_i = R_i/\pi a^2$. Equation (29') could be inserted into Eq. (123), but the derivatives on either side of the resultant equation would be difficult to solve it. The action potential travels along the axon like a wave, and so the wave equation could be used to make the derivatives the same with respect to the variable. This is done by considering a simple sine wave, which can be described by

$$V = \sin(x + \theta t), \quad (124)$$

where θ is the velocity of wave propagation. Differentiation with respect to time and distance gives

$$\frac{\partial V}{\partial x} = \cos(x + \theta t), \qquad \frac{\partial V}{\partial t} = \theta \cos(x + \theta t),$$

$$\frac{\partial^2 V}{\partial x^2} = -\sin(x + \theta t), \qquad \frac{\partial^2 V}{\partial t^2} = -\theta^2 \sin(x + \theta t).$$

IV. Hodgkin–Huxley Equations

Hence it follows that

$$\frac{\partial^2 V}{\partial x^2} = \frac{1}{\theta^2} \frac{\partial^2 V}{\partial t^2}. \tag{125}$$

Thus inserting Eqs. (29′) and (125) into Eq. (123) gives

$$\frac{a}{2R_i\theta^2} \frac{d^2 V}{dt^2} = C_m \frac{dV}{dt} + \bar{G}_K n^4(V - V_K) + \bar{G}_{Na} m^3 h(V - V_{Na}) + \bar{G}_L(V - V_L). \tag{126}$$

This is an ordinary differential equation, which can be solved numerically. But the procedure is involved in that θ is not known in advance; so a value is guessed and inserted in the equation for solution. A numerical solution worked out has been found to agree with the behavior of real nerves. Hodgkin and Huxley (1952d) found that the kinetic equations given above could predict with fair accuracy many of the electrical properties of the giant axon of the squid.

It has become useful to think of excitation as a process mediated by a voltage-sensitive gate and a connected selectivity filter. These are conceptual terms and need not signify the presence of any such physical structures in the membrane. This gate analogy was used to indicate the steep voltage–conductance relation found by Hodgkin and Huxley (1952a,b). Voltage across the membrane controls the opening and closing of gates. When the sodium gates are open (caused by depolarization), sodium ions move down their electrochemical gradient. The resulting depolarization causes more membrane gates to open. The variables m and h of the Hodgkin–Huxley (1952d) theory, since they are voltage sensitive, correspond to a version of the gates. The concept of selectivity filter explains the different permeabilities of the channels to different ions. Both gate and filter together signify the presence of an energy barrier that the ion must surmount to cross the membrane.

The gate structure, since it is voltage sensitive, must be charged. Hodgkin and Huxley (1952d) made an estimate of the amount of charge required to open a single sodium channel. In the model they used, sodium conductance is assumed to be proportional to the fraction of gating "particles" near the inside of the membrane. That is,

$$G_{Na} = \bar{G}_{Na} P_i, \tag{127}$$

$$P_i + P_o = 1, \tag{128}$$

where P_i is the fraction of particles on the inside and P_o is the fraction of particles on the outside. If these particles have independence to move freely, their arrangement in the membrane would follow the Boltzmann distribution. Hence

$$P_i/P_o = \exp[(W + zeE)/\kappa T], \tag{129}$$

where W is the work required for a particle to cross the membrane when $E = 0$ and $\kappa T/e = 25$ mV at 20°C. Combining Eqs. (128) and (129) gives

$$P_i = \frac{1}{1 + \exp[-(W + zeE)/\kappa T]} = \frac{\exp[(W + zeE)/\kappa T]}{1 + \exp[(W + zeE)/\kappa T]},$$

and so Eq. (127) becomes

$$G_{Na} = \bar{G}_{Na} \frac{\exp[(W + zeE)/\kappa T]}{1 + \exp[(W + zeE)/\kappa T]}. \tag{130}$$

When E is large and negative, $\exp[(W + zeE)/\kappa T] \ll 1$, then Eq. (130) becomes

$$G_{Na} \approx \bar{G}_{Na} \exp[zeE/\kappa T] = \exp(zE/25).$$

Hodgkin and Huxley (1952d) found that at large negative potentials when $G_{Na} \ll \bar{G}_{Na}$,

$$G_{Na} \propto \exp(E/4).$$

Thus $z = 6$. This means that six charges must cross the membrane to open one sodium channel, or if a dipole is assumed to move to open the channel, it must have three charges at each end. This movement of charge in the membrane is the so-called displacement or gating current and has been measured in recent years (Armstrong and Bezanilla, 1974; Keynes and Rojas, 1974).

Gating currents are measured by voltage-clamp experiments in which interference from conventional capacitative and ionic currents and noise are reduced or eliminated. The nerve axons are perfused inside and outside with solutions of impermeant ions and suitable agents to block the channels. Matched positive and negative pulses are applied, and the resulting currents are added. In this way symmetrical capacity and leakage currents are canceled. Pulses are repeated several times, and the result is accumulated in a computer. Asymmetrical components of current are attributed to gating current for activation of sodium channels.

The integral of the time course of gating current gives the net charge transferred during a single voltage-clamp step. This can be as large as 1882 $-e/\mu m^2$, where $-e$ is the electronic charge. Assuming 6 charges per channel gives a value of 314 channels per square micrometer of the membrane. The $\bar{G}_{Na}$, according to Hodgkin and Huxley (1952d), is 1200 pS/μm^2. So the sodium single-channel conductance corresponds to 3.82 pS.

References

Archer, W. I., and Armstrong, R. D. (1980). *In* "Electrochemistry" (H. R. Thirsk, ed.) (Specialist Periodical Reports), Vol. 7, p. 158. The Chemical Society, London.
Armstrong, C. M., and Bezanilla, F. (1974). *J. Gen. Physiol.* **63**, 533.
Cole, K. S. (1972) "Membranes, Ions and Impulses." Univ. of California Press, Berkeley.
Danielli, J. F., and Davson, H. (1935). *J. Cell Physiol.* **5**, 495.
Davis, L., Jr., and Lorente de Nó, R. (1947). *Stud. Rockefeller Inst. Med. Res.* **131**, 442.
Ehrenstein, G., and Lecar, H. (1972). *Annu. Rev. Biophys. Bioeng.* **1**, 347.
Hanai, T., Haydon, D. A., and Taylor, J. (1964). *Proc. R. Soc. London. Ser. A* **281**, 377.
Hodgkin, A. L. (1958). *Proc. R. Soc. London, Ser. B.* **148**, 1.
Hodgkin, A. L., and Huxley, A. F. (1952a). *J. Physiol. (London)* **116**, 449.
Hodgkin, A. L., and Huxley, A. F. (1952b). *J. Physiol. (London)* **116**, 474.
Hodgkin, A. L., and Huxley, A. F. (1952c). *J. Physiol. (London)* **116**, 497.
Hodgkin, A. L., and Huxley, A. F. (1952d). *J. Physiol. (London)* **117**, 500.
Hodgkin, A. L., and Katz, B. (1949). *J. Physiol. (London)* **108**, 37.
Hodgkin, A. L., and Rushton, W. A. H. (1946). *Proc. R. Soc. London, Ser. B* **133**, 444.
Horowicz, P., Schneider, M. F., and Begenisich, T. (1980). *In* "Membrane Physiology" (T. E. Andreoli, J. F. Hoffman, and D. D. Fanestil, eds.), p. 185. Plenum, New York.
Katz, B. (1948). *Proc. R. Soc. London, Ser. B* **135**, 506.
Keynes, R. D., and Rojas, E. (1974). *J. Physiol. (London)* **239**, 393.
Lakshminarayanaiah, N. (1979). *Subcell. Biochem.* **6**, 401.
Palti, Y. (1971). *In* "Biophysics and Physiology of Excitable Membranes" (W. J. Adelman, Jr., ed.), p. 168. Van Nostrand-Reinhold, New York.
Rall, W. (1977). *In* "Handbook of Physiology" (J. M. Brookhart and V. B. Mountcastle, eds.), Vol. 1, Sect. 1, Part 1, p. 39 Am. Physiol. Soc. Bethesda, Maryland.
Schanne, O. F., and Ruiz P.-Ceretti, E. (1978). "Impedance Measurements in Biological Cells." Wiley, New York.
Schwan, H. P. (1963). *In* "Physical Techniques in Biological Research" (W. L. Nastuk, ed.), Vol. 6, Part B, p. 323. Academic Press, New York.
Singer, S. J., and Nicolson, G. L. (1972). *Science* **175**, 720.
Tarr, M., and Trank, J. (1971). *J. Gen. Physiol.* **58**, 511.
Taylor, R. E. (1963). *In* "Physical Techniques in Biological Research" (W. L. Nastuk, ed.), Vol. 6, Part B, p. 219. Academic Press, New York.

Chapter **8**

FLUCTUATION ANALYSIS OF THE ELECTRICAL PROPERTIES OF THE MEMBRANE

In Chapter 7, it was shown that there are about 3×10^{10} Na channels per 1 cm^2 of the membrane. In addition there are also K and other channels. When the Na and K channels are open, about 10^{16} or so Na and K ions per second flow through the membrane. The recording techniques used in the measurements of current and voltage give their average values (macroscopic). That is the local potentials and currents associated with individual channels (microscopic), i.e., the unitary parameters, are averaged. On a microscopic level all the constituents of the membrane, viz., ions, water molecules, proteins, and lipids, due to their inherent thermal energies, will be in constant motion, colliding and exchanging energy with one another. Consequently, fluctuations on a microscopic level will occur and exist about the measured (macroscopic) average or mean value. The situation is similar to any random event that occurs in nature. For example, in the case of gas molecules contained in a vessel the macroscopic average number of molecules present in each half of the vessel is the same. But at any given instant the number of molecules in one half of the vessel will be a little more or little less than the other half; but the number of molecules in each half of the vessel will fluctuate about a mean value of one-half of the total number of molecules present in the vessel. Similarly the number of open channels in a biological membrane will also fluctuate about some mean value. Membrane conductance therefore will show fluctuations, and these will appear as noise or fluctuations in current or voltage.

The basic idea underlying fluctuation analysis is that the same molecular processes that govern microscopic events (fluctuations) also govern macroscopic behavior. This means that, taking the example of channels, the same channels are responsible for the mean value as well as fluctuations. Nothing new comes out of noise or fluctuation analysis. But information not available by other methods can be obtained by this method. Resolution of single-channel opening and closing can be realized, and single-channel conductance can be derived conveniently. Also, fluctuation analysis helps in deciding between various models predicting kinetics.

There are several papers and review articles dealing with fluctuation analysis at different levels. The papers are devoted to measurement and analysis of fluctuations in current or voltage occurring in lipid bilayer membranes doped with ionophores (see Chapter 5) or in some biological preparations. Review articles exist at several levels; some are elementary and others are difficult to follow without the required mathematical background. So in this chapter a simple nonmathematical description of fluctuation analysis followed by an outline of the mathematical tools required to understand this special method are presented. The nonmathematical description is simply a summary of the 1975 paper of Stevens in the *Federation Proceedings*. The review articles of interest are those by Verveen and DeFelice (1974), Conti and Wanke (1975), DeFelice (1977), Neher and Stevens (1977), and Lecar and Sachs (1981). The article by Horowicz *et al.* (1980) contains a section that gives a succinct introduction to fluctuation analysis. The text by DeFelice (1981) gives a mathematical treatment of membrane noise with a review of papers devoted to application of noise analysis to transport phenomena in artificial and biological membranes.

I. Nonmathematical Description of Noise Analysis

The parameter that is characteristic of fluctuations is the overall amplitude. The usual procedure is to make a histogram describing the fraction of time a given oscillating quantity has a particular value. If one wants to study membrane current fluctuation around a mean value of 100 μA, the membrane current would be measured a large number of times (e.g., 1000) and a histogram would be constructed showing how many times the membrane current was within ± 1 μA, ± 2 μA, ± 3 μA, etc., of the mean value. The amplitude of fluctuations around the mean value is given by the standard deviation and is indicated by the width of the histogram. The standard deviation is a statistical parameter that is calculated by taking all deviations, plus and minus, from the mean value, squaring them so that nothing is allowed to cancel, and averaging the squared deviations. The square root of this (the averaged squared deviation) is the standard deviation. The amplitude of fluctuations is generally designated as the standard deviation of variations around the mean value. For a normal distribution (Gaussian curve) 68% of the fluctuations fall within ± 1 SD, 95% within ± 2 SD, and 99.7% within ± 3 SD. If a membrane current had a mean value of 100 μA and a standard deviation of 2 μA, then a current larger than 102 μA would be observed about 34% of the time. A current as large as 110 μA will never be seen.

If a biological process is time dependent, then it is important to know the rate at which the deviations from the mean would occur. This may be denoted either by the covariance function (also called autocorrelation or simply correlation function) or the spectrum. Machines are available that compute for a given input signal the covariance function. Also, machines are available for deriving the spectral density functions. With modern computers, because the spectrum can be more easily and rapidly calculated, it has become the most important technique for characterization of the signal.

Most signals can be resolved into a sum of sine and cosine waves of various frequencies and amplitudes. If current emerging from a source (e.g., nerve or muscle membrane, end plate) is recorded for 1 s and this record subjected to Fourier analysis would give weights corresponding to sine and cosine functions of frequencies $1\ s^{-1}$, $2\ s^{-1}$, $3\ s^{-1}$, up to infinity. These sines and cosines multiplied by their weights and added together would reconstruct the original record, and so the weights would characterize the signal. A function of time converted to weights for sine and cosine waves is called a Fourier transform and getting back the original function from weights is termed an inverse Fourier transform. Although in theory the number of sine and cosine waves required to resolve a time function is infinite, in practice the number of sine and cosine frequencies required is small because (1) the signal is recorded for a finite time and (2) the equipment has only a limited bandwidth. The lowest frequency for a 1-s sample would be a sine (and cosine) wave of 1 Hz. For a $\frac{1}{2}$-s sample, the lowest frequency would be 2 Hz. The highest frequency is generally given by the rate at which the original time record is taken and according to the sampling theorem it is given by half the sampling rate. Thus for a signal of 1 kHz, the highest frequency in the Fourier analysis would be 500 Hz. If the sampling frequency is $4\ s^{-1}$, then the highest frequency component would be a sine wave of 2 Hz. So the weights of sines and cosines can be illustrated by the following example. If a 1-s record is taken at a sampling rate of 1 sample/ms, the weights for sine and cosine waves would be $1\ s^{-1}$, $2\ s^{-1}$, $3\ s^{-1}$,..., up to $500\ s^{-1}$ (half of the sampling rate). Similarly, if a $\frac{1}{2}$-s sample is taken at a sampling rate of 4 kHz, the sine wave component would have frequencies $2\ s^{-1}$, $4\ s^{-1}$,..., up to $2000\ s^{-1}$.

In order to calculate the spectrum of an experimental record, the record is Fourier analyzed to weights for sine and cosine functions. These weights are squared and added together and divided by 2 (averaged) for each frequency. The spectrum for the original signal therefore is a collection of average squared weights for various components of different frequencies. The spectrum corresponding to a particular record indicates the rate of fluctuations in the original record. If a measured variable such as current (at constant voltage) or voltage (at constant current) fluctuated at a high rate, its Fourier resolution would contain many high-frequency components, and so its spectrum would have large values for the high frequencies. The opposite would be true if the fluctuations were slowly varying, and the spectrum would have more slow component weights.

Spectra may be calculated in several ways. One simple way is to tape-record the signal and play the tape over many times, first through a bandpass filter with a center frequency of 1 Hz, second with 2 Hz, third with 3 Hz, etc. The output from the bandpass filter is measured with an root-mean-square (rms) meter, and the square of the rms reading gives the amplitude of the spectrum for that particular frequency. Similarly, the amplitudes are obtained for other frequencies. This procedure is too cumbersome and time consuming. Now digital computers are employed in these computations. The time required for calculations is considerably reduced by using an algorithm fast Fourier transform (FFT) to resolve the signal by the Fourier transform.

In calculating spectra two important factors must be kept in mind. Because one is recording fluctuations that are inherently random, estimates of amplitudes may vary from one record to another. To reduce this variation, the spectra derived from a number of records are averaged. The number of spectra taken at an average greatly increases the precision of the spectral estimates.

The second factor that introduces errors is the so called "aliasing." In general, if sine waves of higher frequencies than the highest-frequency limit according to the sampling theorem are present in the signal record, these higher frequency components can produce sample points corresponding to sample points arising from low-frequency sine waves. Fourier analysis by FFT program would treat these high-frequency components as arising from low-frequency components. Thus higher values for some of the lower frequency amplitudes would be obtained. This can be eliminated by the use of analog filters. Before the signal is recorded it is sent through an analog filter to remove the higher frequency components.

The second method used for characterizing the fluctuating signal is to employ the covariance or autocorrelation function if the covariance function can be calculated directly from the signal record. Usually it is more efficient to calculate the covariance function from the spectrum by a Fourier transform. The covariance function is the inverse Fourier transform of the spectrum. So to calculate a covariance function from a particular record, the spectrum is first obtained by the FFT method. Then the spectrum is Fourier transformed by adding together cosine waves of various frequencies weighted in accordance with the corresponding spectral amplitudes. For example, if estimates from the spectrum are at frequencies 1, 2, 3, ..., 500 Hz, cosine waves of 1, 2, 3, ..., 500 Hz are weighted according to their spectral amplitudes and added to yield the covariance function, which would be a function in time. This function on normalization so that its value at zero time is unity is called the autocorrelation function. The covariance function provides in the time domain the same information about the rate of fluctuations as the spectrum does in the frequency domain. The covariance function indicates how well the present values of the fluctuating signal will correlate with its value some t s later. Generally the covariance function for random processes will decay monotonically from a maximum value at time zero to a value of zero for long times.

A decay over a 1-ms period would reveal that the parameter under study has components lasting about a millisecond. A decay over 100 ms would indicate that the signal is made up of slower components. So the correlation function gives information about the rapidity of the signal change in time. In summary fluctuations are characterized by (1) the magnitude of fluctuations given as variance or standard deviation (rms value) and (2) the rapidity of fluctuations expressed by the spectrum or covariance (or correlation or autocorrelation) function.

The next stage is to establish a connection between the two characteristics of the fluctuations described and the macroscopic behavior of the system. This is done, as discussed by Stevens (1975), with the help of two approaches, one based on the fluctuation–dissipation theorem and the other following from explicit mechanistic theories. The fluctuation–dissipation theorem states that the covariance function related to fluctuations around a constant mean in any given system describes the relaxations that arise when that system is perturbed. It is a very general and experimentally untestable method, and its applicability gives little information about the mechanisms governing the behavior of the system. On the other hand, mechanistic theories provide insight into mechanisms. But these mechanisms are as good as the assumptions on which they are based. So inferences drawn about the behavior of the system from analysis of fluctuations are as accurate as the theory used to describe the characteristics of the system.

II. Statistical Concepts

In writing this section books by Boas (1966) and Maksoudian (1969) were used. Some of the numerical examples are taken from Maksoudian.

If there are several equally likely exclusive and exhaustive outcomes of an experiment, then the probability p of an event E is given by

$$p = \frac{\text{number of outcomes favorable to } E}{\text{total number of outcomes}}.$$

A set of all possible mutually exclusive outcomes is called a *sample space*, and each outcome is called a *point* of the sample space. For any given problem there may be several sample spaces. A list of outcomes of equal probability is called uniform space; otherwise, if the outcomes have different probabilities, it is called nonuniform sample space.

Any variable x that has a definite value for each point of the sample space is called a random variable. Thus it follows that a random variable x is a function defined on a sample space, each value of x, for example x_i, has a probability p_i of occurrence. This means the probability that $x = x_i$ may be written as $p_i = f(x_i)$, or $f(x)$ is the probability function for the random variable x. Alternatively, it can be said that x is a random variable if it assumes various values x_i with probabilities $p_i = f(x_i)$; or the probability function is $p = f(x)$. A probability function must satisfy two conditions,

$$f(x) \geq 0, \tag{1}$$

$$\sum f(x) = 1. \tag{2}$$

As an example, let x denote the number of heads when three coins are flipped. The uniform space sample is

hhh   hth   htt   ttt
tht   thh   tth   hht

A table of x and $p = f(x)$ can be constructed thus:

x	0	1	2	3
$p = f(x)$	$\frac{1}{8}$	$\frac{3}{8}$	$\frac{3}{8}$	$\frac{1}{8}$

The probability function $p = f(x)$ of a random variable is also called the probability density, probability distribution, or frequency distribution. This may be represented by a graph of probability function $f(x)$ associated with a random variable x, as shown in Fig. 1. Instead of a line graph, the same represented by a bar graph (see Fig. 2) is commonly called a histogram. In the bar graph the probabilities are represented by areas. It is obvious that the total area represented is equal to unity. If, on the other hand, the areas represented are considered proportional to the probability, the figure would still be similar to a histogram, but the total area would not be unity.

A probability function could be discrete or continuous. This means that the random variable has finite set of values, or assumes values that represent continua.

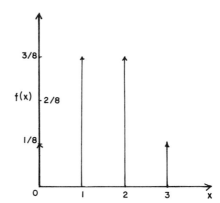

Fig. 1. Probability function $f(x)$ plotted against a random variable x.

Related to each probability function $f(x)$, a related function $F(x_0)$ called the distribution function may be defined. Thus

$$F(x_0) = \sum_{x \leq x_0} f(x).$$

This shows that summation is over all values of random variable that are less than or equal to the specified value of x_0. The probability that the random variable x will take values less than or equal to x_0 is given by $F(x_0)$. On the other hand, $f(x_0)$ gives the probability of x taking the value x_0 only. For example, if $f(x)$ is defined by

$$f(x) = x^3/100, \quad x = 0, 1, 2, 3, 4,$$

is $f(x)$ a probability function? If so, find $F(0)$, $F(1)$, $F(2)$, $F(3)$, $F(4)$. Since $f(0) = 0$, $f(1) = 1/100$, $f(2) = 8/100$, $f(3) = 27/100$, $f(4) = 64/100$ are all greater than or equal to zero and their sum is equal to unity, the two conditions [Eqs. (1) and (2)] of probability are satisfied.

$$F(0) = 0,$$
$$F(1) = f(0) + f(1) = \tfrac{1}{100},$$
$$F(2) = f(0) + f(1) + f(2) = \tfrac{1}{100} + \tfrac{8}{100} = \tfrac{9}{100},$$
$$F(3) = f(0) + f(1) + f(2) + f(3) = \tfrac{9}{100} + \tfrac{27}{100} = \tfrac{36}{100},$$
$$F(4) = f(0) + f(1) + f(2) + f(3) + f(4) = \tfrac{36}{100} + \tfrac{64}{100} = 1.$$

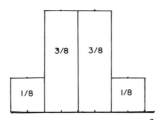

Fig. 2. Histogram.

The two relations applicable to distribution function are

(1) $p(x > x_0) = 1 - F(x_0)$, (3)

(2) $p(x_0 < x \leq x_1) = F(x_1) - F(x_0)$. (4)

These are helpful in computing probabilities.

Three common distribution functions are the binomial distribution, the normal or Gaussian distribution, and the Poisson distribution. Before these are outlined, it is necessary to list the common counting formulas, which are the permutations and combinations of several things taken few at a time.

The permutation of n things taken k at a time denoted as $P(n, k)$ is given by

$$P(n, k) = n!/(n - k)!,$$ (5)

where $n! = n(n - 1)(n - 2) \cdots (2)(1)$ and is read "n factorial." Then $P(n, n) = n!/(n - n)! = n!/0! = n!$ if 0! is defined as equal to 1. It is assumed that all n things are distinguishable. If some things become indistinguishable, then how many distinct permutations Q are possible for n things with some of them being similar is given by

$$Q = \frac{n!}{n_1! \, n_2! \cdots n_k!},$$ (6)

where $n_1, n_2, \ldots, n_k$ represent the number of similar things in subgroups 1, 2, ..., k. Alternatively, Eq. (6) may be written

$$Q = \frac{P(n, n)}{P(n_1, n_1) \, P(n_2, n_2) \cdots P(n_k, n_k)}.$$ (7)

Writing $C(n, k)$ to represent combination of n things taken k at a time,

$$C(n, k) = P(n, k)/k!,$$ (8)

since k things can be arranged in $k!$ ways among themselves. From this definition, it follows that

$$C(n, k) = C(n, n - k).$$ (9)

A binomial $(a + b)$ raised to the power of n ($n = 1, 2, 3, \ldots, n$) is written

$$(a + b)^n = C(n, 0)a^n b^0 + C(n, 1)a^{n-1}b^1 + C(n, 2)a^{n-2}b^2 + \cdots + C(n, n)a^0 b^n$$

$$= \sum_{k=0}^{n} C(n, k) a^{n-k} b^k.$$ (10)

This is called the binomial theorem.

II. Statistical Concepts

In several problems, some things are repeated; at each attempt there are two outcomes of probability, p ("success") and $q = 1 - p$ ("failure"). These separate repeated attempts with constant probabilities p and q are called Bernoulli trials. Consider the function

$$f(x) = C(n, x) p^x q^{n-x} \qquad (x = 0, 1, 2, \ldots, n). \tag{11}$$

Because of the restriction imposed, i.e., $p + q = 1$, $f(x)$ is always greater than or equal to zero, satisfying Eq. (1). Similarly

$$F(x) = \sum_{x=0}^{n} f(x) = \sum_{x=0}^{n} C(n, x) p^x q^{n-x}$$

is given by

$$F(x) = f(0) + f(1) + \cdots + f(x).$$

That is,

$$F(x) = C(n, 0) p^0 q^n + C(n, 1) p^1 q^{n-1} + \cdots + C(n, x) p^x q^{n-x}.$$

This is simply the binomial expansion of $(p + q)^n$, and so

$$(p + q)^n = 1^n = 1,$$

thus satisfying Eq. (2). So Eq. (11) is called the binomial probability function, and $F(x)$ is called the binomial distribution function. It is seen that x is the random variable and n and p are independent parameters (q is related to p and so is not independent). For specific values of n and p, x varies from term to term. The notation $f(x:n, p)$ is adopted. To illustrate the use of the definition, consider an example where a coin is tossed 10 times. What is the probability of getting 2 heads? What is the probability of getting more than 2 heads?

$$p(x = 2) = f(2:10, 0.5) = C(10, 2)(\tfrac{1}{2})^2 (\tfrac{1}{2})^8 = \tfrac{45}{1024} = 0.0439,$$

$$p(x > 2) = F(10:10, 0.5) - F(2:10, 0.5)$$

$$= 1 - \sum_{x=0}^{2} f(x:10, 0.5)$$

$$= 1 - [\tfrac{1}{1024} + \tfrac{10}{1024} + \tfrac{45}{1024}] = 0.9453.$$

The relations useful in computations with binomial function are

$$f(x:n, p) = f(n - x:n, q),$$
$$f(x:n, p) = F(x:n, p) - F(x - 1:n, p),$$
$$F(x:n, p) = 1 - F(n - x - 1:n, q).$$

Again, an illustration is to evaluate $F(9:10, 0.4)$. The solution is to use the third equation thus:

$$F(9:10, 0.4) = 1 - F(0:10, 0.6)$$
$$= 1 - f(0:10, 0.6)$$
$$= 1 - C(10, 0)(0.6)^0(0.4)^{10}$$
$$= 1 - (0.4)^{10} = 0.9999.$$

A great deal of computation is involved in calculating the binomial distribution for any given value of n unless n is very small. Although tables exist, it is useful to consider two approximations to the binomial distribution. These are the normal or Gaussian distribution and the Poisson distribution. Using Stirling's formula, which approximates

$$n! \approx n^n \exp(-n)\sqrt{2\pi n},$$

it can be shown that the binomial probability function when n and np are large approximates to

$$f(x) \approx \frac{1}{\sqrt{2\pi npq}} \exp\left(\frac{-(x-np)^2}{2npq}\right). \tag{12}$$

Equation (12) is the normal (or Gaussian) approximation to the binomial function.

Consider the function

$$f(x, k) = \exp(-k)k^x/x! \quad (k > 0, \quad x = 0, 1, 2, 3, \ldots). \tag{13}$$

Under the conditions of k and x, $f(x, k) \geq 0$ and satisfies Eq. (1). The second condition of Eq. (2) is also satisfied thus

$$\sum_{x=0}^{\infty} f(x, k) = \exp(-k)\left(1 + k + \frac{k^2}{2!} + \frac{k^3}{3!} + \cdots\right) = \exp(-k)\exp(k) = 1.$$

So Eq. (13) is a probability function and is called the Poisson probability function. The binomial function approximately equals Poisson function when $n \to \infty$ and $p \to 0$ such that np remains constant, say k. That is,

$$\lim_{\substack{n \to \infty \\ p \to 0}} C(n, x)p^x q^{n-x} = \frac{\exp(-k)k^x}{x!}. \tag{14}$$

If $f(x)$ is a probability function and $g(x)$ is any function of the random variable x, the expectation of $g(x)$ is defined as the sum of the products of $g(x)$ and $f(x)$ for all x. This in symbols is written

$$E[g(x)] = \sum g(x)f(x).$$

Some rules to compute expectations are as follows.

(1) Expectation of a constant k is k. Thus $E(k) = k$.
(2) Expectation of $kg(x)$, where $g(x)$ is a random variable, is k times the expectation of $g(x)$. That is,

$$E[kg(x)] = kE[g(x)].$$

(3) The expectation of the sum of two random variables is the sum of the respective expectations. Thus

$$E\left[\sum_{j=1}^{n} g_j(x)\right] = \sum_{j=1}^{n} E[g_j(x)].$$

An example of how to compute expectation is to find the expectation of x^2 when $f(x)$ is the binomial probability function $f(x:4, 0.5)$. The binomial function for $x = 0, 1, 2, 3, 4$ is computed. These values are $\frac{1}{16}, \frac{4}{16}, \frac{6}{16}, \frac{4}{16}$, and $\frac{1}{16}$; each of these are multiplied by x^2 ($0, 1, 2^2, 3^2, 4^2$). Thus

$$E(x^2) = \tfrac{1}{16}[(0 \times 1) + (1 \times 4) + (4 \times 6) + (9 \times 4) + (16 \times 1)] = \tfrac{80}{16} = 5.$$

A family of expectations can be written $E[(x - a)^k]$, where a is a constant and $k = 0, 1, 2, \ldots$. If

$$m_{k,a} = E[(x - a)^k] = \sum (x - a)^k f(x), \tag{15}$$

then $m_{k,a}$ is called the "kth moment about the point a." Two illustrations are to find the third moment about point 2 and second moment about point 0 for the binomial probability function $f(x:4, 0.5)$:

$$m_{3,2} = E[(x - 2)^3] = \sum_{x=0}^{4} (x - 2)^3 C(n, x)(\tfrac{1}{2})^x (\tfrac{1}{2})^{4-x}$$

$$= -\tfrac{8}{16} - \tfrac{4}{16} + 0 + \tfrac{4}{16} + \tfrac{8}{16} = 0,$$

$$m_{2,0} = E[(x - 0)^2] = E(x^2).$$

This is the same as the illustration shown above, and so the second moment about the origin $m_{2,0}$ (or simply m_2) is $m_2 = 5$. The first moment m_1 about the origin of the binomial probability function $f(x:4, 0.5)$ is

$$m_1 = \sum_{x=0} xf(x) = \tfrac{1}{16}[(0 \times 1) + (1 \times 4) + (2 \times 6) + (3 \times 4) + (4 \times 1)] = 2.$$

The histogram of the probability function $f(x:4, 0.5)$ is shown in Fig. 3. The symmetry shows that the center of gravity of the histogram is at $x = 2$. The calculations show that the first moment $m_1 = 2$. Thus the first moment about the origin m_1 gives the average or mean, generally represented by μ. From this definition it can be shown that the mean of the binomial probability

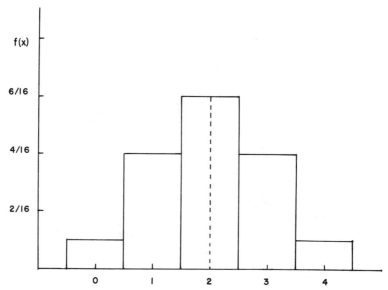

Fig. 3. Histogram of binomial function $(x:4, \frac{1}{2})$.

function $f(x:n,p)$ is given by

$$\mu = np, \tag{16}$$

and the mean of the Poisson probability function $f(x:k)$ is given by

$$\mu = k. \tag{17}$$

If a is replaced by μ, then Eq. (15) becomes

$$m_{k,\mu} = E[(x - \mu)^k] = \sum (x - \mu)^k f(x).$$

As the primary concern is related to moments about the mean, $m_{k,\mu}$ (the kth moment about the mean) may be represented by μ_k. The mean of the binomial probability function $f(x:4, 0.5)$ is np, and so $\mu = 2$. First, second, and third moments about this mean are given by

$$\mu_1 = E[(x-2)] = \sum_{x=0}^{4} (x-2)C(4,x)(\tfrac{1}{2})^x(\tfrac{1}{2})^{4-x} = 0,$$

$$\mu_2 = E[(x-2)^2] = 1, \tag{18}$$

$$\mu_3 = E[(x-2)^3] = 0.$$

The second moment about the mean is called the variance. The square root of variance is called the standard deviation (σ). So σ and Eq. (18) can be equated to give

$$\sigma^2 = E[(x - \mu)^2] = E[(x^2 - 2\mu x + \mu^2)].$$

Since $\mu = E(x)$ is a constant,

$$\begin{aligned}\sigma^2 &= E(x^2) - 2\mu E(x) + \mu^2 \\ &= E(x^2) - 2\mu^2 + \mu^2 \\ &= E(x^2) - [E(x)]^2.\end{aligned} \quad (19)$$

So the variance of a probability function is given by the second moment minus the square of the first moment (mean). Another form of Eq. (19) that is useful is given by

$$E(x^2) - [E(x)]^2 = E[x(x - 1)] + E(x) - [E(x)]^2. \quad (20)$$

Furthermore, it can be shown that the variance of the binomial probability function $f(x:n, p)$ is given by

$$\sigma^2 = npq \quad (21)$$

and that of the Poisson probability function by

$$\sigma^2 = k. \quad (22)$$

If a probability function shows asymmetry about its mean, then it is said to be skewed, positively skewed if $\mu_3 > 0$ and negatively skewed if $\mu_3 < 0$.

III. Mathematical Preliminaries

In the case of a membrane containing only the simplest channel, a channel that simply switches randomly in the steady state from one state (open) to another (closed), the rate of transition from open state to closed state does not change with time. When the number of channels are few, the individual events can be seen clearly. But when the number of channels becomes very large, the individual events are no longer resolvable. The many-channel records can, however, be analyzed as a form of noise having random properties related to the unitary events underlying the fluctuations. The simplest way to derive an expression for the amplitude of a single-channel contribution to membrane conductance is by the application of statistical principles. Let the membrane have a total of N channels of which n channels open with a probability of p. For just a single channel following binomial probability, n can be either 0 (closed) or 1 (open), and so according to Eq. (16), the mean is given by

$$\langle n \rangle = p,$$

and according to Eq. (21) the variance is given by

$$\sigma^2 = p(1 - p).$$

So, if there are m independent channels, the mean number of channels is given by

$$\langle n \rangle = Np, \tag{23}$$

and the variance by

$$\sigma^2 = Np(1 - p). \tag{24}$$

If the conductance of a single channel is γ, then the average membrane conductance is given by

$$\langle G \rangle = \gamma \langle n \rangle = \gamma Np \tag{25}$$

and its variance by

$$\sigma_G^2 = \gamma^2 \sigma^2 = \gamma^2 Np(1 - p). \tag{26}$$

The ratio of variance to mean gives an expression for γ. Thus Eqs. (24) and (25) give

$$\sigma_G^2 / \langle G \rangle = \gamma(1 - p). \tag{27}$$

Equation (27) becomes

$$\sigma_G^2 / \langle G \rangle = \gamma/2 \tag{28}$$

when $p = \tfrac{1}{2}$ and

$$\sigma_G^2 / \langle G \rangle = \gamma \tag{29}$$

when $p \ll 1$.

Equation (29) is frequently used in the estimation of the conductance of a single channel. This is valid only when the channels are independent and $p \ll 1$.

When a single channel opens, the current may emerge as a rectangular impulse, or it may have a complex shape. Trains of impulses of complex shape can emerge from one or several channels. If they arrive independently with a mean rate v, the statistical parameters can be derived with the help of Campbell's theorem. The average properties of noise signals are related to the unitary events underlying the noise without postulating any specific mechanism. If N impulses of some function $F(t)$ arrive randomly in time interval T (see Fig. 4a), the sum of these will result in a random noise signal $x(t)$ or conductance signal $G(t)$. The average of one train of these impulses is

$$\langle F(t) \rangle = \frac{1}{T} \int_{-\infty}^{\infty} F(t) \, dt. \tag{30}$$

III. Mathematical Preliminaries

(a)

Fig. 4. Schematic representation of (a) impulses of form $F(t)$ arriving randomly in time interval T; (b) solid line is the sum of impulses and the dashed line is the average $\langle x(t) \rangle$.

(b)

For N trains,

$$\langle x(t) \rangle = N \langle F(t) \rangle = \frac{N}{T} \int_{-\infty}^{\infty} F(t)\, dt. \tag{31}$$

When N and T are large, v is the average rate of arrival of elementary impulses; then $N/T = v$. Thus the average value of $x(t)$ is given by

$$\langle x(t) \rangle = v \int_{-\infty}^{\infty} F(t)\, dt. \tag{32}$$

The variance of $x(t)$ can be calculated as follows.
The mean-square value by definition is given by

$$\langle F^2(t) \rangle = \frac{1}{T} \int_{-\infty}^{\infty} F^2(t)\, dt. \tag{33}$$

At time $t \leq 0$, $\langle F(t) \rangle = 0$. This means $\langle F^2(t) \rangle$ is the variance of $F(t)$. N impulses would give the total variance in $x(t)$ as

$$\sigma_x^2 = N\sigma_F^2 = v \int_{-\infty}^{\infty} F^2(t)\, dt, \tag{34}$$

where N/T tends to v when N and T are very large. Dividing Eq. (34) by Eq. (31) gives

$$\frac{\sigma_x^2}{\langle x(t) \rangle} = \left(\int_{-\infty}^{\infty} F^2(t)\, dt \right) \Big/ \left(\int_{-\infty}^{\infty} F(t)\, dt \right). \tag{35}$$

Equation (35) contains only quantities related to the elementary event. If a square pulse function for $F(t)$ is inserted into Eq. (35), it leads to Eq. (29) if σ_x^2 and $\langle x(t) \rangle$ represent σ_G^2 and $\langle G \rangle$, respectively. However, if an exponential function is considered, that is, $F(t) = \gamma \exp(-t/\tau)$ (for $t > 0$ and 0 for $t \leq 0$), then

$$\frac{\sigma^2}{\langle G \rangle} = \left(\int_0^{\infty} \gamma^2 \exp(-2t/\tau)\, dt \right) \Big/ \int_0^{\infty} \gamma \exp(-t/\tau)\, dt$$

$$= \frac{\left[-\gamma^2 \tau \dfrac{\exp(-2t/\tau)}{2} \right]_0^{\infty}}{[-\gamma\tau \exp(-t/\tau)]_0^{\infty}} = \frac{\gamma^2 \tau/2}{\gamma \tau} = \frac{\gamma}{2}.$$

Thus the estimate of γ depends on the assumption made about the shape of the underlying unitary event. There is no unique relation between the shape and the exact number of γ. The shape of the elementary event has been integrated and replaced by a number. The temporal characteristics of the random fluctuation are unavailable and are not combined in the mean and the variance of $x(t)$. Furthermore, different impulse trains of varied shape can give the same $\langle x \rangle$ and $\langle x^2 \rangle$. So one has to resort to other means to extract the temporal information from the fluctuations. There are two ways of doing this—one is by spectrum analysis and the other is by the covariance or correlation or autocorrelation function. These two methods are different representations of the same information since they are related to each other by a Fourier transform.

A. Fourier Series

This subsection and the next are summaries derived from the chapter by Franz (1982) on the introduction to signals and systems.

Periodic functions can be represented by an infinite series of sinusoids known as Fourier series. Periodic signals can be decomposed into well-understood sinusoids. Periodic signal of arbitrary shape can be resolved into its sinusoidal components. This is specially true in the case of linear systems where the superposition theorem allows reconstruction of the original signal from the system by adding together the individual sinusoidal components of the signal. Two versions of the Fourier series are given below.

(1) $\quad F(t) = a_0 + \sum_{n=1}^{\infty} [a_n \cos(n\omega t) + b_n \sin(n\omega t)],$ \hfill (36)

$$a_0 = \frac{1}{T} \int_0^T F(t)\,dt = \frac{1}{T} \int_{-T/2}^{T/2} F(t)\,dt, \tag{37}$$

$$a_n = \frac{2}{T} \int_0^T F(t)\cos(n\omega t)\,dt = \frac{2}{T} \int_{-T/2}^{T/2} F(t)\cos(n\omega t)\,dt, \tag{38}$$

$$b_n = \frac{2}{T} \int_0^T F(t)\sin(n\omega t)\,dt = \frac{2}{T} \int_{-T/2}^{T/2} F(t)\sin(n\omega t)\,dt, \tag{39}$$

where a_0, a_n, and b_n are the Fourier coefficients, T is periodic time, and radian frequency ω is given by $\omega = 2\pi f = 2\pi/T$.

(2) $\quad F(t) = \sum_{n=-\infty}^{\infty} C_n \exp(jn\omega t),$ \hfill (40)

$$C_n = \frac{1}{T} \int_0^T F(t)\exp(-jn\omega t)\,dt = \frac{1}{T} \int_{-T/2}^{T/2} F(t)\exp(-jn\omega t)\,dt, \tag{41}$$

and
$$C_0 = a_0. \tag{42}$$

For $n \neq 0$ ($n = \pm 1, \pm 2, \ldots$), the C_n coefficients are complex numbers:

$$C_n = (a_n - jb_n)/2, \tag{43}$$

$$C_n = |C_n|\exp(j\phi_n) = (r_n/2)\exp(j\phi_n), \quad n = \pm 1, \pm 2, \ldots, \tag{44}$$

where a_n and b_n are given by Eqs. (38) and (39),

$$r_n = \sqrt{a_n^2 + b_n^2}, \tag{45}$$

$$\tan \phi_n = b_n/a_n. \tag{46}$$

a_0 of Eq. (37) or C_0 of Eq. (42) represent the dc level of the signal $F(t)$. The components with radian frequency ω ($n = 1$) having the same repetition frequency as $F(t)$ are called fundamentals of the series expansion; others which are integral multiples of the fundamental ($\omega_n = n\omega$) are called nth-order harmonics of the periodic signal. The series expansion of Eq. (36) could be in the form of cosines and sines with Fourier amplitude coefficients (a_n, b_n) defined by Eqs. (38) and (39). All the coefficients need not be present in any given case. In the case of signals without dc component $a_0 = 0$. For even functions, i.e., replacing $F(t)$ by $F(-t)$ does not change the value, all $b_n = 0$. Similarly, for odd functions, i.e., value changed by replacement of $F(t)$ by $F(-t)$, all $a_n = 0$. The second version [Eq. (40)] requires a magnitude coefficient given by Eqs. (44) and (45) and a phase given by Eq. (46).

An example of Fourier series for square waves may be considered. When $F(t) = E$ in Eq. (37) gives

$$a_0 = \frac{1}{T}\left[\int_0^{T/2} E\,dt - \int_{T/2}^{T} E\,dt\right] = \frac{1}{T}[(Et)_0^{T/2} - (Et)_{T/2}^{T}] = 0.$$

Equation (38) gives the cosine amplitude coefficients a_n ($n \neq 0$):

$$a_n = \frac{2}{T}\left[\int_{-T/2}^{0} E\cos(n\omega t)\,dt + \int_{0}^{T/2} E\cos(n\omega t)\,dt\right]$$

$$= \frac{2}{T}\left[\frac{E\sin n\omega t}{n\omega}\right]_{-T/2}^{0} + \frac{2}{T}\left[\frac{E\sin n\omega t}{n\omega}\right]_{0}^{T/2} = 0,$$

and all values of a_n are equal to zero. This is true of all waveforms (sine wave) that are odd. Cosine waveforms are even. So the sine amplitude coefficients given by Eq. (39) are given by

$$b_n = \frac{2E}{T}\int_{-T/2}^{0}\sin(n\omega t)\,dt + \frac{2E}{T}\int_{0}^{T/2}\sin(n\omega t)\,dt.$$

Integration gives

$$b_n = -\frac{2E}{n\omega T}\{[\cos(n\omega t)]_0^{T/2} + [\cos(n\omega t)]_0^{T/2}\} = -\frac{4E}{n\omega T}[\cos(n\omega t)]_0^{T/2}.$$

Substituting $[n\omega T/2] = n\pi$ gives

$$b_n = -\frac{2E}{n\pi}(\cos n\pi - 1) = -\frac{2E}{n\pi}[(-1)^n - 1].$$

When n is even (2, 4, 6, ...), $b_n = 0$, and when n is odd, $b_n = 4E/n\pi$. Alternatively, substituting $n = 2k + 1$, where $k = 0, 1, 2, \ldots$, the Fourier series is completely given by

$$F_1(t) = \frac{4E}{\pi} \sum_{k=0}^{\infty} \frac{1}{2k+1} \sin[(2k+1)\omega t] \tag{47}$$

or

$$F_1(t) = \frac{4E}{\pi}[\sin \omega t + \tfrac{1}{3}\sin(3\omega t) + \tfrac{1}{5}\sin(5\omega t) + \cdots].$$

Fourier square waves with and without dc bias level are shown in Fig. 5.

For the complex Fourier series given by Eq. (40), the parameters can be calculated for the square wave shown in Fig. 5b. Equation (41) becomes

$$C_n = \frac{1}{T}\int_{-T/2}^{T/2} F_2(t)\exp(-jn\omega t)\,dt = \frac{2E}{T}\int_{-d/2}^{d/2} \exp(-jn\omega t)\,dt.$$

Integration gives

$$C_n = \frac{2E}{T}\left[\frac{\exp(-jn\omega t)}{-jn\omega}\right]_{-d/2}^{d/2},$$

$$C_n = \frac{2E}{T}\left[\frac{\exp(jn\omega d/2) - \exp(-jn\omega d/2)}{jn\omega}\right] = \frac{2Ed}{T}\frac{\sin(n\omega d/2)}{n\omega d/2}. \tag{48}$$

Since

$$\frac{\sin(n\omega d/2)}{n\omega d/2} = \frac{n\omega d/2 - (n\omega d/2)^3/3 + \cdots}{n\omega d/2} = 1$$

for $n = 0$ and so $C_0 = a_0 = 2Ed/T$. The dc level is proportional to d/T, i.e., (pulse duration)/period. In the present case, $d = T/2$ and so $C_0 = a_0 = E$. The broken line in Fig. 5b indicates the level of E.

When n is not equal to zero, Eq. (48) gives zero for even values of n and nonzero values for odd n. These nonzero values alternate in sign. With the

III. Mathematical Preliminaries

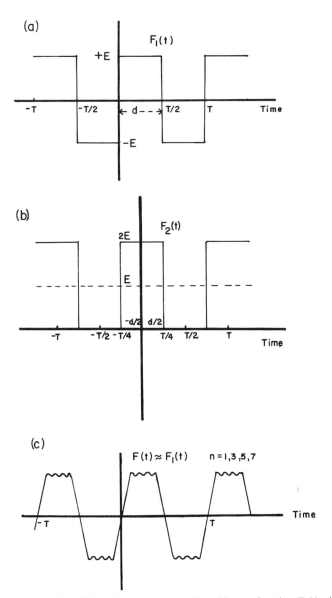

Fig. 5. (a) Odd function $F_1(t)$ with zero average value; (b) even function $F_2(t)$ with average value E; (c) approximation of square wave in (a) by the first four terms of the Fourier series. (After Franz, 1982.)

help of Eqs. (43) and (48) one can write

$$a_k = \frac{(-1)^k 4E}{(2k+1)\pi} \quad \text{for} \quad k = 0, 1, 2, \ldots .$$

Thus the function $F_2(t)$ of Fig. 5b is given by

$$F_2(t) = E + \frac{4E}{\pi} \sum_{k=0}^{\infty} \frac{(-1)^k}{2k+1} \cos[(2k+1)\omega t] \tag{49}$$

or

$$F_2(t) = E + \frac{4E}{\pi}[\cos \omega t - \tfrac{1}{3}\cos(3\omega t) + \tfrac{1}{5}\cos(5\omega t) - \cdots].$$

The identification of a periodic function by the Fourier series is determined uniquely by the two numbers. These depend on the form of the Fourier expansion chosen. For the two forms chosen, the two sets of numbers are the magnitudes of the amplitude coefficients a_n, b_n and the phases r_n, $|C_n|$, ϕ_n. Graphic representation of the Fourier parameters as a function of frequency ($n\omega$) or harmonic multiple (n) is called a spectrum, Fourier spectrum, or frequency spectrum of the periodic function. The spectrum thus specifies the signal in the frequenc domain as shown in Fig. 6a and b for the functions $F_1(t)$ and $F_2(t)$, which appear in Fig. 5a and b in the time

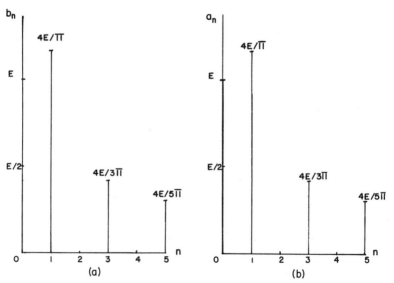

Fig. 6. Discrete spectra for the square waves shown in Figs. 5a and b: (a) odd-function spectrum; and (b) even-function spectrum with $a_0 = E$. (After Franz, 1982.)

domain. In terms of magnitudes, both the signals have the same frequency response. But the phases differ, as shown by Eqs. (47) and (49). The spectra are discrete because the occurrence of Fourier components is in multiples of the fundamental frequency, and they are one-sided because of the summation extending over positive values of n only [see Eq. (36)]. On the other hand, the complex Fourier series [Eq. (40)] extends over both positive and negative values of n, and so would give a two-sided spectrum for $|C_n|$ and would be symmetric to the ordinate at half the magnitude of the corresponding one-sided spectrum [see Eq. (44)]. An example is shown in Fig. 7a.

The response of a linear system to a periodic signal of arbitrary shape can be deduced from the frequency response function of the system in two steps. This is the basis of the spectrum concept. The two steps are (1) the frequency-response function is used to compute the response to each sinusoidal Fourier component, and (2) these individual component responses are added to synthesize the complete response. This is not done in practice. Generally the frequency spectrum of a signal is used to specify the frequency-response properties of a system. The spectrum concept and the resolution of signals into steady-state sinusoids can be extended to nonperiodic signals. This involves three processes. (1) Finite period t is extended to cover $-\infty$ to $+\infty$; (2) instead of summation of the Fourier series [see Eq. (40)], integration is used; and (3) the discrete point spectra is converted to continuous spectral densities (see Fig. 7). These form the basis for the Fourier integral or transformation or just transform.

B. Fourier Transform

The Fourier transformation of any function of time $V(t)$ is defined by the relation

$$\bar{V}(f) = \int_{-\infty}^{\infty} V(t)\exp(-j\omega t)\,dt. \tag{50}$$

The transformed function depends on frequency f. Equation (50) may be regarded as an integral equation with $\bar{V}(f)$ known and $V(t)$ to be evaluated. Then

$$V(t) = \int_{-\infty}^{\infty} \bar{V}(f)\exp(j\omega t)\,df. \tag{51}$$

This is called the inverse Fourier transform. The two equations (50) and (51), which are the counterparts of Eqs. (41) and (40) of the Fourier series, form a transform pair representing two equivalent formulas of the same signal. Equation (51) can be expressed in terms of ω. Since $df = d\omega/2\pi$, Eq. (51)

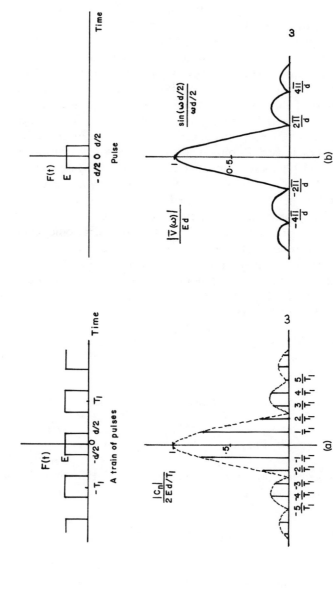

Fig. 7. Transition of a periodic signal to a aperiodic one: (a) discrete point spectrum; (b) continuous density spectrum. (After Franz, 1982.)

becomes

$$V(t) = \frac{1}{2\pi} \int_{-\infty}^{\infty} \bar{V}(j\omega) \exp(j\omega t)\, d\omega. \tag{52}$$

The Fourier transformation $\bar{V}(j\omega)$ is generally a complex function of radian frequency and is given by

$$\bar{V}(j\omega) = |\bar{V}(j\omega)| \exp[j\phi(\omega)],$$

where $|\bar{V}(j\omega)|$ is the amplitude or magnitude density (spectrum) defined over the entire range $-\infty < \omega < \infty$ with the quantity $|\bar{V}(j\omega)|\, d\omega$ indicating the contribution of oscillations in the very small frequency band $d\omega$. For a unique description of a signal, the phase density spectrum $\phi(\omega)$ must be included. Many practical methods of spectral analysis give only a magnitude spectrum. This is enough if one is interested in the relative contributions of the component frequencies without requiring unique signal identification.

Fourier coefficients computed with the help of Eq. (48) giving a pulse train are shown in Fig. 7a. The same figure also contains the discrete magnitude spectrum plotted after normalization. Discrete spectral points become denser with increase in T_1. The pulse train at the limit is reduced to a single pulse (see Fig. 7b), which produces a continuous density spectrum. The envelope of the discrete spectrum is equivalent to the continuous density spectrum. In other words, a periodic signal is made aperiodic. Quantitative demonstration follows from Eq. (50):

$$\bar{V}(f) = \int_{-\infty}^{\infty} E \exp(-j\omega t)\, dt. \tag{53}$$

If the duration of the pulse is d, then integration of Eq. (53) gives

$$\bar{V}(f) = E \int_{-d/2}^{d/2} \exp(-j\omega t)\, dt = -\frac{E}{j\omega}\left[\exp\left(\frac{-j\omega d}{2}\right) - \exp\left(\frac{j\omega d}{2}\right)\right].$$

Substitution of Eqs. (54) and (55) into Eq. (53),

$$\exp(-j\omega d/2) = \cos(-\omega d/2) + j\sin(-\omega d/2)$$
$$= \cos(\omega d/2) - j\sin(\omega d/2), \tag{54}$$
$$\exp(j\omega d/2) = \cos(\omega d/2) + j\sin(\omega d/2), \tag{55}$$

gives

$$\bar{V}(f) = -\frac{E}{j\omega}\left[-2j\sin\left(\frac{\omega d}{2}\right)\right] = \frac{2E}{\omega} \sin\left(\frac{\omega d}{2}\right),$$

i.e.,

$$\bar{V}(f) = Ed\, \frac{\sin(\omega d/2)}{(\omega d/2)}. \tag{56}$$

Comparing Eq. (56) to Eq. (48) shows that normalization should lead to identical curves. The main portion of the spectrum is in the interval $(-\omega_d, +\omega_d)$ and $(\pm\omega_d = \pm 2\pi/d)$. As the pulse width d is made smaller, the frequency interval ω_d is increased. To capture the correct signal, a system with a large bandwidth should be used.

Some important properties of the Fourier transform are the following:

(1) The Fourier transformation is a linear operation obeying the superposition theorem:

$$\bar{V}[aV_1(t) + bV_2(t)] = a\bar{V}_1(f) + b\bar{V}_2(f). \tag{57}$$

(2) Amplitude modulation of sinusoids by a signal $V(t)$ corresponds to a shift of the Fourier spectrum representing the signal

(a) $\bar{V}[V(t)\exp(j\omega_0 t)] = \bar{V}[j(\omega - \omega_0)]$,
(b) $\bar{V}[V(t)\exp(-j\omega_0 t)] = \bar{V}[j(\omega + \omega_0)]$, \hfill (58)
(c) $\bar{V}[V(t)\cos\omega_0 t] = (\bar{V}/2)[j(\omega - \omega_0)] + (\bar{V}/2)[j(\omega + \omega_0)]$.

(3) Convolution in the time domain is equivalent to multiplication in the frequency domain:

$$\bar{V}\left\{\int_{-\infty}^{\infty} V_1(\tau)V_2(t-\tau)d\tau\right\} = \bar{V}\left\{\int_{-\infty}^{\infty} V_1(t-\tau)V_2(\tau)d\tau\right\}$$

$$= \bar{V}_1(j\omega)\bar{V}_2(j\omega). \tag{59}$$

(4) Differentiation in the time domain is equivalent to multiplication with $(j\omega)$ in the frequency domain:

$$\bar{V}\left\{\frac{d^n}{dt^n}V(t)\right\} = (j\omega)^n \bar{V}(j\omega). \tag{60}$$

Here linear differential equations are reduced to linear algebraic equations.

(5) Integration in the time domain is equivalent to division in the frequency domain:

$$\bar{V}\left\{\int_{-\infty}^{t}\cdots\int_{-\infty}^{t} V(t)(dt)^n\right\} = \frac{1}{(j\omega)^n}\bar{V}(j\omega). \tag{61}$$

In connection with property (2), it is useful to consider another function called the delta (δ) function, whose property is generally defined by

$$\int_{-\infty}^{\infty} \delta(x - x_0)F(x)\,dx = F(x_0) \quad \text{and} \quad \int_{-\infty}^{\infty} \delta(x)\,dx = 1, \tag{62}$$

where $F(x)$ is an arbitrary function. The significance of the δ function is revealed by considering the integrals $\int F(x)f_n(x - x_0)\,dx$, where functions

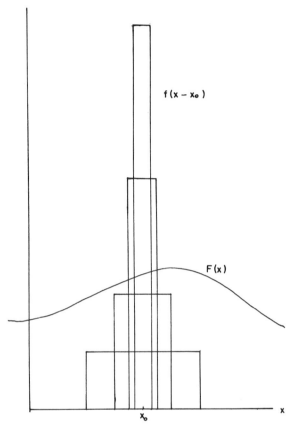

Fig. 8. Functions $f_n(X - X_0)$ peaking at X_0 as n is increased keeping the area constant (i.e., δ function). (After Boas, 1966.)

$f_n(x - x_0)$ peak higher and higher at x_0 as n is increased, since the area under each curve $f_n(x - x_0) = 1$ (see Fig. 8). When $f_n(x - x_0)$ becomes very very narrow, $F(x)$ becomes constant $[F(x_0)]$ over the width $f_n(x - x_0)$ and the integral becomes

$$F(x_0) \int f_n(x - x_0)\, dx = F(x_0) \times 1 = F(x_0).$$

In essence it corresponds to imposing an instantaneous impulse. A sine wave gets compressed to an impulse (δ function) at the frequency of oscillation when it is Fourier transformed. This can be expressed as

$$\delta(f) = \int_{-\infty}^{\infty} \exp(-j\omega t)\, dt. \tag{63}$$

This means the δ function is the Fourier transformation of $V(t) = 1$.

A sine wave can be expressed as

$$V(t) = V_0 \sin \omega_0 t. \tag{64}$$

The common relations of complex quantities are given by Eqs. (54) and (55), and so it follows that

$$\sin \omega_0 t = \frac{1}{2j} [\exp(j\omega_0 t) - \exp(-j\omega_0 t)], \tag{65}$$

$$\cos \omega_0 t = \frac{1}{2} [\exp(j\omega_0 t) + \exp(-j\omega_0 t)]. \tag{66}$$

Substituting Eq. (65) into Eq. (64) and taking the Fourier transform give

$$\bar{V}(f) = \frac{V_0}{2j} \int_{-\infty}^{\infty} [\exp(j\omega_0 t) - \exp(-j\omega_0 t)] \exp(-j\omega t)\, dt. \tag{67}$$

In terms of Eq. (63), Eq. (67) becomes

$$\bar{V}(f) = (V_0/2j)[\delta(f - f_0) - \delta(f + f_0)].$$

Similarly, it can be shown that the Fourier transform of a cosine wave is

$$\bar{V}(f) = (V_0/2)[\delta(f - f_0) + \delta(f + f_0)].$$

It follows therefore that a sine wave in the frequency domain is represented by two imaginary δ functions symmetric about zero at the frequency of oscillation. Similarly a cosine wave has two δ functions on the real axis. The former function is odd, and the latter is even.

In recent years Fourier transformation for spectral analysis of recorded signals has increased mainly because of the elimination of tedious calculations by digital computers. But digital computers require approximations of continuously recorded signals in the time domain to be reduced to discrete sequence of points representing the value of the signal sampled every few seconds (say ΔT seconds, which is the sampling interval). Then these are summed by the computer, and the summation is confined to a finite time interval.

If N is the total number of signal samples sampled every ΔT seconds, then $T = N \Delta T = 1/\Delta f$ is the total time interval over which the signal has been sampled (Δf is the smallest frequency interval to be resolved). Analogous to Eqs. (50) and (51), one can define the discrete Fourier transform (DFT) and its inverse by

$$X(n) = \Delta T \sum_{k=0}^{N-1} x(k\, \Delta T) \exp\left(\frac{-j2\pi nk}{N}\right), \quad n = 0, 1, 2, \ldots, N-1,$$

and

$$x(kT) = \Delta f \sum_{n=0}^{N-1} X(n) \exp\left(\frac{j2\pi nk}{N}\right), \quad k = 0, 1, 2, \ldots, N-1,$$

where $X(n)$ is the spectral component at frequency n (Δf) and $x(k\Delta T)$ is the signal sample at $k\Delta T$.

The fast Fourier transform is a particularly efficient algorithm for computing DFT on digital computers. Many computer centers have programs for FFT; however, one should be familiar with the peculiarities and shortcomings of the DFT/FFT before they are used.

IV. Spectral Density and Rayleigh's Theorem

The several definitions given here follow the descriptions provided in DeFelice's text (1981), which should be consulted for details.

Any arbitrary waveform from a noise source restricted to the time domain $0 < t < T$ (zero elsewhere) can be represented in the frequency domain to extend to infinity. Description of the signal $e(t)$ in the time domain may not be convenient, but its frequency composition can be determined and used to describe the signal in the frequency domain. This is the basis of noise analysis.

The relation between the parameters of the time domain to those in the frequency domain is given by Rayleigh's theorem, expressed as

$$\int_{-\infty}^{\infty} |e(t)|^2 \, dt = \int_{-\infty}^{\infty} |\bar{e}(f)|^2 \, df. \tag{68}$$

This means that the area under the square modulus of a function equals the area under the square modulus of its Fourier transform. For an arbitrary signal which exists between $t = 0$ and T, Rayleigh's theorem gives

$$\int_{0}^{T} e^2(t) \, dt = \int_{-\infty}^{\infty} |\bar{e}(f)|^2 \, df. \tag{69}$$

The average value of $e^2(t)$ is defined as

$$\langle e^2(t) \rangle = \frac{1}{T} \int_{0}^{T} e^2(t) \, dt. \tag{70}$$

If the average $\langle e(t) \rangle$ is zero at $t = 0$, then variance σ^2 is given by $\langle e^2(t) \rangle$ [see Eq. (19)]. So Eq. (69) can be written

$$\sigma^2 = \frac{1}{T} \int_{0}^{T} e^2(t) \, dt = \frac{2}{T} \int_{0}^{\infty} |\bar{e}(f)|^2 \, df \tag{71}$$

(2 appears because only positive values are considered). The integrand on the right-hand side of Eq. (71) is called the spectral density. Thus

$$\hat{S}(f) = (2/T)|\bar{e}(f)|^2. \tag{72}$$

$\hat{S}(f)$ is for a finite time only. If a large sample of the signal is taken, then spectral density for a signal going on forever can be obtained. Thus one can write

$$S(f) = \lim_{T \to \infty} \hat{S}(f).$$

The spectral density $S(f)$ has units of volts squared seconds if the units of the signal $e(t)$ are in volts. Since $S(f)$ is represented as a frequency density, the units become volts squared per hertz.

The variance of a signal in an infinitesimal range of frequency df is given by $S(f)\,df$, and so the total variance is given by

$$\sigma^2 = \int_0^\infty S(f)\,df. \tag{73}$$

The signal coming from a noise source may be well defined (say, $\sin \omega t$ or $\cos \omega t$) or random. In the former case, both the spectral density and the variance assume definite values. In the latter case, the values differ depending on each time interval T. But in the steady state both the spectral density and the variance of a random signal will attain a mean value represented by σ^2 and $S(f)$.

V. Spectral Density and Source Impedance

Since the elements of noise sources are mostly resistors R, capacitors C, and inductors L, their behavior in response to a voltage or current signal may be described either in the time domain or in the frequency domain.

The time domain responses of an inductor and a capacitor are given by

$$V_L(t) = L\frac{d}{dt}[i_L(t)], \tag{74}$$

$$i_C(t) = C\frac{d}{dt}[V_C(t)]. \tag{75}$$

For the circuit shown in Fig. 9, one can write

$$(E - V)/R = i \quad \text{or} \quad V = E - iR$$

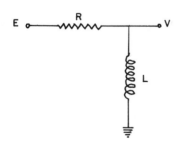

Fig. 9. Resistor R in series with an inductor L. E is applied voltage (input), and V is the output voltage across the inductor L. (After DeFelice, 1981.)

V. Spectral Density and Source Impedance

or

$$L\frac{di}{dt} = E - iR.$$

Solving this equation gives

$$V(t) = E\exp(-t/\tau), \qquad (76)$$

where $\tau = (L/R)$.

Similarly for the circuit shown in Fig. 10, one can write

$$i = \frac{dq}{dt} = C\frac{dV}{dt} = \frac{E-V}{R}$$

or

$$RC\frac{dV}{dt} = E - V.$$

Solving this equation gives

$$V(t) = E[1 - \exp(-t/\tau)], \qquad (77)$$

where $\tau = RC$.

To describe the responses of the circuits of Figs. 9 and 10 in the frequency domain, the Fourier transformation of Eqs. (74) and (75) should be taken. So applying Eq. (60) to Eqs. (74) and (75) gives

$$\bar{V}_L(f) = j\omega L \bar{i}_L(f), \qquad (78)$$

$$\bar{i}_C(f) = j\omega C \bar{V}_C(f). \qquad (79)$$

Impedances of inductor and capacitor are given by

$$Z_L = j\omega L, \qquad (80)$$

$$Z_C = 1/j\omega C. \qquad (81)$$

Now consider the RC circuit of Fig. 10. In the frequency domain, output for any input is written as

$$(\bar{E} - \bar{V}_C)/R = \bar{V}_C/Z_C.$$

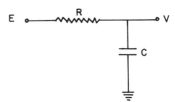

Fig. 10. Voltage E applied to capacitor C through resistance R. (After DeFelice, 1981.)

Rearrangement gives

$$\bar{V}_C \left[\frac{1}{Z_C} + \frac{1}{R} \right] = \frac{\bar{E}}{R},$$

$$\bar{V}_C = \frac{\bar{E}}{R} \bigg/ \left[\frac{1}{Z_C} + \frac{1}{R} \right] = \frac{\bar{E}}{1 + R/Z_C},$$

or

$$\bar{V}_C = \frac{\bar{E}}{1 + j\omega RC} = \frac{\bar{E}}{1 + j\omega\tau}, \qquad (82)$$

where $\tau = RC$. Similarly, for the circuit of Fig. 9, one can write

$$\bar{V}_L = \frac{\bar{E} j\omega\tau}{1 + j\omega\tau} \qquad (83)$$

where $\tau = L/R$.

Because of the impedance of the source, an internal signal gets distorted as seen at an external point. The rectangular signal coming out of an *RC* circuit looks distorted, as shown in Fig. 11. The output voltage in the frequency domain is given by Eq. (82). Taking the square modulus gives

$$|\bar{V}_C(f)|^2 = \frac{|\bar{e}(f)|^2}{1 + \omega^2\tau^2}$$

or

$$\frac{2}{T} |\bar{V}_C(f)|^2 = \frac{2/T |\bar{e}(f)|^2}{1 + \omega^2\tau^2}.$$

In the limit of a high value of *T*,

$$S_v(f) = \frac{S_e(f)}{1 + \omega^2\tau^2}.$$

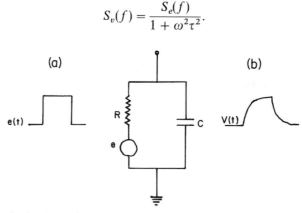

Fig. 11. Qualitative shape of rectangular pulse (a) measured across the parallel circuit looks as in (b). (After DeFelice, 1981.)

VI. Filters

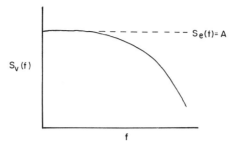

Fig. 12. Voltage spectral density as a function of frequency. $S_e(f) = A$ is flat, indicating white noise. (After DeFelice, 1981.)

$S_v(f)$ is the voltage spectral density of the signal $V(t)$ observed at the output and $S_e(f)$ is the voltage spectral density of the internal source of noise $e(t)$. The spectral density observed across the RC circuit is shown in Fig. 12.

If the emf in the resistor R is a source of white noise (all frequencies present equally), then $S_e(f) = A$ is flat. If current flows in the external branch, the frequency domain current is given by

$$\bar{I}(f) = \bar{e}(f)/R.$$

The current spectral density is

$$S_I(f) = S_e(f)/R^2.$$

The power spectral density may be represented by

$$S_w(f) = \frac{2}{T}|\bar{I}\bar{V}| = \frac{2}{T}\frac{|\bar{e}|^2}{|Z|}$$

Thus

$$S_w(f) = \frac{S_v(f)}{|Z|} = S_I(f)|Z| \quad (\text{W/Hz}).$$

Sometimes both voltage and current spectra are loosely called power spectra.

VI. Filters

Measurement of spectral density as pointed out in the very beginning involves use of filters to remove unwanted frequencies. An example of a low-pass filter is the circuit shown in Fig. 10. Its output-to-input ratio (called the transfer function Y_{lp}) is given by Eq. (82) as

$$\frac{\bar{V}}{\bar{E}} = Y_{lp} = \frac{1}{1 + j\omega\tau} \quad \text{and} \quad \tau = RC.$$

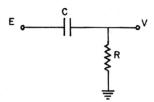

Fig. 13. High-pass RC-filter circuit. (After DeFelice, 1981.)

When $\omega = 0$, $Y_{lp} = 1$, and $\omega \to \infty$, $Y_{lp} = 0$. Thus the circuit allows only low frequencies to go through. In Fig. 13 is given the high-pass filter circuit, which allows only high frequencies to go through. Its transfer function Y_{hp} is given by

$$\frac{\bar{V}}{\bar{E}} = Y_{hp} = \frac{j\omega t}{1 + j\omega t}.$$

An example of a buffered RC bandpass filter is shown in Fig. 14. It is a combination of a low-pass and a high-pass filter with an operational amplifier in between acting as a buffer. Its transfer function is given by

$$Y = Y_{hp} Y_{lp} = \frac{j\omega \tau_2}{(1 + j\omega \tau_1)(1 + j\omega \tau_2)}.$$

The special case when $\tau_1 = \tau_2 = \tau$ is called an L filter, and its transfer function is given by

$$Y_L = \frac{\bar{V}(f)}{\bar{E}(f)} = \frac{j\omega \tau}{(1 + j\omega \tau)^2}.$$

Since $\bar{V}(f) = Y_L \bar{E}(f)$, the spectral density of the output is given by

$$S_v(f) = \frac{2}{T} |Y_L \bar{E}(f)|^2$$

$$= |Y_L|^2 \left[\frac{2}{T} |\bar{E}(f)|^2 \right]$$

$$= |Y_L|^2 S_E(f)$$

in the limit of a high value of T. $|Y_L|^2$ is given by

$$|Y_L|^2 = \left| \frac{\omega^2 \tau^2}{(1 + \omega^2 \tau^2)^2} \right|.$$

In general one can write

$$S_v(f) = |Y|^2 S_E(f).$$

Spectra at the output V is equal to the input spectra coming out of the internal noise source E multiplied by the square of the transfer function of

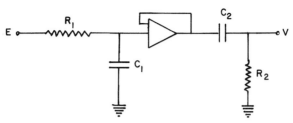

Fig. 14. Combination of low-pass and high-pass RC filters with an operational amplifier in between (buffered RC bandpass filter). (After DeFelice, 1981.)

the filter. This relation is useful in that $|Y|^2$ can be determined by injecting noise of a known spectra $[S_E(f)]$ and measuring the spectra of the noise at the output. If the input noise is white (spectra constant at all frequencies = A), then

$$|Y|^2 = S_v(f)/A.$$

Thus the output spectrum is identical in shape to the square of the modulus of the transfer function of the filter.

The transfer function of a network of resistances and capacitors is related to the impedance of the circuit. Consider the circuit shown in Fig. 15. The closed circuit current is given by

$$\bar{I} = \frac{\bar{E} Z_2}{Z_1 Z_2} = \frac{\bar{E}}{Z_1}.$$

Circuit impedance is given by $Z = \bar{V}/\bar{I}$ and the transfer function by $Y = \bar{V}/\bar{E}$ or $Z = YZ_1$. The transfer function of the circuit multiplied by the impedance of the input branch is equal to the impedance of the circuit.

If the operational amplifier acting as a buffer in the circuit of Fig. 14 is removed, the two stages of the circuit interact. For the first stage, the impedance is given by

$$Z = \frac{R_1}{1 + j\omega\tau_1}$$

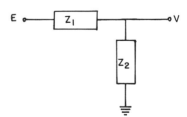

Fig. 15. General input E and output V circuit. Z_1 and Z_2 are impedances. (After DeFelice, 1981.)

and

$$\frac{\bar{E} - \bar{E}'}{1/j\omega C_1} = \frac{\bar{E}'}{R_1},$$

and so the transfer function is given by

$$Y = \frac{\bar{E}'}{\bar{E}} = \frac{j\omega\tau_1}{1 + j\omega\tau_1}.$$

The open-circuit voltage is given by

$$\bar{V} = \frac{\bar{E}'(1/j\omega C_2)}{Z + R_2 + (1/j\omega C_2)}.$$

Substituting the values of Z and $\bar{E}'$ gives after simplification the transfer function for the unbuffered RC bandpass filter as

$$\frac{\bar{V}}{\bar{E}} = Y = \frac{j\omega\tau_1}{(1 + j\omega\tau_1)(1 + j\omega\tau_2) + j\omega R_1 C_2}.$$

The circuit of Fig. 16 represents a filter called a Q filter. The impedance of the parallel L and C is given by

$$j\omega L/(1 - \omega^2 LC).$$

By inspection one can write

$$\frac{\bar{E} - \bar{V}}{R} = \frac{\bar{V}(1 - \omega^2 LC)}{j\omega L}.$$

On simplification, the transfer function Y_Q of the Q filter is given by

$$Y_Q = \frac{\bar{V}}{\bar{E}} = \frac{j\omega L}{R(1 - \omega^2 LC) + j\omega L},$$

$$|Y_Q|^2 = \frac{\omega^2 L^2}{R^2(1 - \omega^2 LC)^2 + \omega^2 L^2} = \frac{1}{1 + (R^2/\omega^2 L^2)(1 - \omega^2 LC)^2}.$$

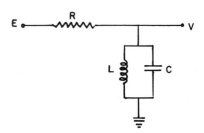

Fig. 16. Representation of the Q filter. (After DeFelice, 1981.)

This equation may be written

$$|Y_Q|^2 = \left(1 + \frac{R^2 C^2}{LC}\left[\frac{1}{\omega\sqrt{LC}} - \omega\sqrt{LC}\right]^2\right)^{-1}.$$

This simplifies to

$$|Y_Q|^2 = \left(1 + Q^2\left[\frac{\omega_0}{\omega} - \frac{\omega}{\omega_0}\right]^2\right)^{-1},$$

where $\omega_0^2 = 1/LC$ and $Q = \omega_0 RC = R/\sqrt{L/C}$. When $\omega = \omega_0$, $|Y_Q|^2$ has a maximum value. The corresponding frequency $f_0 [\omega = 2\pi f; \omega_0 = 2\pi f_0]$ is called the center frequency. In terms of frequency

$$|Y_Q|^2 = \left[1 + Q^2\left(\frac{f_0}{f} - \frac{f}{f_0}\right)^2\right]^{-1}. \quad (84)$$

Filters with high values of Q are called narrow-band filters. f_0 and Q are adjusted independently. f_0 may be set at any particular value by adjusting L and C. R may be adjusted either to keep Q constant or to keep the area under $|Y_Q|^2$ constant. Such filters are called constant-Q filters or constant-bandwidth filters.

Spectral density $S(f)$ of a signal may be evaluated by passing the signal through a battery of filters and measuring the rms value of the output from each filter. The filter selects a range of frequencies from any arbitrary input signal. The selection depends on the center frequency f_0 and the filter bandwidth Δf.

If a noise signal $x_i(t)$ is put into a Q filter with center frequency f_0, let $x_0(t)$ be the signal coming out of the filter. Then the frequency domain output is given by

$$\bar{x}_0(f) = Y_Q \bar{x}(f).$$

So one can write

$$S_0(f) = |Y_Q|^2 S(f).$$

$S(f)$ is the spectral density to be measured. At frequency $f = f_0$, the spectral density of the input signal can be defined as

$$S(f_0) = \left(\int_0^\infty S_0(f)\,df\right)\bigg/\left(\int_0^\infty |Y_Q|^2\,df\right)$$

$$= \left(\int_0^\infty |Y_Q|^2 S(f)\,df\right)\bigg/\left(\int_0^\infty |Y_Q|^2\,df\right). \quad (85)$$

As the Q filter becomes narrow, $|Y_Q|^2$ tends towards $\delta(f - f_0)$, and so according to Eq. (62), $S(f)$ becomes equal to $S(f_0)$, thereby validating Eq. (85).

The denominator of Eq. (85) is called the bandwidth of the filter, and so

$$\Delta f = \int_0^\infty |Y_Q|^2 \, df. \tag{86}$$

Substituting for $|Y_Q|^2$ from Eq. (84), Eq. (86) on integration gives

$$\Delta f = (\pi/2) f_0/Q. \tag{87}$$

As Q is constant, Δf shows a linear rise with increase in f_0.

The integral in the numerator of Eq. (85) according to Eq. (73) is the variance, and so it becomes

$$\sigma_0^2 = \int_0^\infty S_0(f) \, df. \tag{88}$$

Substituting Eqs. (88) and (86) into Eq. (85) gives the spectral density of $x(t)$ at $f = f_0$. Thus

$$S(f_0) = \sigma_0^2/\Delta f. \tag{89}$$

As σ_0 is the rms value of the output signal, Eq. (89) on substituting Eq. (87) becomes

$$S(f_0) = (2Q/\pi f_0)(\text{rms})^2. \tag{90}$$

If a noise signal is passed through a battery of Q filters of various values of f_0 (2, 5, 10, 20 Hz), the points of the spectra $S(f)$ are obtained as described by Eq. (90). So the procedure is as follows: (1) measure the rms value of the output from each Q filter of frequency f_0; (2) square the rms value; (3) multiply it by the factor $2Q/\pi$; and (4) divide it by the center frequency f_0 of the filter. For each value of f_0, a value of the right-hand side of Eq. (90) is obtained. These several points plotted generally on log–log paper constitute a plot of voltage spectral density of the noise source (V^2/Hz versus Hz).

VII. Correlation Function and Spectra

The Wiener–Khintchine theorem relates the power spectrum to the correlation or autocorrelation function. The correlation function $C(\tau)$ is defined as the average value of the product of a stationary random function multiplied by a delayed version of itself. If $e(t)$ is a stationary random process (mean value constant; no drift with time), then

$$C(\tau) = \lim_{T \to \infty} \frac{1}{2T} \int_{-T}^{T} e(t) e(t + \tau) \, dt$$

or

$$C(\tau) = \lim_{T \to \infty} \frac{1}{T} \int_0^T e(t)e(t+\tau)\,dt. \tag{91}$$

All the available information about a random event is contained in the correlation function. Using the inverse Fourier transform [see Eq. (51)], i.e.,

$$e(t) = \int_{-\infty}^{\infty} \bar{e}(f) \exp(j\omega t)\,df,$$

Eq. (91) can be written (see Kittel, 1967)

$$C(\tau) = \lim_{T \to \infty} \frac{1}{T} \iiint \bar{e}(f) \exp(j\omega t) \bar{e}(f') \exp[j\omega(t+\tau)]\,dt\,df\,df'$$

$$= \lim_{T \to \infty} \frac{1}{T} \int_{-\infty}^{\infty} \bar{e}(f) \exp(j\omega t) \bar{e}(-f) \exp[-j\omega(t+\tau)]\,df$$

$$= \lim_{T \to \infty} \frac{1}{T} \int_{-\infty}^{\infty} df\, \bar{e}(f)\bar{e}(-f) \exp(-j\omega\tau).$$

This can be written

$$C(\tau) = \lim_{T \to \infty} \frac{2}{T} \int_0^{\infty} |\bar{e}(f)|^2 \exp(-j\omega\tau)\,df,$$

i.e.,

$$C(\tau) = \lim_{T \to \infty} \frac{2}{T} \int_0^{\infty} |\bar{e}(f)|^2 \cos(\omega\tau)\,df. \tag{92}$$

But the spectral density $S(f)$ is given by [see Eqs. (71)–(73)]

$$S(f) = \lim_{T \to \infty} \frac{2}{T} |\bar{e}(f)|^2,$$

and so Eq. (92) becomes

$$C(\tau) = \int_0^{\infty} S(f) \cos(\omega\tau)\,df. \tag{93}$$

The Fourier cosine inversion yields

$$S(f) = 4 \int_0^{\infty} C(\tau) \cos(\omega\tau)\,d\tau. \tag{94}$$

Equation (94) together with Eq. (93) is called the Wiener–Khintchine theorem. In practice it is convenient to measure the spectrum of electrical noise. Equation (93) of the Wiener–Khintchine theorem can be used to calculate the correlation function, which is easier to relate to theoretical models that describe the stochastic process.

Lecar and Sachs (1981) have shown how a power spectrum is derived from a random process, the process being the fluctuating current $X(t)$ through a

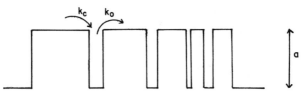

Fig. 17. Open and closed two-state channel as a random process. Rate constants K_o and K_c for opening and closing, and a is the modulus. (After Lecar and Sachs, 1981.)

single channel whose opening and closing is random. As this simple example illustrates the ideas pertaining to the analysis of channel noise, an outline of their derivation is given.

The opening and the closing of a channel is shown schematically in Fig. 17. When there are odd number of transitions during the interval t and $(t + \tau)$, the product $X(t) X(t + \tau) = 0$. When there are even number of transitions, the product is equal to a^2, the square of the modulus. So to get the correlation function the probability of observing, an even number of transitions in a given time must be determined. The current X (random variable) through the channel assumes values 0 and a. The transition probability for the closed to open state per unit time is k_o and k_c is the transition probability for the open to closed state. The rate of opening of the channel is equal to the rate constant k_o multiplied by the probability p_c of being closed. Similarly the rate of closing of the channel is given by $k_c p_o$, and so $k_o p_c = k_c p_o$ and $p_o + p_c = 1$. Combining these two equations gives

$$p_o + \frac{k_c}{k_o} p_o = 1 \quad \text{or} \quad p_o = \frac{k_o}{k_o + k_c}. \tag{95}$$

If the mean value of the random variable X is $\langle X \rangle$, then

$$\langle X \rangle = \lim_{T \to \infty} \frac{1}{2T} \int_{-\infty}^{\infty} X(t)\, dt = \frac{k_o a}{k_o + k_c}. \tag{96}$$

The correlation function is given by

$$C(\tau) = \lim_{T \to \infty} \frac{1}{2T} \int_{-T}^{T} [X(t) - \langle X \rangle][X(t + \tau) - \langle X \rangle]\, dt$$

$$= \lim_{T \to \infty} \frac{1}{2T} \int_{-T}^{T} \{X(t)X(t + \tau) - \langle X \rangle[X(t) + X(t + \tau)] + \langle X \rangle^2\}\, dt. \tag{97}$$

In view of Eq. (96), integrals $X(t)$ and $X(t + \tau)$ can be replaced by $\langle X \rangle$, and so Eq. (97) becomes

$$C(\tau) = \lim_{T \to \infty} \frac{1}{2T} \int_{-T}^{T} [X(t)X(t + \tau) - \langle X \rangle^2]\, dt. \tag{98}$$

The integrand in Eq. (98) can have values 0 and a^2; a^2 is attained when the channel is open at t and $t + \tau$. The integral is dependent further on the probability at any time of the channel being open and on the conditional probability that an initially open channel will undergo an even number of transitions in time τ. Thus

$$\lim_{T \to \infty} \frac{1}{2T} \int_{-T}^{T} X(t)X(t + \tau)\,dt = a^2 \frac{k_o}{k_o + k_c} p_e(\tau). \tag{99}$$

Given that the channel is open at $t = 0$, $p_e(\tau)$ represents the conditional probability that in time τ the channel undergoes even number of transitions. If $d\tau$ is the infinitesimal interval during which no more than one transition occurs and $p_{od}(\tau)$ is the probability of an odd number of jumps in time τ, then

$$p_e(\tau + d\tau) = p_e(\tau) \times \text{(probability of 0 jumps in } d\tau\text{)}$$
$$+ p_{od}(\tau) \times \text{(probability of 1 jump in } d\tau\text{).} \tag{100}$$

But

$$p_e(\tau) + p_{od}(\tau) = 1,$$

and in $d\tau$, the probability of an additional transition from the closed to open state is $k_o d\tau$, and similarly the probability of an additional jump from open to closed state is $k_c\,d\tau$. Thus Eq. (100) becomes

$$p_e(\tau + d\tau) = p_e(\tau)(1 - k_c d\tau) + [1 - p_e(\tau)] k_o\,d\tau.$$

This equation may be written

$$dp_e(\tau) = k_o d\tau - p_e(\tau)(k_o + k_c)\,d\tau$$

or

$$\frac{dp_e(\tau)}{k_o - p_e(\tau)(k_o + k_c)} = d\tau.$$

Integration gives

$$\ln[k_o - p_e(\tau)(k_o + k_c)]_{p_e(\tau) = 1}^{p_e(\tau)} = [-(k_o + k_c)\tau]_{\tau = 0}^{\tau}$$

or

$$\frac{k_o - p_e(\tau)(k_o + k_c)}{-k_c} = \exp\left(-\frac{\tau}{\tau'}\right),$$

where $\tau' = 1/(k_o + k_c)$. Rearrangement gives

$$p_e(\tau)(k_o + k_c) = k_c \exp(-\tau/\tau') + k_o$$

or

$$p_e(\tau) = \frac{k_o}{k_o + k_c} + \frac{k_c}{k_o + k_c}\exp\left(-\frac{\tau}{\tau'}\right). \quad (101)$$

Combining Eqs. (96) and (98)–(101) gives

$$C(\tau) = a^2 \frac{k_o}{k_o + k_c}\left[\frac{k_o}{k_o + k_c} + \frac{k_c}{k_o + k_c}\exp\left(-\frac{\tau}{\tau'}\right)\right] - \left(\frac{k_o a}{k_o + k_c}\right)^2.$$

Simplification results in

$$C(\tau) = \frac{k_o k_c}{k_o + k_c} a^2 \exp\left(-\frac{\tau}{\tau'}\right). \quad (102)$$

The probability that a channel is open is given by $p = k_o/(k_o + k_c)$, and so

$$1 - p = \frac{k_c}{k_o + k_c}.$$

The modulus $a = \gamma$, the single channel conductance. Substituting these into Eq. (102) gives

$$C(\tau) = \gamma^2 p(1-p)\exp(-\tau/\tau'). \quad (103)$$

The power spectrum is obtained by substituting Eq. (103) into Eq. (94), yielding

$$S(f) = 4\gamma^2 p(1-p)\int_0^\infty \exp\left(-\frac{\tau}{\tau'}\right)\cos(\omega\tau)\,d\tau.$$

Completing the integration gives

$$S(f) = \frac{4\gamma^2 p(1-p)\tau'}{1 + \omega^2(\tau')^2}. \quad (104)$$

The correlation function $C(\tau)$ for the process described is given in Fig. 18a and the power spectrum is shown in Fig. 18b. The function $\tau'/(1 + \omega^2\tau'^2)$ in Eq. (104) is called a Lorentzian, and the corresponding spectrum is called a Lorentzian spectrum. Equation (104) represents the spectrum for the simplest kinetic process and is useful in the analysis of channel noise. More complex kinetic schemes can be postulated. Channels may undergo transitions among several states. In these cases the correlation function will be the weighted sum of exponentials. The spectrum will be a weighted sum of Lorentzians. Some of the complex schemes are discussed by DeFelice (1981) and by Neher and Stevens (1977).

Conductance fluctuation spectra need not always be Lorentzian. Actually the power spectrum depends on the shape of the underlying unit pulse. For

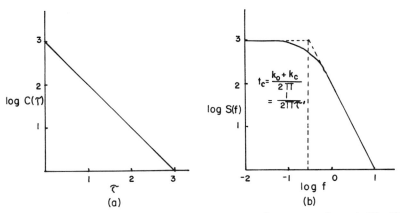

Fig. 18. (a) Autocorrelation function for the two-state random process shown in Fig. 17 as a function of transition time constant τ. (b) Power spectrum for the random process shown in Fig. 17. The cutoff frequency f_c is shown in the figure at the intersection of the high- and low-frequency asymptotes. (After Lecar and Sachs, 1981.)

the case of the unit pulse of exponential shape, the spectrum is Lorentzian but pulses of other shapes give other spectra.

VIII. Types of Noise Sources

Mean values of membrane current I, conductance G, and potential E are related by

$$I = G(E - E_r), \tag{105}$$

where E_r is the nonfluctuating resting membrane potential. If small fluctuations of ΔI, ΔG, and ΔE are inserted, then

$$\begin{aligned} I + \Delta I &= (G + \Delta G)(E + \Delta E - E_r) \\ &= G(E - E_r) + G\,\Delta E + \Delta G(E - E_r) + \Delta G\,\Delta E. \end{aligned} \tag{106}$$

Subtracting Eq. (105) from Eq. (106) gives

$$\Delta I = G\,\Delta E + \Delta G(E - E_r) + \Delta G\,\Delta E.$$

Ignoring the last term (the product of two small quantities) gives

$$\Delta I = G\,\Delta E + \Delta G(E - E_r). \tag{107}$$

Experiments can be carried out under constant-current (current-clamp) or under constant-voltage (voltage-clamp) conditions. In the former case voltage fluctuates, and in the latter case current fluctuates. Under current-clamp conditions, $\Delta I = 0$, and so Eq. (107) becomes

$$\Delta G = -G\,\Delta E/(E - E_r). \tag{108}$$

Since spectral density is related to the square of the amplitude, mean-square fluctuations are to be computed. So squaring and averaging the terms of Eq. (108) gives

$$\langle \Delta G^2 \rangle = \langle G^2 \rangle \frac{\langle \Delta E^2 \rangle}{(E - E_r)^2}.$$

If the frequency dependence of voltage fluctuations [$\langle \Delta E^2 \rangle$] is followed, it will give information about the frequency dependence of conductance fluctuations. But unfortunately this relation is not straightforward, because conductance also will change with frequency and calculations become complicated. Instead, if current fluctuations under voltage-clamp conditions ($\Delta E = 0$) are followed, Eq. (107) becomes

$$\Delta I = \Delta G(E - E_r).$$

Squaring and averaging both sides of the equation give on rearrangement

$$\langle \Delta G^2 \rangle = \langle \Delta I^2 \rangle/(E - E_r)^2.$$

In this case current fluctuations are directly related to conductance fluctuations. Current fluctuation noise spectra occur very commonly. It is found in electrical resistors. Its fluctuation spectrum $S_I(f)$ (A² s) is independent of frequency and is given by

$$S_I(f) = 4\kappa T/R,$$

where R is the value of the resistance. Its voltage spectrum is given by

$$S_E(f) = 4\kappa TR \quad (\text{V}^2/\text{Hz}).$$

The two are related by the general expression

$$S_E(f) = S_I(f)|Z|^2,$$

where Z is the impedance of the system.

The noise that is frequency independent is generally called white, thermal, or Johnson noise. It arises from the electrical fluctuations due to thermal motion of charges in the conductors.

$1/f$ noise has the spectral density

$$S(f) = B/f^\alpha,$$

where B is a constant and α is close to 1. This kind of noise is also known as flicker noise and is seen in nonequilibrium systems, small openings between conductive solutions and artificial membranes (DeFelice and Michalides,

1972). This spectral characteristic is interesting but is not observed in linear RC circuits. However, early work of Verveen and Derksen (1965) showed that $1/f$ noise spectra was obtained from the node of Ranvier. Derksen and Verveen (1966) found the $1/f$ noise to disappear leaving a white noise spectrum when the node was bathed in isotonic KCl. Later, when Verveen et al. (1967) applied artificial currents to the membrane, the noise signal was the smallest at the potassium equilibrium potential, suggesting that $1/f$ noise was due to passive movement of potassium ions across the membrane. Similarly, in the voltage-clamped lobster axon, Poussart (1971) found potassium current fluctuations to conform to $1/f$ noise spectra. However, when Siebenga and Verveen (1972) examined the potential fluctuations in the node of Ranvier at more positive potentials than they had previously used, they found a component of spectral density function with the form

$$S'(f) = \frac{C}{1 + (f/f_c)^2}, \tag{109}$$

where C is a constant and f_c is the corner frequency at which $S'(f) = C/2$. Thus a Lorentzian spectral component was observed. But later work by Siebenga et al. (1974) showed that application of tetraethylammonium ion (tetrodotoxin used to block Na channels) reduced this Lorentzian component indicating that the Lorentzian noise arose by the opening and closing of active potassium channels.

The conventional voltage-clamp experimental arrangement of the Hodgkin–Huxley type gives no noise fluctuations due to large membrane area and low membrane resistance. In 1973, Fishman used a small area of squid axon membrane (patch) in which potentials and currents were independent of those in the surrounding membrane. With this patch (voltage-clamped) preparation, a combination of $1/f$ and Lorentzian noise in the fluctuations of potential was observed. The $1/f$ noise was abolished when 50-mM tetraethylammonium ion was applied. The difference between the two spectra obtained conformed to the Lorentzian of Eq. (109). Thus the entire voltage noise spectrum realized from the patch clamped squid axon membrane can be represented by

$$S_E(f) = a + \frac{b}{f} + \frac{C}{1 + (f/f_c)^2},$$

where a, b, and C, are the appropriate constants for the white noise, $1/f$ noise, and Lorentzian noise, respectively.

In 1974 Wanke et al. observed a "hump" in the voltage spectrum but not in the current spectrum in the squid giant axon. They showed that this arose from a peak in the impedance–frequency curve for the membrane, the known theoretical relation between voltage spectrum $S_E(f)$, current spectrum $S_I(f)$, and impedance $Z(f)$ being

$$S_E(f) = S_I(f)|Z(f)|^2.$$

This shows that voltage and current spectra will have the same form only when the impedance is independent of frequency. Further work by Fishman et al. (1975) showed that perfusion of the squid axon with 100-mM cesium ions or low potassium ions (substituted with isotonic sucrose) eliminated the Lorentzian noise. These results and the TEA effect already referred to support the idea that the source of Lorentzian noise is the opening and closing of potassium channels.

Fluctuation analysis applied to several other biological and lipid bilayer preparations has been reviewed by DeFelice (1977, 1981), Neher and Stevens (1977), and Lecar and Sachs (1981), which should be consulted for details.

There are two methods that are used to extract values for the single-channel conductance (γ) from measurements of noise spectra. In the first method, with γ as a parameter, the experimental noise data are fitted to the equation of a kinetic theory. For example, consider the probabilistic version of the Hodgkin–Huxley equations (see DeFelice, 1981). For the K channel, it is assumed that there are N_K independent channels per unit area, each of which consists of four subunits. The subunits can be either in the open state or in the closed state. When all four subunits are in the open state, the channel conductance is γ_K. When one or more subunits are in the closed state, the conductance is zero. The assumption of four subunits gives rise to four Lorentzians in the spectral density function (see Conti et al., 1975; DeFelice, 1981). Thus

$$S_K(f) = \frac{4I_K^2 \tau_n}{N_K} \sum_{i=1}^{4} C(4,i) \left(\frac{1-n}{n}\right)^i \frac{1}{i} \frac{1}{1+(\omega\tau_n/i)^2}.$$

Expansion gives

$$S_K(f) = \frac{4I_K^2 \tau_n}{N_K}\left[\left(\frac{1-n}{n}\right)^4 \frac{4}{16+\omega^2\tau_n^2} + \left(\frac{1-n}{n}\right)^3 \frac{12}{9+\omega^2\tau_n^2}\right.$$
$$\left.+ \left(\frac{1-n}{n}\right)^2 \frac{12}{4+\omega^2\tau_n^2} + \left(\frac{1-n}{n}\right) \frac{4}{1+\omega^2\tau_n^2}\right], \tag{110}$$

where I_K is the steady-state K current, n is the Hodgkin–Huxley parameter, τ_n is the time constant for the n process, $n = \alpha_n/(\alpha_n + \beta_n)$, and $\tau_n = 1/(\alpha_n + \beta_n)$ (see Chapter 7).

Equation (110) predicts the spectral density for the squid axon potassium channels. Conti et al. (1975) showed that TTX-treated axon (Na currents blocked) $S_K(f)$ was approximated by a single Lorentzian of the form

$$S_K(f) = \frac{b}{f} + \frac{C}{1 + (f/f_c)^2},$$

where $b = 1.3 \times 10^{-20}$ (at 1 Hz) and $C = 2.7 \times 10^{-21}$ ($f_c = 38$ Hz). However, when $f = 0$, Eq. (110) becomes

$$S_K(0) = \frac{4I_K^2}{N_K}\tau_n\left[\left(\frac{1-n}{n}\right)^4\frac{1}{4} + \left(\frac{1-n}{n}\right)^3\frac{4}{3} + \left(\frac{1-n}{n}\right)^2 3 + \left(\frac{1-n}{n}\right)4\right]. \quad (111)$$

I_K, τ_n, and n were measured under voltage clamp as functions of membrane potential. By fitting Eq. (111) to a curve of $S_K(0)$ versus potential, a value for N_K giving the best fit was found. This value of N_K was found to be $60\mu\text{m}^{-2}$. With the value of $\bar{G}_K$ measured for the axons, single-channel conductance ($\gamma = \bar{G}_K/N_K$) of 12 pS was determined.

A similar analysis was performed on the spectral density function for the Na channel by blocking K current with injected TEA. The current noise spectral density for the Hodgkin–Huxley Na channel is given by (Conti et al., 1975; DeFelice, 1981)

$$S_{Na}(f) = \frac{4I_{Na}}{N_{Na}}\left\{\frac{1-h}{h}\frac{\tau_h}{1 + \omega^2\tau_h^2} + \tau_m\sum_{i=1}^{3}\frac{1}{i}C(3,i)\left(\frac{1-m}{m}\right)^i\right.$$

$$\left.\times\left[\frac{1}{1 + (\omega\tau_m/i)^2} + \frac{1-h}{h}\frac{\tau_h}{1 + i\tau_h}\cdot\frac{1}{1 + [\omega\tau_h\tau_m/\tau_m + i\tau_h]^2}\right]\right\}, \quad (112)$$

where

$$C(3, i) = \frac{3!}{(3-i)!i!}, \quad \tau_h = \frac{1}{\alpha_h + \beta_h}, \quad h = \frac{\alpha_h}{\alpha_h + \beta_h},$$

$$\tau_m = \frac{1}{\alpha_m + \beta_m}, \quad m = \frac{\alpha_m}{\alpha_m + \beta_m}.$$

When $f \geq 100$ Hz, Conti et al. (1975) obtained reliable data, and at these frequencies Eq. (112) approximates to

$$S_{Na}(f) = \frac{4I_{Na}^2}{N_{Na}}\frac{\tau_m}{h}\sum_{i=1}^{3}\frac{1}{i}C(3,i)\frac{[(1-m)/m]^i}{1 + (\omega\tau_m/i)^2}. \quad (113)$$

Again, when $f = 0$, Eq. (113) becomes

$$S_{Na}(0) = \frac{4I_{Na}^2}{N_{Na}}\frac{\tau_m}{h}\left[3\frac{1-m}{m} + \frac{3}{2}\left(\frac{1-m}{m}\right)^2 + \frac{1}{3}\left(\frac{1-m}{m}\right)^3\right]. \quad (114)$$

Equation (114) was fitted to the experimental curve of $S_{Na}(0)$ versus membrane potential yielding a value of 330 μm^{-2} for N_{Na}. This corresponds to a single-channel conductance of 4 ps. This theory applied to the single node of Ranvier gave values of 2–5 pS (van den Berg et al., 1975) and 8 pS (Conti et al., 1976) for γ_{Na}.

The second method for determining the values for γ does not depend on a model. The measured noise parameter is the mean-square membrane current fluctuation $\langle \Delta I^2 \rangle$. This is obtained from the mean-square fluctuation of the current through a single channel $\langle \Delta i^2 \rangle$ by multiplying by the total number of channels N, i.e.,

$$\langle \Delta I^2 \rangle = N \langle \Delta i^2 \rangle.$$

Horowicz et al. (1980) have given a derivation for the single-channel conductance where the channel has three states, γ, δ, and 0 instead of only, say, 1 state γ (open) and 0 state (closed). If the probabilities are p and q for the channel being in γ and δ states, then the zero-conductance state will have a probability $1 - p - q$. The average current through a single channel is given by

$$\langle i \rangle = \sum(\text{probability of the state}) \times (\text{conductance of the state})$$
$$= p\gamma(E - E_r) + q\delta(E - E_r) + 0 \text{ (zero for the state 0).}$$

Similarly

$$\langle i^2 \rangle = p\gamma^2(E - E_r)^2 + q\delta^2(E - E_r)^2.$$

But

$$\langle \Delta i^2 \rangle = \langle i^2 \rangle - \langle i \rangle^2, \tag{115}$$

$$\langle \Delta i^2 \rangle = (E - E_r)^2 [p\gamma^2 + q\delta^2 - (p\gamma + q\delta)^2]. \tag{116}$$

Relevant equations for single-channel conductance can be written

$$\langle G \rangle = p\gamma + q\delta \quad \text{and} \quad \langle G^2 \rangle = p\gamma^2 + q\delta^2.$$

Substituting these in Eq. (116) gives

$$\langle \Delta i^2 \rangle = (E - E_r)^2 [\langle G^2 \rangle - \langle G \rangle^2]$$

or

$$\langle \Delta i^2 \rangle = (E - E_r)^2 \langle G^2 \rangle [1 - \langle G \rangle^2 / \langle G^2 \rangle].$$

Rearrangement gives

$$\langle G^2 \rangle = \frac{\langle \Delta i^2 \rangle}{(E - E_r)^2 [1 - \langle G \rangle^2 / \langle G^2 \rangle]}. \tag{117}$$

Since
$$\langle \Delta I^2 \rangle = N \langle \Delta i^2 \rangle, \tag{118}$$
$$\langle I \rangle = N \langle G \rangle (E - E_r),$$

or
$$\langle G \rangle = \langle I \rangle / N(E - E_r). \tag{119}$$

Using Eqs. (118) and (119) in Eq. (117) gives
$$\frac{\langle G^2 \rangle}{\langle G \rangle} = \frac{\langle \Delta I^2 \rangle}{\langle I \rangle [1 - \langle G \rangle^2 / \langle G^2 \rangle](E - E_r)}. \tag{120}$$

For only one conductance state γ of probability p ($\delta = 0$ and $q = 0$, i.e., the other two states are nonconducting)
$$\langle G \rangle = p\gamma \quad \text{and} \quad \langle G^2 \rangle = p\gamma^2$$

and so Eq. (120) becomes
$$\gamma = \frac{\langle \Delta I^2 \rangle}{\langle I \rangle (1 - p)(E - E_r)}. \tag{121}$$

Begenisich and Stevens (1975) used Eq. (121) to estimate the value of γ (4 pS) for the K channels of the frog node. The mean current was measured, p was estimated from the relation $[G(E)/\bar{G}]$ (conductance of membrane relative to its value at high depolarizations), and $\langle \Delta I^2 \rangle$ was obtained either from the zero-time value of the correlation function or from the integration of the one-sided spectrum over all frequencies, i.e.,

$$\langle \Delta I^2 \rangle = 2 \int_0^\infty S_I(f) \, df.$$

Sigworth (1980) used a method whereby the noise in the sodium channel of the frog node was studied. Na channel noise, because it is fast and inactivates quickly, is difficult to measure. Using internal cesium ions and external TEA, K currents were blocked. Sets of 80–512 identical depolarizing pulses of 20-ms duration were applied. The pulses were repeated at intervals of 300–600 ms. The resulting membrane current records were processed in groups. The group mean I_j and variance σ_j^2 were calculated from the jth group of n records. With the help of equations

$$I_j(t) = \frac{1}{n} \sum_{k=1}^n Y_{jk}(t)$$

and
$$\sigma_j^2(t) = \frac{1}{n-1} \sum_{k=1}^n [Y_{jk}(t) - I_j(t)]^2,$$

where Y_{jk} is the kth membrane current record and the terms in the brackets represent the deviation of each current sample from its group mean value. From the theory of channel noise [see Eqs. (23)–(27)], it follows that

$$\sigma_I^2 = Np(1 - p)i^2,$$

where N is the total number of membrane channels. Substituting $I = Npi$ gives

$$\sigma_I^2 = iI - I^2/N.$$

This equation was used to fit the experimental data (plot of corrected σ_I^2 and mean current I) by choosing appropriate values for i and N. Typical values were $N = 20,400$ and $i = -0.55$ pA. The single-channel conductance was calculated from $\gamma = i/(E - E_{Na})$ and was found to be $\gamma = 6.4$ pS.

Artificial lipid bilayers doped with gramicidin A and extra junctional acetyl choline receptors of frog muscle are two systems in which single channels can be observed and also noise measurements can be made. Neher and Zingsheim (1974) measured the conductance and average lifetime of single channels in gramicidin A–doped bilayer membranes. These values agreed with those derived for the same system from the autocorrelation function of the current noise. Similarly Kolb et al. (1975) determined γ again for the same system by both methods and found good agreement. Also Neher and Sakmann (1976) measured the conductance and average lifetime of extra junctional acetyl choline channels in frog muscle (channels opened by the action of suberyldicholine) and found the value of γ to agree with that derived from noise analysis by Colquhoun et al. (1975). These findings from two different techniques (noise and relaxation experiments) yielding similar results show that noise analysis is a reliable technique that could be applied with confidence to derive single-channel parameters in membrane systems in which single channels cannot be seen directly.

From the foregoing, it is obvious that noise theory by itself is abstract and sterile and raw experimental data are nonilluminating. But the two together with appropriate kinetic models illuminate the dark and unfathomable crevices of the cell membrane.

References

Begenisich, T. B., and Stevens, C. F. (1975). *Biophys. J.* **15**, 843.
Boas, M. L. (1966). "Mathematical Methods in the Physical Sciences." Wiley, New York.
Colquhoun, D., Dionne, V. E., Steinbach, J. H., and Stevens, C. F. (1975). *Nature (London)* **253**, 204.
Conti, F., and Wanke, E. (1975). *Q. Rev. Biophys.* **8**, 451.
Conti, F., DeFelice, L. J., and Wanke, E. (1975). *J. Physiol. (London)* **248**, 45.

References

Conti, F., Hille, B., Neumcke, B., Nonner, W., and Stämpfli, R. (1976). *J. Physiol. (London)* **262**, 699.
DeFelice, L. J. (1977). *Int. Rev. Neurobiol.* **20**, 169.
DeFelice, L. J. (1981). "Introduction to Membrane Noise." Plenum, New York.
DeFelice, L. J., and Michalides, J. P. L. M. (1972). *J. Membr. Biol.* **9**, 261.
Derksen, H. E., and Verveen, A. A. (1966). *Science* **151**, 1388.
Fishman, H. M. (1973). *Proc. Natl. Acad. Sci. U.S.A.* **70**, 876.
Fishman, H. M., Moore, L. E., and Poussart, D. J. M. (1975). *J. Membr. Biol.* **24**, 305.
Franz, G. N. (1982). *In* "Electronics for the Modern Scientist" (P. B. Brown, G. N. Franz, and H. Moraff, eds.), p. 165. Elsevier, New York.
Horowicz, P., Schneider, M. F., and Begenisich, T. (1980). *In* "Membrane Physiology" (T. E. Andreoli, J. F. Hoffman, and D. D. Fanestil, eds.), p. 185. Plenum, New York.
Kittel, C. (1967). "Elementary Statistical Physics." Wiley, New York.
Kolb, H. A., Läuger, P., and Bamberg, E. (1975). *J. Membr. Biol.* **20**, 133.
Lecar, H., and Sachs, F. (1981). *In* "Excitable Cells in Tissue Culture" (M. Lieberman and P. G. Nelson, eds.), p. 137. Plenum, New York.
Maksoudian, Y. L. (1969). "Probability and Statistics with Applications." International Text-Book Co., Scranton, Pennsylvania.
Neher, E., and Sakmann B. (1976). *Nature (London)* **260**, 799.
Neher, E., and Stevens, C. F. (1977). *Annu. Rev. Biophys. Bioeng.* **6**, 345.
Neher, E., and Zingsheim, H. P. (1974). *Pfluegers. Arch.* **351**, 61.
Poussart, D. J. M. (1971). *Biophys. J.* **11**, 211.
Siebenga, E., and Verveen, A. A. (1972). *Biomembranes* **3**, 473.
Siebenga, E., de Goede, J., and Verveen, A. A. (1974). *Pfluegers Arch.* **351**, 25.
Sigworth, F. (1980). *J. Physiol. (London)* **307**, 97.
Stevens, C. F. (1975). *Fed. Proc., Fed. Am. Soc. Exp. Biol.* **34**, 1364.
van den Berg, R. J., de Goede, J., and Verveen, A. A. (1975). *Pfluegers Arch.* **360**, 17.
Verveen, A. A., and DeFelice, L. J. (1974). *Prog. Biophys. Mol. Biol.* **28**, 189.
Verveen, A. A., and Derksen, H. E. (1965). *Kybernetik* **2**, 152.
Verveen, A. A., Derksen, H. E., and Schick, K. L. (1967). Nature *(London)* **216**, 588.
Wanke, E., DeFelice, L. J., and Conti, F. (1974). *Pfluegers Arch.* **347**, 63.

INDEX

A

Action potential, 328
Activated complex, 43, 44
Activation, standard free energy of, 44
Activation energy, 53, 179, 197, 247
Active transport, 147, 148
Activity, 9, 12–15
Activity coefficient, 12, 130
   Davies equation, 73
   Guggenheim equation, 73
   Guntelberg equation, 73
   mean, 14, 73
   molal, 12
   molar, 12
   practical, 12
   rational, 12, 71
   stoichiometric, 12
Admittance, 331
Adsorption isotherm, *see* Langmuir isotherm
Aliasing, 372
Arrhenius equation, 52
Avogadro number, 51, 70
Axon
   crayfish, 120, 122
   lobster, 411
   Myxicola, 122
   squid, 122, 179, 412

B

Barrier energy, 177
Behn equation, 58
Bernoulli trials, 377
Binomial probability function, 377, 380, 381
Binomial theorem, 376
Boltzmann equation, 50, 66, 83
Born energy, 189

C

Cable equation, 337, 339
Cable theory, 334–352
Campbell's theorem, 382
Capacitance, 29, 30
   double-layer, 92
Capacitor, 396, 397
   combination of, 30, 31
   energy stored in, 31
   parallel plate, 29, 31
Carrier model, 210–229
Cell constant, 48
   electrical, 344
Center frequency, 403, 404
Channel
   alamethicin, 239–254
      micellar model, 240
      oligomer model, 240, 244–247
   amphotericin B, 265
   black widow venom, 265
   closing, 369
   excitability-inducing material (EIM), 259–263
   gramicidin A, 229–239
   hemocyanin, 263, 264

monazomycin, 254-258
nystatin, 265
opening, 369
potassium, 369
sodium, 369
single-ion, 234
subunits, 412
two-barrier model, 162
two ions, 253
Channel formers, *see* Channel-forming ionophores
Channel-forming ionophores, 229-265
Characteristic frequency, 334
Charge density, 66
  surface, 209
    evaluation of, 122, 123
  volume, 126
Charge pulse method, 201, 217
Chemical potential, 9
Combination, 376
Compact layer, *see* Stern layer
Complex quantity, 394
Concentration, 38
  molal, 12
  molar, 12
  mole fraction, 12
  polarization, 195
Conductance, 48, 382
  electrical, 276, 279-282
  equivalent, 49
  fluctuation spectra, 408
  infinite dilution, 49, 75, 76
  integral, 49
  mechanical, 279, 280
  molar, 49
  single-channel, 412, 414
    potassium, 413, 415
    sodium, 416
  single-ion, 49, 148
  specific, 32, 48, 102, 103
  surface, 102, 103
  total, 49
Conductor, 25
Constant field equation, *see* Goldman equation
Convolution, 392
Core conductor model, 334, 335
Core resistance, 335
Correlation function, 371, 384, 404, 405, 406, 408

Coulombic force, 73
Coulomb's law, 17, 19, 23
Coupling coefficient, 270, 285
  L, 270
  R, 270
Coupling ratio, 144
Covariance function, *see* Correlation function
Current, 32
  capacitative, 337
  clamp, 409
  definition, 25
  density, 32, 275
Curve peeling, 62

D

Debye-Hückel limiting law, 71
Debye-Hückel theory, 66-78
  activity coefficient, 70-73
  electrolyte conductance, 75-78
Debye length, 69, 86, 90, 116, 152, 204
Delta function, 392-394
Dialysis, 3
Diasolysis, 3
Dielectric, 25-27
  polarized, 26, 27
Dielectric constant, 25, 28, 30, 66, 70
Diffuse double layer, 88, 99, 103
  characteristics, 93
Diffusion
  rate theory of, 43
  time-dependent, 58
  zone, 54
Diffusion coefficient, 38, 42, 52, 177
  restricted, 315
Diffusion potential, 54-57
Diffusional flow, 315
Dipole moment, 26, 27, 84-86
Dipole potential, *see* Potential, surface
Displacement vector, 27
Dissipation function, 270, 278, 283, 296, 297, 322
Dissociation constant, 153, 237
Distribution coefficient, *see* Partition coefficient
Distribution function
  binomial, 376
  Gaussian, 376, 378
  Poisson, 376, 378

Donnan ratio, 109, 111, 112, 114, 115, 138
Double layer
  theory of, 87–98
    applications of, 119–122
    thickness of, 54, 80, 83, 116
Drift velocity, 47
Dufour effect, 3
Dwell time, 259

E

Einstein relation, 51
Einstein–Smoluchowski equation, 42
Electric field strength, *see* Electric intensity
Electric flux, 23
Electric intensity, 20, 21
Electrochemical potential, 38
Electrodecantation, 4
Electrodialysis, 3
Electrodiffusion model; *see* Flux, equation, Nernst–Planck
Electrokinetic phenomena, 98–107, 277–283
Electrolyte conductance
  electrophoretic effect, 76
  relaxation effect, 76, 77
Electromotive force, *see* Emf
Electron
  charge, 17
  mass, 17
Electro-osmosis, 3, 100–103, 106, 324
  flow, second, 280
  permeability, 306
  pressure, 103, 104, 279, 280
    second, 280
  velocity, 102
Electrophoresis, 4, 106, 107
  forced flow, 4
Electrostatic capacity, 25
Electrostatic energy, 22
Electrostatic field, 33
Electrostatic interaction energy, 188
Electrostatic potential, 20, 67, 85
Electrostatics, 17
El-Sharkawy–Daniel equation, 145
Emf, 32, 33, 36
Emu, 18, 19
Energy barrier, 174, 175
  trapezoidal, 201
Energy of transfer, 274

Energy well, 178, 179
Enthalpy, 7, 53, 197
Entropy, 7, 197
  production, 269, 271, 273
Enzyme kinetics
  equations of, 165–174
  mediated transport, 173, 174
Equilibrium, 11
  between different phases, 37
  conditions for, 8
  constant, 44, 113, 114, 132, 168, 244–246
  heterogeneous, 211
  Donnan, 107–111
  charged membrane, 114–117
Equipotential, 22, 338
Equivalent circuit, 330
Esu, 18, 19
Even function, 385
Expectation, 378, 379

F

Farad, 29
Fast Fourier transform (FFT), 372
Faxen equation, 315
Fick's law, 315
  first, 38, 39
  second, 40
Filter
  bandpass, 400–402
  bandwidth, 403, 404
  high-pass, 400
  L, 400
  low-pass, 397, 400
  narrow-band, 403
  Q, 402, 403
Filtration coefficient, 279, 281, 291
Fluctuation
  macroscopic, 369
  microscopic, 369
Fluctuation analysis, 369
  nonmathematical, 370–372
Fluctuation–dissipation theorem, 373
Flux, 53
  definition, 38
  diffusional, 50
  equation
    Nernst–Planck, 53, 54, 129, 198
    integration of, 137–139
Fourier analysis, 371

Fourier coefficients, 384, 385, 391
  parameters, 388
  series, 384–389
    complex, 384, 389
  square wave, 386
  transform, 371, 389, 392
    complex function, 391
    discrete, 394
      inverse, 394
    fast, 395
    inverse, 371, 389, 405
Free energy, 7, 9, 11
  activation, 43
  partial molar, 9
Frequency
  density, 396
  domain, 395, 395
Friction coefficient, 288–293, 307, 308
  urea–Visking dialysis tubing, 294

G

Galvanic cell, 32, 36
  potential of, 37
Gating current, 366
  particles, 365, 366
Gauss's theorem, 23, 24
Gibbs–Duhem equation, 10, 16, 284, 303
Gibbs equation, 272, 273
Gibbs–Helmholtz equation, 9
Goldman equation, 139
Goldman–Hodgkin–Katz equation, 142, 180

H

Haldane equation, 168
Heat content, see Enthalpy
Heat of transfer, 275
Helmholtz layer, 88
Henderson equation, 56
Henry's law, 206
Histogram, 370
Hodgkin–Huxley equations, 352–365
  h parameter, 360, 363, 413
  m parameter, 360–363, 413
  n parameter, 358, 359, 361, 412
Hydraulic permeability, 4
Hydrolysis constant, 112
Hyperfiltration, 316

I

Ideal solution, 11
Impedance, 401
  capacitor, 397
  complex plane, 330
  imaginary component of, 329, 330
  inductor, 397
  input, 344
  locus, 333
  real component of, 329, 330
  source, 396–399
Impulse, see Action potential
Inactivation, 253
Independence principle, 149
Induced dipoles, 25–27
Inductor, 396
Inhibition
  competitive, 169
  complete, 169
  mixed, 169
  noncompetitive, 169
  partial, 169
  uncompetitive, 169
Inhibitor, 168
  irreversible, 168
  reversible, 168
Input resistance, 340, 341
Insulator, see Dielectric
Intercept discrepancy method, 219
Ion atmosphere, 66, 68, 70
  model, 92
  thickness, 69
Ion cloud, see Ion atmosphere
Ion hydration, 74, 75
  primary, 65
  secondary, 65
Ion–ion interaction, 65, 70, 71, 75
Ion size parameter, 72–74
Ion–water interaction, 63, 75
Ionic hypothesis, 328, 356
Ionic strength, 69
Isoelectric point, 113, 114
Isotope interaction, 309–311

J

Jump distance, 174, 175
  relation to diffusion coefficient, 43

K

Kinetics
  channel formation, 230
  exchange, 62
Kirchhoff's laws, 36, 339
Kohlrausch's law, 75

L

Laminar flow, 103
Langmuir isotherm, 94, 152, 153
Laplace's equation, 23, 25
Law of mass action, 44
Length constant, *see* Space constant
L'Hospital's rule, 140
Liquid junction, 54
  constrained type, 56
  continuous mixture type, 56
  flowing type, 56
  free diffusion type, 56
  potential, 54, 55
Long pore effect, 147
Lorentzian, 408
  spectrum, 408

M

Mechanical permeability, *see* Hydraulic permeability
Membrane
  bilayer, 188, 189
    equivalent circuit, 345
    doped, 370
    potential energy, 189
    transport of lipophilic ion, 188–210
  biological
    Danielli–Davson model, 327
    equivalent circuit, 353
    Singer–Nicolson model, 327
  charge density, evaluation of, 123–126
  charged, 299, 301
    effective charge density, 123–126
  classification, 1, 2
  composite, 46, 305, 306
  conductance, 146, 198
    initial, 194
    single-ion, 454
  definition, 1
  dielectric constant, 189
  impedance, 328–334
  liquid, 155–159
  mosaic, 327
  permeability, 45, 46, 177, 178, 238
    ratio, 183, 184, 186
  permeation, Eyring model, 174–178
  permselectivity, 1, 118, 124
  pore radius, 318, 319
  porosity, 317
  rate constant, 197
  selectivity, 178
    evaluation of, 149
  semipermeable, 1, 15, 283, 285, 286
  time constant, 335, 337, 344
  transport
    kinetic models, 165
  transport number, 1
  transport processes, 3
Membrane potential, 117, 118
  kinetic approach, 160–163
  resting, 327, 355
  thermal, 277
  thermodynamic approach, 159, 160
Mean square value, definition of, 383
Michaelis–Menten constant, 165, 171
Michaelis–Menten equation, 165, 169
  reversible reaction, 166
Mobility
  absolute, 47
  electrical, 47, 53, 77
Molality, mean, 14
Moment
  first, 379
  second, 379
  third, 379
Moreton equation, 145, 146
Muscle
  barnacle, 125
  frog, 416
  sartorius, 125
  semitendinosus, 125

N

Na–K pump, 144, 145
Negative anomalous osmosis, 287
Nernst–Einstein relation, 51
Nernst layer, 47
Newton, 19
  relation to dyne, 19
Node
  Ranvier, 122, 411, 414
  Xenopus, 122

Noise, 409–416
  analysis, 395
  current, 370
  flicker, 410
  Johnson, 410
  Lorentzian, 411
  spectra, 411
  voltage, 370
  white, 410
Nonpolar molecule, 26
Normal distribution, 370, 371
Northop–Anson equation, 61

O

Odd function, 385
Ohm's law, 32, 102, 146
Onsager relation, 271, 274, 279, 285
Osmoionosis, 3
Osmosis, 3
Osmotic coefficient, 75, 111
  definition, 15
  practical, 16
  rational, 15
  relation
    to activity coefficient, 16, 17
    to osmotic pressure, 16
Osmotic pressure, 109–111, 283, 284

P

Partial molar quantity, 9
  pressure, 11
  volume, 10, 11, 108
Partition coefficient, 46, 80, 178, 191
Patch clamp, 411
Periodic function, 384
Permeability coefficient, 46
Permitivity, 20, 28
Permutation, 376
Pervaporation, 4
Phase angle, 329, 333
Phase plane method, 62
Phenomenological coefficients, 270, 274, 275
  second-order, 312, 313
  equation, 270, 279, 285, 297, 299
Piezodialysis, 4
Piezoelectricity, 3
Planck constant, 43, 44
  equation, 57
Poiseuille law, 104, 304

Poisson–Boltzmann equation, linearized, 67
Poisson probability function, 380, 381
Poisson's equation, 23, 24, 66, 87, 105
Polar molecule, 26
Potential
  asymmetry, 152
  bi-ionic, 130, 131–134
  concentration, 151–154
  contact, 79, 83
  diffusion, 276, 277
  distribution, 79–81
  Donnan, 109, 110, 112, 116, 117, 133, 151
  electric, 129
  electrochemical, 147
  electrotonic, 335
  equilibrium for Na, 356
  Galvani, 79, 80
  interface, 79, 82
    Gouy–Chapman theory of, 87–93
    Stern theory of, 92–97
  multi-ionic, 131
  reversal, 150, 151
  surface, 79, 82, 84, 120
  zeta, 98–100, 120
Potential energy
  ion, 50, 66
  unit charge, 20
Potential energy barrier
  with independence, 179–182
  without independence, 182–185
Probability distribution function, 374, 375
Probability function, 374, 378
  continuous, 374
  discrete, 374
  histogram, 379
  skewed, 381
Proton
  charge, 17
  mass, 17

R

Radian frequency, 384
Random variable, 374
Raoult's law, 12
Rate constants, 60, 161, 162, 180, 214–217, 221–223, 228, 233, 237, 243, 252, 264
  α and β, 358–364
Rate equation, schematic method of King and Altman, 170–173

Rayleigh's theorem, 395
Reactance, 333
Reaction rate, 167
Relaxation current, 194, 196, 200, 213, 214, 247
Resistance, 32
  electrical, 279, 280, 309
  electrotonic, 341, 342, 348
  integral, 49
  internal, 33, 34, 37
  mechanical, 279, 280
  phenomenological, 309
  specific, 32
Resistors, 396, 397
  parallel, 34, 35
  series, 34, 35
Reverse osmosis, 4
Rms, 373
Rothmund-Kornfeld equation, 132

S

Sample space, 374
  nonuniform, 374
  uniform, 374
Saxen's relations, 106, 279
Scatchard equation, 159
Selectivity
  filter, 365
  kinetic view, 185–188
Selectivity coefficient, 130, 134
  determination of, 134–137
Shear plane, 98, 99
Single-file transport, 238, 321
Sjodin-Ortiz equation, 145
Solute flux interactions, 293
Solute-solute interaction, 293, 295
Solvent drag, 147, 287
Soret effect, 4
Space charge density, 23
Space constant, 335, 337, 339, 343, 344
Spectral density, 395, 396
  current, 399
  power, 399
  voltage, 399, 404
Spectrum, 371, 384, 388, 390
  continuous, 389–391
  current, 412
  discrete, 388, 389
  even-function, 388

input, 400
odd-function, 388
one-sided, 389
output, 400, 401
two-sided, 389
voltage, 412
Square wave analysis, 349
Standard deviation, 370
Standard state, 12
Statcoulomb, 19
Statistical concepts, 373–381
Statvolt, 30
Staverman equation, 159
Staverman reflection coefficient, 286, 292, 293
  definition of, 1
Stefan-Maxwell diffusivity coefficients, 311
Steric hindrance, 315
Stern layer, 88, 94, 99, 204
Stern model, 88
Stern plane, 98, 99
Stirling formula, 378
Stokes-Einstein relation, 51, 52
Streaming current, 4, 105, 279, 280
  second, 280
Streaming potential, 4, 105, 106, 277, 280, 324
Superposition theorem, 384
Surface charge density, 26, 27
Susceptance, 331
Susceptibility, 27, 28
Synapse, 150

T

Temperature coefficient, 179
Teorell equation, 138
Teorell-Meyer-Sievers (TMS) theory, 117, 118, 130
Terminal voltage, 34
Tetrodotoxin, 356, 411, 412
Theories of membrane behavior, 129
Theory of rate processes, 161, 176
Thermodynamics
  concepts, 7
  first law, 8
  irreversible processes, 269, 273, 277
Thermo-emf, 4
Thermo-osmosis, 4
Three-capacitor model, 203–208

Time constant, 60, 332
Time domain, 395, 396
Tortuosity, 301
Transfer function, 399–402
Transition probability, 406
Transmission coefficient, 44, 176, 180
Transport
 mediated, 148
 water, 129
Transport depletion, 4
Transport number, 53, 118, 123, 124, 301, 303, 305, 306–308, 354
 mass, 276
Two-barrier model, 183

U

Ultrafiltration, *see* Reverse osmosis
Unit of charge, definition of, 20
Units of conversion, 18
 system of, 18

Ussing flux ratio, 309
 equation, 147, 234

V

van't Hoff equation, 109, 284
 factor, 16
Variance, 380–382, 395, 396, 404
Vibration frequency, 44, 161
Viscous flow, 315
Volta potential, *see* Potential, distribution
Voltage clamp, 355, 357, 409
Voltage jump method, 194
Voltage-sensitive gate, 365
Volume flow, 278, 285, 304

W

Wiener–Khintchine theorem, 404, 405
Work
 expansion, 8
 maximum, 8
 net, 8